Instabilities in Multiphase Flows

Instabilities in Multiphase Flows

Edited by

G. Gouesbet and A. Berlemont

Laboratoire d'Energétique des Systèmes et Procédés
URA CNRS 230–INSA de Rouen
Mont Saint Aignan, France

Springer Science+Business Media, LLC

Library of Congress Cataloging-in-Publication Data

Instabilities in multiphase flows / edited by G. Gouesbet and A.
Berlemont.
p. cm.
"Proceedings of an International Symposium on Instabilities in
Multiphase Flows, held May 11-14, 1992, in Rouen, France"--Galley.
Includes index.

ISBN 978-1-4899-1596-2

1. Multiphase flow--Mathematical models--Congresses.
I. Gouesbet, Gérard. II. Berlemont, A. III. International
Symposium on Instabilities in Multiphase Flows (1992 : Rouen,
France)
TA357.5.M84I52 1993
620.1'064--dc20 93-20770
CIP

Proceedings of an International Symposium on Instabilities in Multiphase Flows, held May 11–14, 1992, in Rouen, France

ISBN 978-1-4899-1596-2 ISBN 978-1-4899-1594-8 (eBook)
DOI 10.1007/978-1-4899-1594-8

PREFACE

The study of multiphase flows is of utmost interest for engineers who are more or less inevitably faced with them when handling various industrial processes or when dealing with environmental problems such as the dispersion of pollutants. It is also a large kingdom assembling many beautiful and weird landscapes in which wandering researchers may be caught by the fascination of precious stones or mysterious insets to deep and obscure caverns.

Unfortunately, it is also an historically disconnected field of research, as testified by any textbook contents or by the scientific programs of conferences devoted to multiphase flows. For instance, is there a relation between fluidization and the study of interfacial waves, or between the behaviour of an annular film of liquid and the one of a free surface heated from below ? The answer is indeed : yes.

To help reveal some unity behind the avatars of multiphase flow behaviours, it has been decided to focus the interest on the instability phenomena. This book therefore provides the reader with most of the papers which have been accepted and/or presented at the international symposium on "instabilities in multiphase flows" held at the National Institute for Applied Science (INSA) in Rouen, France, from the 11th of May to the 14th of May 1992. The topic of the conference has produced a strong emphasis on instability theory and nonlinear dynamics, including chaotic phenomena. It has been a great pleasure to gather experts in multiphase flows and others in nonlinear dynamics, many of them being not used to getting together and to observe how they shared their common or disconnected experiences for fruitful discussions.

Such an aim was somewhat of a challenge and the organizers appreciated very much the commitments of colleagues and friends who, by accepting to join the scientific committee, gave a boost to the enterprise. Scientific committee members were namely:

J.M. Delhaye	*CENG, Grenoble, France*
J.J. Dorning	*University of Virginia, Charlottesville, USA*
G. Gouesbet	*CNRS, INSA de Rouen, France*
G. Hetsroni	*Technion, Haifa, Israel*
M. Maeda	*University of Keio, Japan*
H. Mori	*University of Kyushu, Japan*
G. Nicolis	*University of Bruxelles, Belgium*
Y. Pomeau	*ENS, Paris, France*
H.B. Stewart	*Brookhaven National Laboratory, USA*

Many thanks also to the advisory committee members who, in collaboration with the scientific committee members, had to select the abstracts, namely :

G.I. Barenblatt	*Sciences Academy, Moscou, USSR*
J.A. Bouré	*CENG, Grenoble, France*
P. Cerisier	*Provence University, Marseille, France*

A. Clausse	*Instituto Balsiro, Bariloche, Argentina*
C.T. Crowe	*Washington State University, Pullman, USA*
F. Durst	*LSTM, Erlangen, Germany*
F. Ebert	*Kaiserslautern University, Germany*
K. Hishida	*Keio University, Yokohama,Japan*
G. Iooss	*Nice University, France*
R.T. Lahey, Jr.	*Rennselaer Polytechnic Institute, Troy, USA*
M. Lance	*LMF, Lyon University, France*
J. March Leuba	*ORNL, Oak Ridge, USA*
G. Matsui	*Tsukuba University, Japan*
T. Mullin	*Oxford University, United Kingdom*
M. Rizwan Uddin	*University of Virginia, Charlottesville, USA*
J.H. Whitelaw	*ICST, London, United Kingdom*

The conference would have been impossible to hold without the aid of many local collaborators who had to take care of every detail to the best of their ability, including coffee breaks with cakes. This task has been particularly uncomfortable because new buildings were under construction in our institute during the time of the conference, a fact which could not be forecast three years before, when the decision of holding the conference was made. As a result, two scenarii with two different locations had to be simultaneously prepared and it is only two weeks before D day that the final choice was made. To avoid giving here an endless number of names, let us be satisfied with thanking the official members of the local organizing committee, namely :

A. Berlemont	**H. Cavalier**	**R. Darrigo**	**P. Desjonquères**
G. Gouesbet	**B. Maheu**	**J. Maquet**	**C.Rozé**

A French idiomatic expression tells us that money is the nerve of war. It is then a pleasure to also thank the organizations which helped in that respect, namely :

Conseil Régional de Haute Normandie

Centre National de la Recherche Scientifique

INSA de Rouen

Université de Rouen

Syndicat d'Initiative de Rouen

Banque Nationale de Paris

AXA Assurances

Some papers are absent from these proceedings. First, this is the case of the plenary lectures which, following an agreement with Prof. G. Hetsroni, will be published in the International Journal of Multiphase Flows. The plenary lecturers were as follows :

J.J. Dorning	*University of Virginia, Charlottesville, USA*
D.D. Joseph	*University of Minnesota, Minneapolis, USA*
R.T. Lahey, Jr.	*Rennselaer Polytechnic Institute, Troy, USA*
D. Riley	*University of Bristol, United Kingdom*
H.B. Stewart	*Brookhaven National Laboratory,, USA*
M. Velarde	*UNED, Facultad de Ciencas, Madrid, Spain*

and it is obviously a nice opportunity to warmly thank the plenary lecturers who gave their time and their enthusiasm to present up-to-date accounts of some topics relevant to the conference.

Also, some papers could not be incorporated due to the poor technical quality of the manuscripts, although the scientific quality was possibly excellent. We are sad and we must say ashamed that rejections could result from such casual circumstances. But it has been impossible for us to provide a local retyping. Another cause of sadness is the big number of no-shows mainly due to scientists from eastern countries of Europe. Due to this fact, we unfortunately missed opportunities to meet excellent people, and to make friends, and we are praying for a fast and positive improvement of the economic situation of such countries. Science is by essence something international which should help unifying humankind. And, not having easy opportunities to meet all of our colleagues is like being a body with dead legs or arms.

In any case, optimism is a matter of will, and pessimism a matter of mood, wrote one philosopher named Alain. So, let us be optimistic and look forward to a second conference on "instabilities in multiphase flows" in which all causes of sadness would have disappeared. At least, we believe that the attendants of the first conference found it enjoyable and useful and we would like to end this preface by thanking them because, by their presence, they certainly performed the best and most essential roles.

The Editors
G. Gouesbet, A. Berlemont

CONTENTS

CHAOTIC BEHAVIOR OF AN ANNULAR FILM OF LIQUID UNSTABILIZED BY AN INTERFACIAL SHEAR STRESS

Boujema Izrar and Francois Lusseyran

LEMTA - CNRS-URA 875
2 Avenue de la Fôret de Haye, B.P. 160 54504 Vandoeuvre lès Nancy Cedex
France

In this work, two methods are used to analyze the interface in vertical annular counter current gas liquid flow in the vicinity of the flooding point. More generally, both methods may be applied to the case of free boundary flows.

In the first section we recall the motivations for this study. In the second section we develop the wavelet transform. This transform makes it possible to map a signal into its scale components.

In the third section we compute the fractal dimension of the original signal as well as its different scales. Then we show that this dimension abruptly grows above a critical scale.

INTRODUCTION

For some flow conditions, the interface between two non miscible fluids is unstabilized. Its shape is of great significance in the exchanges between the phases and also in the mechanisms of transition (from stratified to bubbly regimes in horizontal flows and from bubbly to annular in vertical and horizontal flows). However there is a lack of tools that fits the waves characteristics. The Fourier transform does not distinguish between different scales;

Instabilities in Multiphase Flows, Edited by G. Gouesbet
and A. Berlemont, Plenum Press, New York, 1993

and therefore underlying physical phenomena like turbulence viscosity effects, gas liquid coupling, capillarity, transition, inertial effects, etc.

On the other hand global measurements are too poor to quantify loss of stability; it is necessary to measure parameters which are suitable for this kind of approach. Among them the most common method, for weakly non linear models, is the equation of dispersion, to connect wave length velocity, and scale.

Melville made the first interesting attempt in that direction (Melville, 1983).The signal is represented in terms of phase and amplitude by means of Hilberts Transform. The wave number k can be determined with the help of two well located probes. The study concerned the Stokes waves the equation of dispersion of which is known to the second order. Nevertheless this method is limited for evident reasons :

- it seems to be efficient only for well defined modes. As a matter of fact when a fundamental frequency and its harmonics are identified or the equation of dispersion is known, one has a reliable selection criterion.
- even if the signal is expanded with a band pass filter for example and if the central frequencies are rapidly varying, the related instantaneous parameters (amplitude and phase) are as complicated as the whole signal.

A tool, in our conviction, that seems to be without the caveats of the Hilbert transform is the wavelet transform technique. The mode selection does not depend any longer on the signal, but is inherent to the scale-expansion method. This expansion is unique under certain conditions and keeps both time and frequency variables.

At last the localization in time - and - frequency space is scale dependent. The validation of models become much easier if you can have from the experimental data at least one qualitative piece of information about the dynamical aspect, for example the minimum number of equations revealing the complete or partial description of the interface.

This leads directly to the estimation of the dimension of the attractor, associated with the dynamical system in question. The first dimension calculations, in gas-liquid two face flow, were done for the bubble regime (Lusseyran, 1987) and the plane falling film (Biage and al. 1989). In this last case the analysis of the film thickness yields to very high values (9-12 depending on the probe position relatively to the inflow). Recently, the methods of deterministic chaos are applied to a free falling film at low Reynolds number (Lacy C.E. et al. 1991).

As we shall see, wavelets make it possible to split the signal into two components that complement one another; the high scale one having at least the dimension of the signal, the other is significantly reduced.

THE WAVELET TRANSFORM : A TIME - SCALE REPRESENTATION

The earliest works on join distributions goe back to the thirties. The Wigner - Ville transform gives a time - frequency map of a temporal signal or its spectrum in signal analysis (Ville 1948) and a phase space representation of a wave function in quantum mechanics

(Wigner 1932). The W-V transform has the properties of a probability distribution except invertibility and non negativity (Feix et al, 1983; Izrar, 1986).

Among the optimum smoothing kernels which guarantiee positive energies one corresponds to the short times Fourier transform (STFT), the other to the wavelet transform (WT) .

If W(t) is an admissible wavelet the analysis of a signal s(t) gives the time-scale distribution :

$$C(b,a) = \langle s \mid W_{a,b} \rangle = \int_{R} s(t)\, W_{a}^{*}{}_{,b}(t)\, dt$$

where $W_{a,b}(t)$ is obtained from a translation 'b' in time and a dilation of a factor 'a' of the wavelet mother W(t) (Grossmann and Morlet, 1984) ; a is the scale of analysis :

$$W_{a,b}(t) = W\left(\frac{t-b}{a}\right) / \sqrt{a}$$

The only condition that must be satisfied by a function to be an admissible wavelet is :

$$C_W = \int \frac{|\hat{W}(\omega)|^2}{\omega}\, d\omega < \infty$$

In practice the wavelet should also be well localized an have cancellation properties (Meyer, 1990). The coefficients C(b,a) give information on the behavior of the signal both at time t and pulsation $\Omega = \overline{\omega} / a$, where :

$$\overline{\omega} = \frac{1}{\langle W \mid W \rangle} \int |\hat{W}(\omega)|^2 \omega\, d\omega$$

The orders of magnitude of the time-scale resolution, adapted to each scale, are :

$$\sigma_t^2 = \frac{1}{C_W} \int_{R} (t-b)^2\, |S(t)|^2\, dt$$

$$\sigma_\omega^2 = \frac{1}{C_W} \int_{R} (\omega-\Omega)^2\, |\hat{S}(t)|^2\, d\omega$$

In Fourier space , this analysis is equivalent to a bank of band pass filters keeping $d\omega/\omega$ constant . Once the C(b,a) are known, the synthesis of the signal is achieved with :

$$s(t) = \frac{1}{C_w} \int C(b,a)\, W_{ab}(t) \frac{da\, db}{a^2}$$

By discretization the a-axis, the compounds of the signal for each scale may be defined with (Goupillaud et al., 1985):

$$s_a(t) = \frac{\Delta a}{C_w} \int C(b,a)\, W_{ab}(t) \frac{db}{a^2}$$

Notice that the implemented algorithm uses the dyadic transform where the dilation parameter $a=2^j$ and $b=\pm k\, a$ with j,k positive integers.
The WT satisfies the Plancherel's identity :

$$\|s\|^2 = \frac{1}{C_w} \int |C(b,a)|^2 \frac{da\, db}{a^2}$$

and allows a scale by scale allocation of the energy of the signal :

$$E(a) = \int |C(b,a)|^2\, db = \frac{a}{2a} \int |\hat{S}(\omega)|^2\, |\hat{g}(a\omega)|^2\, d\omega$$

This last point is of principal interest in the study of the film, because the gas-liquid interface is perturbed by a wide variety of waves covering a very large span of scales, from the small one of the substrata to the large rolling waves. The wavelet transform allows the separation in groups of scales, which can then be considered separately. This approach was already used (Farge 1990) to extract the coherent vorticity structures, from the turbulent substrata of a numerical simulated turbulent bounded flow.

A large variety of wavelet functions has been proposed in the literature. We have selected the most common one: Morlet's wavelet. It has a Gaussian envelope shape and therefore minimizes the product of the RMS in time and frequency $\sigma_t \sigma_\omega$ (Lemarié, 1990).

Figures 1a to 1c exemplify the behavior of the wavelet coefficient for an academic signal with a frequency doubling. The enlargement in time towards the large scales at the frequency change is a consequence of the singularity. The sudden change of wave length, both sides of each singularity, is well detected and quantified.

EXPERIMENTAL SETUP

The test section is a pipe of plexiglass (diameter 0.0406m). The gas is nitrogen and the liquid is an aqueous solution of KCl (0.33 mol/l), potassium ferrocyanide (2.10^{-3} mol/l) and ferricyanide (4.10^{-3} mol/l). The film is injected and withdrawn through two porous sec-

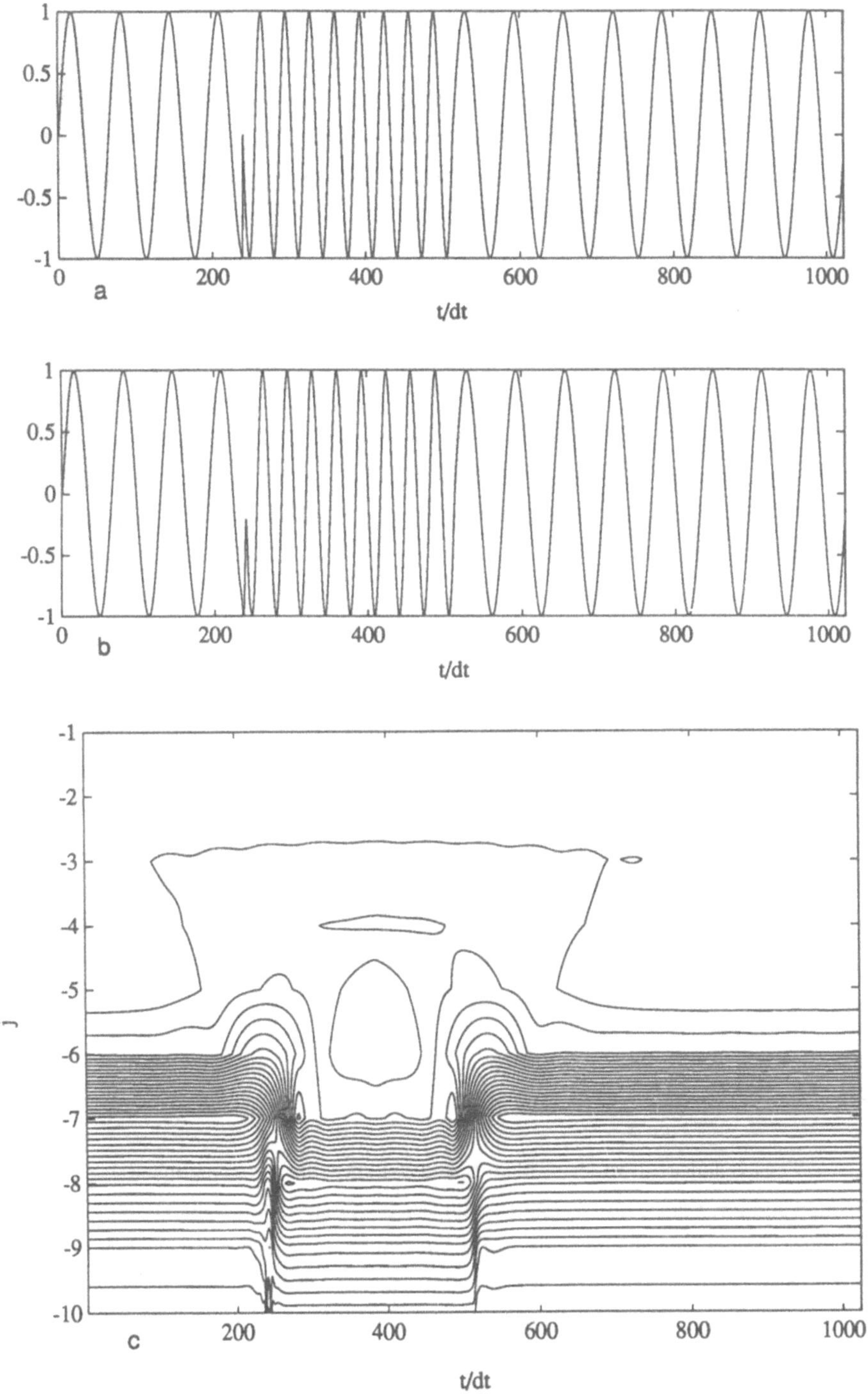

Figures 1 . The signal versus time (a) ; synthesis (b) ; level contours of the distribution of $|C(t,j)|^2$ (c).

tions separated by a variable distance from 0.6 to 2.72m. Before entering the test section, the gas flows through a pipe 2.5m of length connected to the test section. This enable the gas to pass through the test section with an already developed and steady velocity profile.

The measured quantities are local instantaneous liquid film thickness (conductive method) and wall shear stress (electrochemical method). Two film thickness probes were used located at the distance 580, 1640mm from the injection point of the liquid and three wall shear probes were located at the distance 580, 1110 and 2160mm from this injection point (Indarto, 1991).

All measurements were carried out at atmospheric pressure and 20°C temperature. Analog data from the transducers were digitized for 4 flow rates of the liquid ($\mathcal{R}e_l$: 2000, 5000, 7500, 11000) at 512Hz, and at least 3min. of record was stored for subsequent processing. The results presented here concern only the wall shear stress and one value of the liquid flow rate ($\mathcal{R}e_l=4Q_l/\pi D\nu=5000$) and for two gas flow rates : j_g = 0.32m/s and j_g = 0.44m/s ; the last one is chosen just at the flooding point. The sampling of each data set is done at the frequency of 512Hz and during 176s (i.e. 90112 points) with a low noise acquisition system (HP).

WAVELET ANALYSIS OF THE SIGNAL

The results of the analysis are summarized by a map of the $|C|^2$ versus time and scales (figure 2). The distribution of the time average of the energy indicates the predominance of the scale 7 and 8 (figure 3). Obviously a large wave structure emerges in the figure 2b, corresponding to the flooding transition. This high solitary wave always occurs at the transition in our experiments (Izrar et al, 1991). Energy is then spanned over the scales 4 to 8 and with a maximum at 5. Notice that this kind of situation is never caught in Fourier spectrum due to the lack of resolution in time.This difference is still visible in the partial synthesis. The 'best' approximation is obtained when summing the first 7 scales (figure 4).

PHASE SPACE ANALYSIS

The existence of two levels of complexity in the film dynamic, linked respectively to the high and low scales, is investigated with two relatively simple methods. Among the different techniques available to study non-linear temporal sequences, we use the phase space portrait and the estimation of the attractor dimension of correlation. The phase space is built using the delays method applied to the one variable space. Other variables are formed by translating the first one of a quantity τ in such a way that the vector $\{x(t),x(t+\tau), x(t+2\tau), ...\}$ represents the embedment space. After Takens (1981) the so derived attractor conserves the same topological properties as the canonical variables based one. One of the practical difficulties met here, relates to the determination of the delay interval suitable to the problem.

The first zero crossing of the signal autocorrelation function is obtained for an approximate value of 50 points (figure 5). The stationarity of the computed dynamic parameters is then checked for different values of τ inside the interval. Another construction algorithm for the pseudo phase space is proposed by Eckmann and Ruelle (1985). It consists in simultaneous measurements of the signal at several points of the experimental loop. This approach cannot be attempted in the case of falling films because of the sharp influence of the distance to the inflow on the evolution of the gas liquid interface.

The dimension is computed as usual with the Grassberger - Procaccia's algorithm (1983). The correlation summation is given by :

$$C(r) = \lim_{N \to \infty} \frac{1}{N^2} \sum_{ij} H[r - |\vec{X}i - \vec{X}j|]$$

or with a slight modification (Atten & alii 1984).

$$C(r) = \frac{1}{N'} \sum_{j=1}^{N'} \frac{1}{N} \sum_{i=1}^{N} H[r - | X_i - X_j |]$$

with $r \in \mathbf{R}$, $p \in \mathbf{N}$ and $| \vec{X_i} - \vec{X_j} | = \sup(| X_i - X_j | ,..., | X_{i+p\tau} - X_{j+p\tau} |)$ the sup norm.

For N' between 50 and 250, relation (III-2) converges rapidly to the general definition. Index j of the osculating spheres is randomly generated inside an interval not exceeding 75% of the mean diameter of the attractor. The dimension reflects the rate of growth of C(r) versus the length scale r, that is to say : $C(r) \alpha\, r^d$. The calculations are repeated for increasing values of p till d becomes independent of p. For a non random process this happens when $p > 2m + 1$, the dimension of the embedment space m being the first integer number greater than d. In what follows, we take N'=100 and r is varied exponentially with 100 steps.

PHASE SPACE PORTRAIT

Maps in figures 5a,b,c (j_g=.32) and 5d,e,f (j_g=.44) represent 16384 points (i.e. 32s) in a three dimension space for t = 30 points. The complete signal for the two superficial velocities appears in (a & d). Large and small scales are represented respectively in (b &e) and (c & f). The contrast between these plots leads to the following remarks which stay valid for gas flow rates less or equal to the flooding Jg :

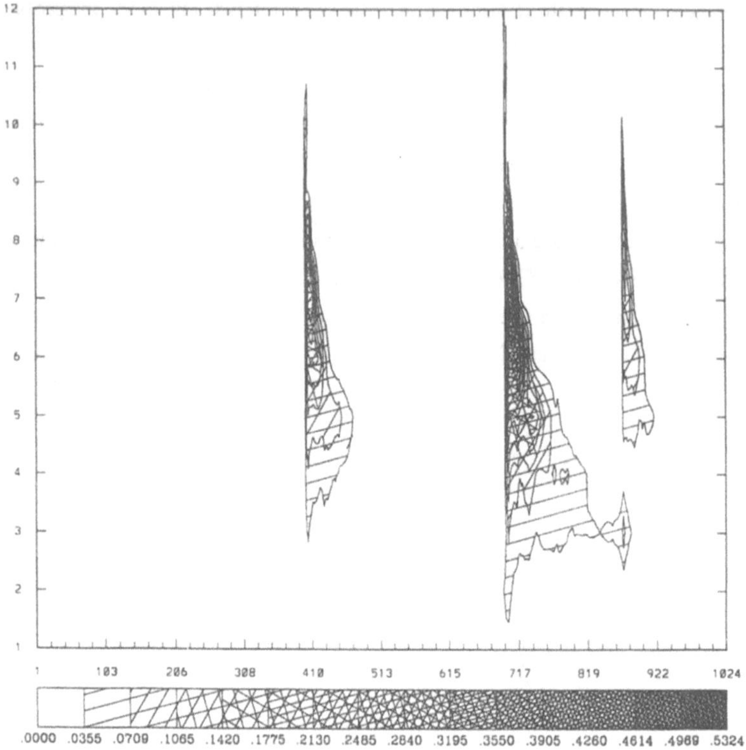

Figure 2a (see caption below)

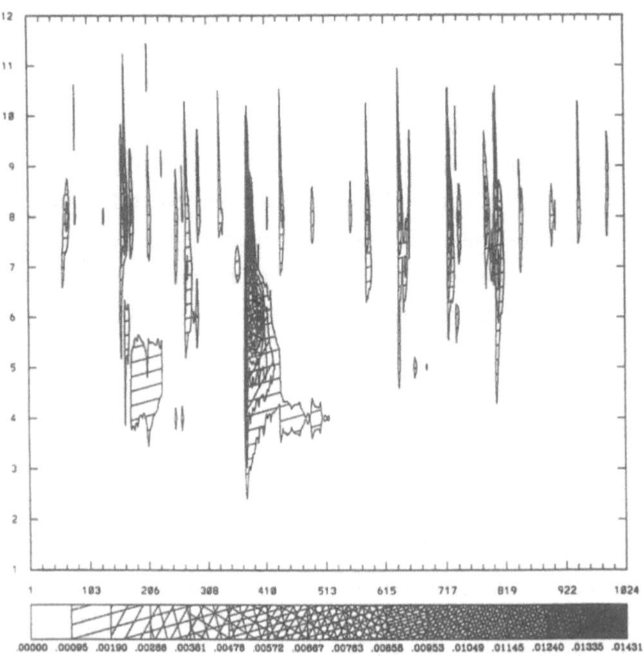

(b)

Figure 2 . $|C|^2$ for j_g=0.32m/s (a) and j_g=0.44m/s (b).

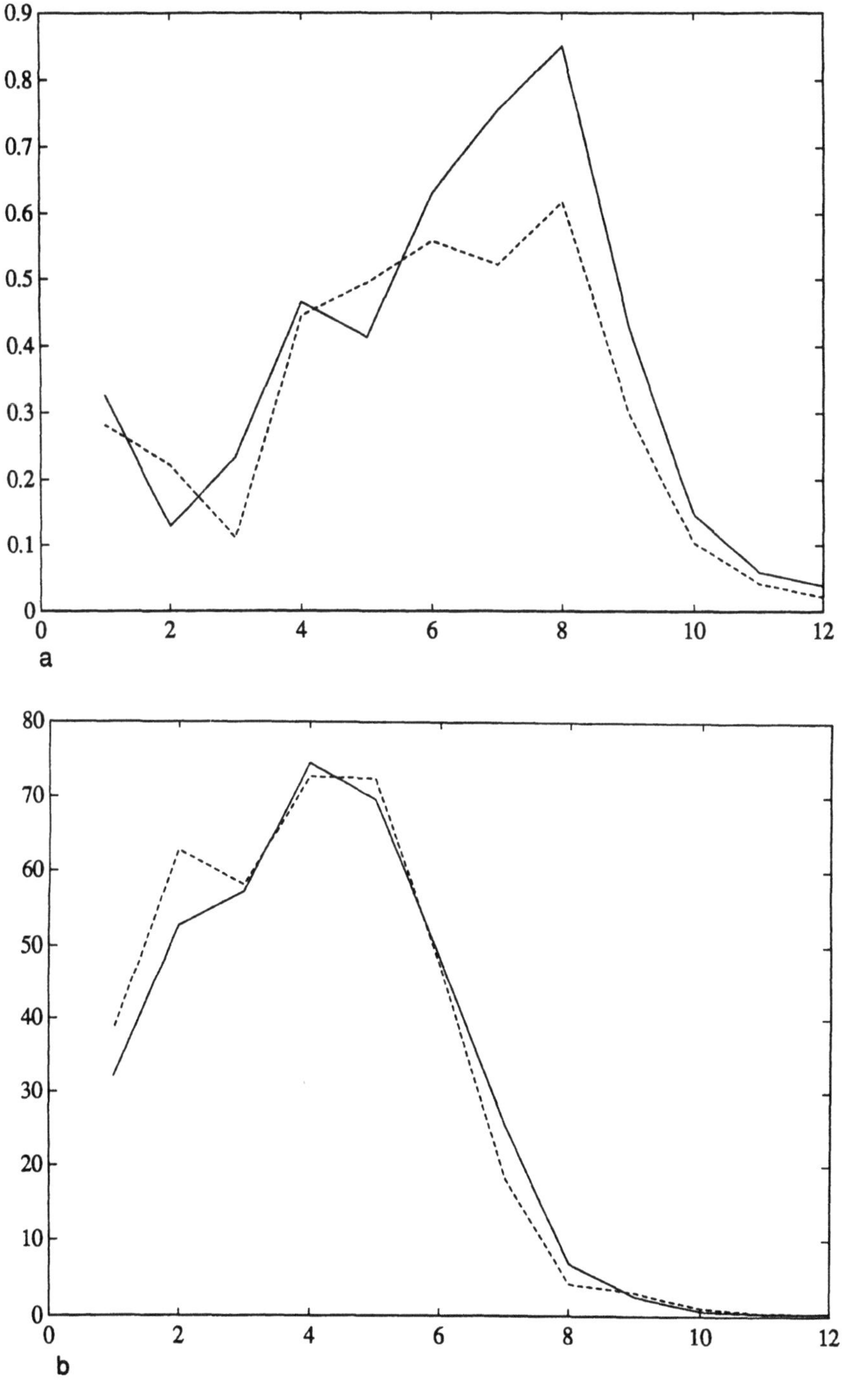

Figure 3 . distribution of the time average of the energy versus the index scales j. j_g=0.32m/s (a) and j_g=0.44m/s (b).

- the first reconstruction of the signal from scales 1 to 7 presents well organized quasi cycles around the zero mean value while the second from 8 to 12 is a constellation of points and is typical of strongly random processes with high number of degrees of freedom. The behavior breaking off between scales 7 and 8 reveals that the structure of the interface outcomes from the superposition of two distinct system dynamics.
- in the phase portrait of the whole signal the most random components hide the phenomena governed by a small number of degrees of freedom. It is probably a general property, knowing the strong effect of noise on the measured dimension of correlation (Atten et al, 1984). Furthermore the energies of the scales 7 and 8 are nearly equal as shown in figure 3.
- increasing of the gas flow rate up to the flooding point do not modify greatly the topology of the two compounds of the signal, in particular the rupture of the dynamics between the scales 7 and 8.

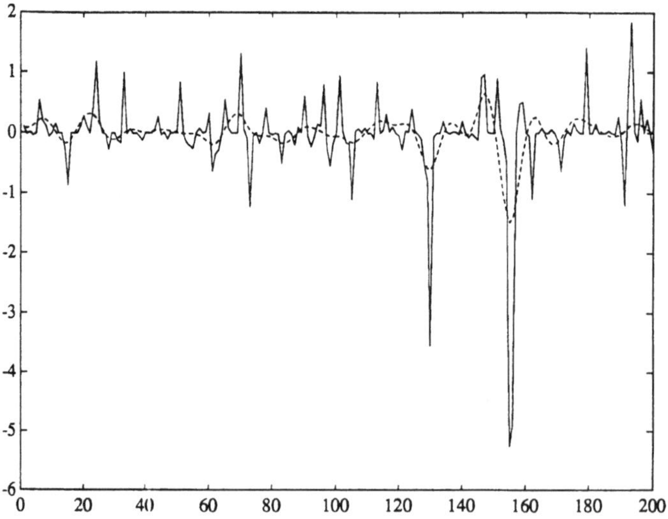

Figure 4 . Fitting of the signal with the first 7 components (zoom on the first 200 points).

THE CORRELATION DIMENSION

The preceding results are confirmed and indicated by the measure of the attractor dimension of every component and of the complete signal .The evolution of the correlation integrals and the increment of the embeding space dimension with a single flow rate are given as illustration (figure 6). The figure 7 combines the dimension measures of the whole attractor and of its components for $j_g = 0.32$ m/s.

The correct increment saturation of the correlation integral at the value of $p = 2m+1$ speaks for the existence of a deterministic chaos for the attractor, which is based on the complete signal as well as for those, which are based on its two components. The relatively weak

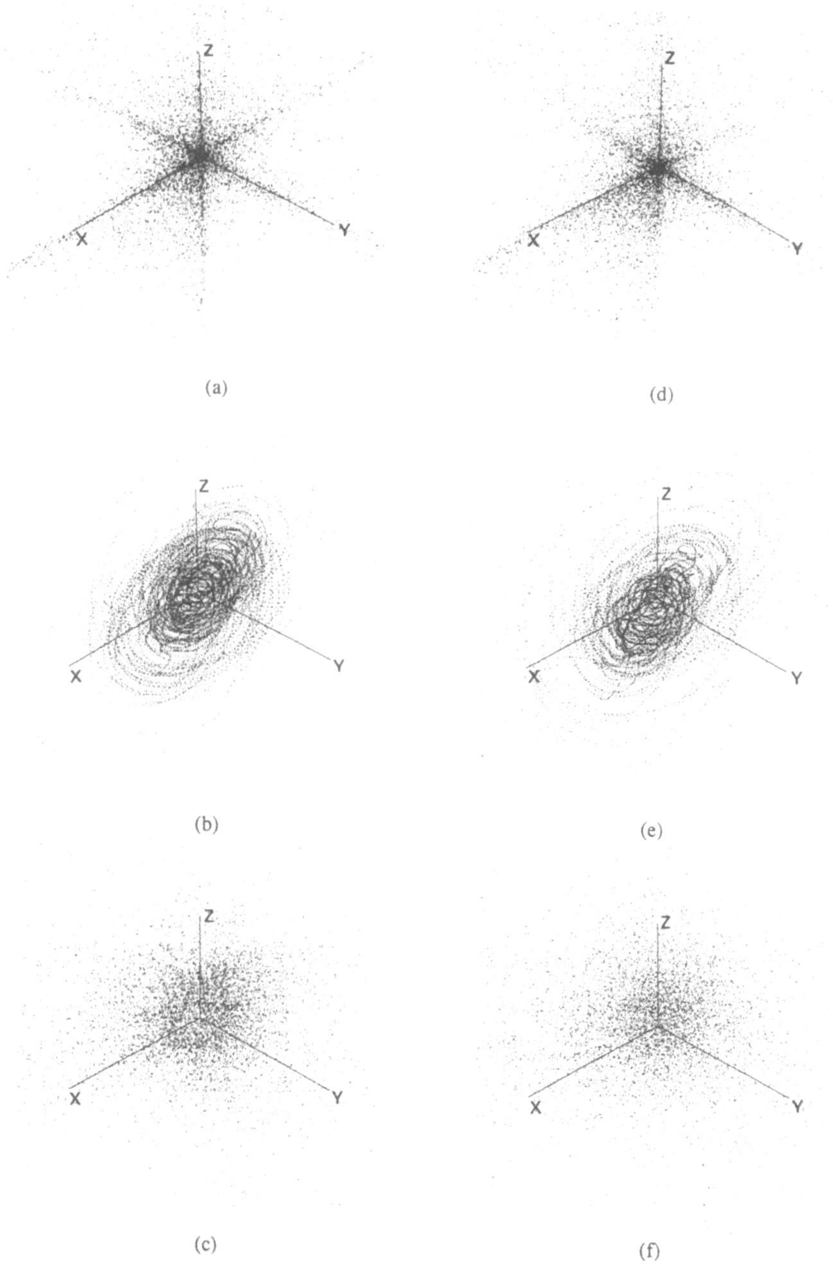

Figure 5 . Phase portraits :
- whole signals (a) ; large (b), small (c) scales components for j_g=0.32m/s.
- whole signals (d) ; large (e), small (f) scales components for j_g=0.44m/s.

dimension, order of magnitude 7, obtained for the complete signal, is unexpected compared with the values, which are higher than 10 and measured for a vertical turbulent plane film (Biage et al, 1989). The reason why might be sought in the increment of the spatial confinement induced by annular geometry. But the most important result is the evidence of the signal partition in two components, the one conforms to high scales of very weak dimension, order of magnitude 3, and the other to small scales, whose dimension is equal to the one of the complete signal.The high scale component represents the signal energy and much more corresponds to the waves of large amplitudes, which are induced by the interaction between the liquid and the gas flow.

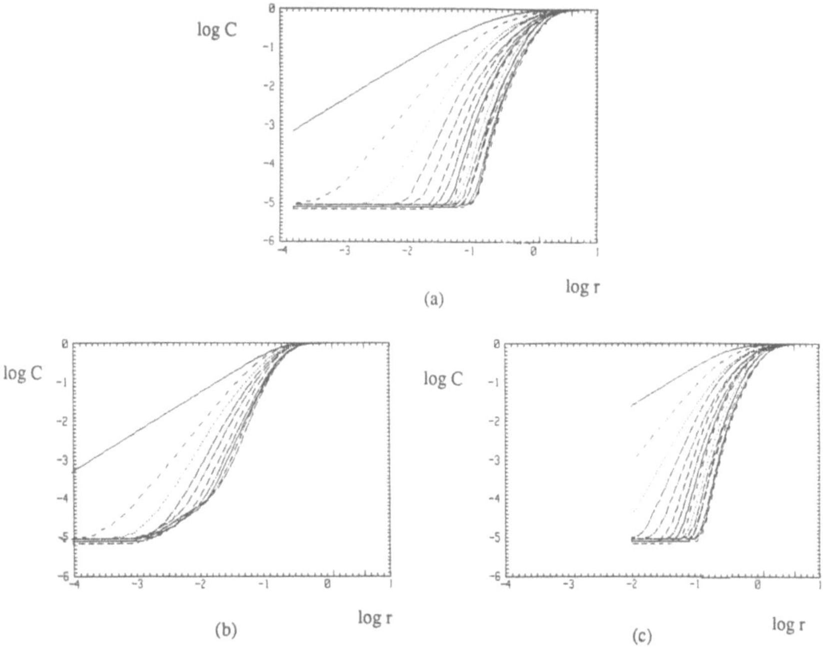

Figure 6 . Correlation function: a) whole signal, b) large scales, c) small scales for jg=0.32 m/s.

These large structures lead to the apparition of one or several solitary waves, when you reach the flooding point (Izrar and Indarto, 1991).

We therefore make the following conjecture : the big waves, which are related to the film unstabilisation are governed by a reduced set of three differential equations at most. The small scales component probably includes the contribution of the substrata waves and the hydrodynamic turbulence of the film ; both contributions weakly depend on the liquid- gas interaction. The analysis of the results for the flow at the flooding point (jg = 0.42 m/s) leads to the same conclusions; there is still a small decrease in the attractor dimension related to small scales. This persistence was not predictible from the spectral energy density which suffers a sudden and global translation towards the low frequencies at the flooding point.

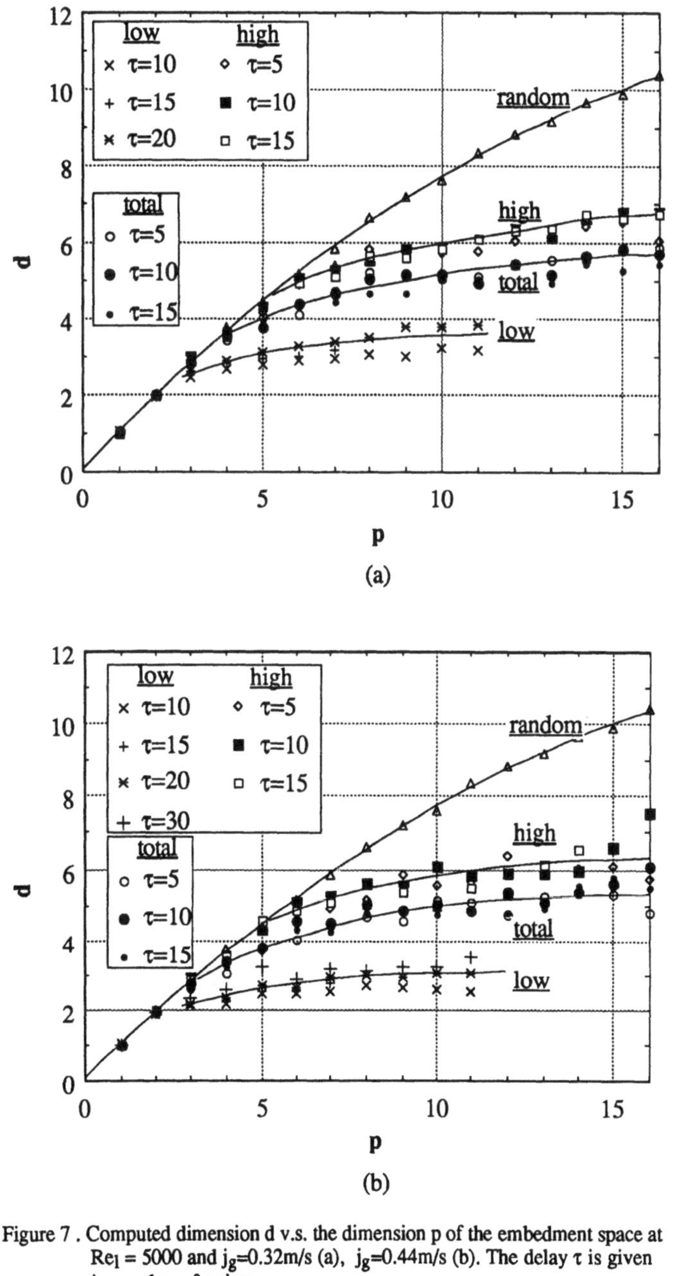

Figure 7 . Computed dimension d v.s. the dimension p of the embedment space at Re_l = 5000 and j_g=0.32m/s (a), j_g=0.44m/s (b). The delay τ is given in number of points.

CONCLUSION

We introduced a method of analysis that completes the traditional techniques . The analysis of the averaged energy in time by scale is in itself unable to split the signal . The dynamics of each component demonstrates completely different behaviors and enabled us to isolate a maximal part in the meaning of the signal analysis with a very small dimension. Moreover, the extraction of pertinent scales provides a way to eliminate the noise from experimental data, which constitute a strong limitation for the use of the methods of deterministic chaos.

REFERENCES

Atten P., Caputo J.G., Malraison B., Gagne Y., 1984.
Détermination de Dimension d'Attracteurs pour Différents Ecoulements. *Journal de Mécanique Théorique et Appliquée* , n° spécial 133.

Biage M., Delhaye J.M., Vernier P.H., 1989.
The Flooding Transition : a Detailed Experimental Investigation of the Liquid Film Flow Before the Flooding Point. *National Heat Transfer Conference, ANS Proceedings* , ANS 53-60.

Eckmann J.P. and Ruelle D., 1985.
Ergodic theory of chaos and strange attractors. Rev. Mod. Phys. 57, 617.

Farge M., 1992.
The Wavelet Transform and its Application to Turbulence, to appear in "Annual Review of Fluids Mechanics".

Feix M.R., Mourgues J., Bertrand P., Izrar B., 1983.
Lissage des fonctions de Wigner-Ville, *Colloque sur le signal et ses applications*, Vol.1, pp.57-62, Nice.

Goupillaud A., Grossmann A. and Morlet J., 1984-1985.
4° Cycle-Octave and related Transforms in Seismic Signal Analysis", *Geoexploration*, Vol.23, p.85-102.

Grossmann A., Morlet J., 1984.
Decomposition of Hardy Funcitons into Square Integrable Wavelets of Constant Shape", *SIAM J. Mater. Anal.*, Vol.15, n°4, p.723.

Indarto I. 1991.
Transition Contre-Courant Co-Courant d'un Ecoulement de Gaz et d'un Film Liquide en Conduite Verticale, *Thèse INPL* Nancy, Juin.

Izrar B., 1986.
Modèles classiques et quantiques pour la simulation dans l'espace des phases, Thèse d'Etat, Nancy.

Izrar B., Indarto I., and Cognet G., 1991.
Caractérisation d'une onde solitaire en frottement dans le flooding, *10ème Congrès Français de Mécanique*, Vol. 2, p.37, 2-6 Septembre 1991, Paris.

Lacy C.E., Sheintuch M. and Dukler A. E., 1991.
Methods of deterministic chaos applied to the flow of thin wavy films, AIChE Journal, vol 37, n° 4, p481.

Lemarié P.G., 1990.
Analyse multi-échelles et ondelettes à suppot compact, in "les Ondelettes en 1989", Springer Verlag, Vol.1438.

Lusseyran F. , 1987.
Portraits de phases en écoulements gaz)liquide ascendants verticaux. EuroMech 234 (Toulouse).

Lacy C.E., Sheintuch M. and Dukler A.E., 1991.
Methods of deterministic chaos applied to the flow of thin wavy films. AIChE J. vol 37, N°4, pp481-489.

Meyer Y. , 1990.
Ondelettes et opérateurs, Actualites Mathématiques, Hermann ed., Paris.
Melville W.K., 1983.
Wave modulation and breadown, *J. Fluid Mech.*, Vol.128, pp.489-506.
Grassberger, P. and Proccaccia, I., 1983.
Measearing the strangeness of strange attractors, Physica 9D, p189.
Takens F., 1981.
Detecting Strange Attractors in Turbulence. *Symp.Dynamical Systems and Turbulence,Rand D.A. et Young LS., Eds, Lecture, Notes in Mathématics, Springer-Verlag*, Vol.898, pp.366-381.
Ville J., 1948.
Théorie et application de la notion de signal analytique, *Câbles et Transmissions*, Vol.1, n°2A, p.61.
Wigner E.P., 1932.
On the quantum correction for thermodynamic equilibrium, *Physical Review*, Vol.40, p.749.

EXPERIMENTAL OBSERVATION OF TYPE-I AND TYPE-II INTERMITTENCIES

Eric Ringuet, Claude Roze, and Gérard Gouesbet

Laboratoire d'Energétique des systèmes et procédés
URA CNRS n° 230, INSA de Rouen, CORIA
BP 08, 76131 Mont Saint Aigan Cedex

INTRODUCTION

Intermittencies have been extensively studied since 1980 when Pomeau and Manneville[1] found that kind of behaviour in the Lorenz model where a limit cycle becomes unstable near R = 166. They proposed three types of intermittencies related to the destabilization of a limit cycle. The first and the third ones are known both theoretically and experimentally[2-4]. However, if the type-II intermittency has been found theoretically[5], on the other hand the only experimental report was found in a coupled nonlinear oscillator circuit[6].

In this paper, we present two different intermittent behaviours. The first one is analysed as being type-I intermittency and the second one seems to be a good candidate for the first observation of type-II intermittency in a hydrodynamic system.

EXPERIMENTAL SETUP

These results have been found in so-called hot-wire experiments with a new experimental setup. Modifications with respect to Rozé[7] concern the length of the wire (now 6 cm instead of 3) and the fact that the wire is current controlled instead of being temperature controlled. The wire diameter is equal to 20 μm and the temperature coefficient of the material (platinum) is $\alpha = 3.9\ 10^{-3}\ K^{-1}$. The wire resistance can be written as : $R(T) = R_0 (1 + \alpha (T - T_0))$ in which R_0 ($\simeq 18\ \Omega$) is the wire resistance at the reference temperature (T_0 = 25 °C) and T is the spatially averaged wire temperature. The wire is placed in a silicon oil (Prandtl number = 130) which is contained in a tank (17*12*10 cm^3). Tank walls are kept at a constant temperature (T_0) by using thermostated water. The tank is in a thermostated box which keeps the ambient temperature at T_0 with a precision of 0.01 K. There are two control parameters in these experiments. The first one is the distance (d) from the wire to the free surface (with a precision of 20 μm). The second one is a constant current (I) supplying the wire by means of a perfect current source driven by a micro-computer. I is controlled with high accuracy and can be adjusted by steps of 0.3 mA.

The studied signal is the wire voltage which is related to the spatially averaged temperature of the wire (using Ohm's law), simply called "temperature" in this paper. When the free surface is motionless,

Instabilities in Multiphase Flows, Edited by G. Gouesbet
and A. Berlemont, Plenum Press, New York, 1993

the temperature is constant. When oscillatory instability occurs, a temperature gradient appears moving along the wire; in this situation, the temporal signal shows oscillatory variations which are studied by a discrete FFT program on a sequence of 8192 sampled values, and also by other tools to be described. Signals are recorded and digitalized with three adjustable parameters: sampling frequency (from 10 Hz to 300), offset (up to 10 V) and amplification ratio (from 1 to 32). Up to 160,000 points (on 14 bits) may be recorded and the cut-off frequency of the time signal is about 100 Hz.

TYPE-I INTERMITTENCY

First example

In order to characterize type-I intermittency in our experiment, we study two consecutive time series of 160,000 sampled values. The time

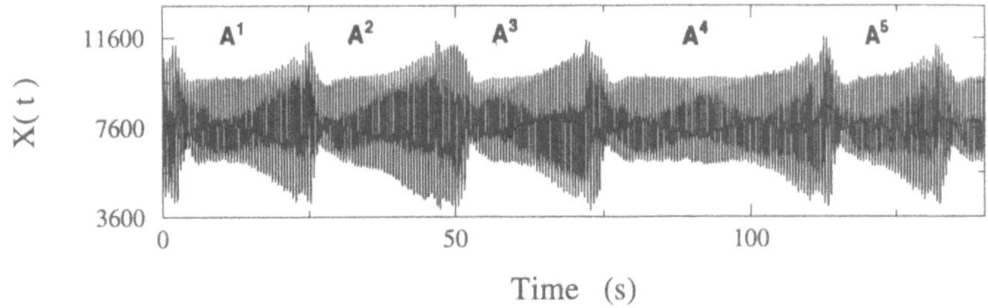

Figure 1. Time signal showing 5 laminar periods.

signal (Fig. 1) is composed of fundamental patterns (A^i, which are called laminar periods): the length of each one have been measured and one can see the probability distribution (number of laminar periods for a given length) where 164 laminar periods are considered (Fig. 2). That sort of histogram is one of the characteristics of type-I intermittency[2] because there are no very long lengths (that is to say very long in comparison with l_{max} = 20 s which is the most probable value) and there are few short lengths (that is to say less than l_{max}). One can also study the statistical features with the number of laminar periods which are longer than a certain length (Fig. 3).

Such features are well illustrated by theoretical results[2] and experimental data in the Belousov-Zhabotinsky reaction[8], in agreement with our own results in figures 2, 3.

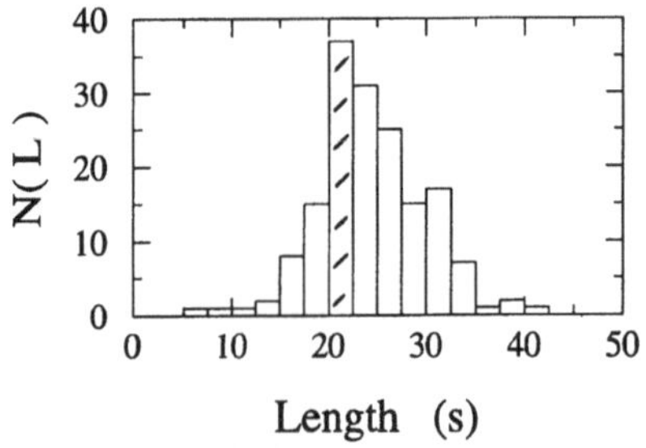

Figure 2. Probability distribution for the laminar periods.

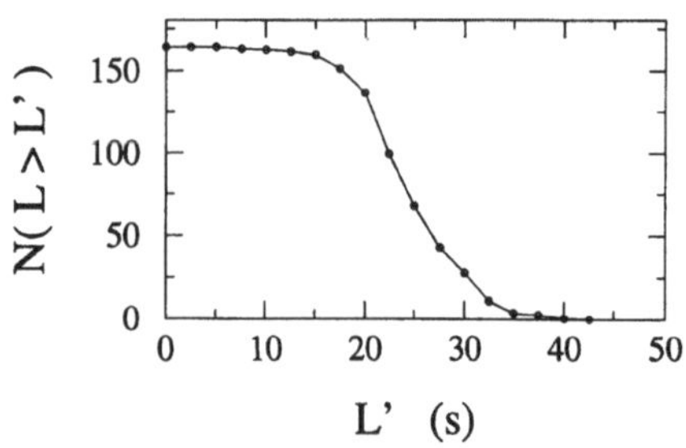

Figure 3. Number of laminar periods such that $L \geq L'$.

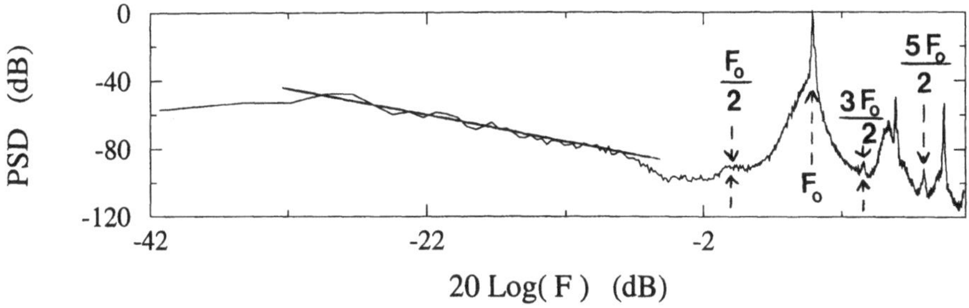

Figure 4. Averaged power spectrum (log-log scales).

Moreover, these rather arbitrarily lengths are pointed out by the average power spectrum (Fig. 4). 38 power spectra have been computed; then, the power spectral density is added for each frequency and the result is divided by 38. This average spectrum exhibits a power law behaviour for the small frequencies over one decade. We get $PSD(F) \propto F^a$ in which $a = -1.65 \pm 0.04$. Notice that this divergence is not available for the smallest frequencies, as it is pointed out by Ben-Mizrachi et al.[9]. The presence of peaks near $F_0/2$, $3F_0/2$ and $5F_0/2$ which was not expected may be explained as below.

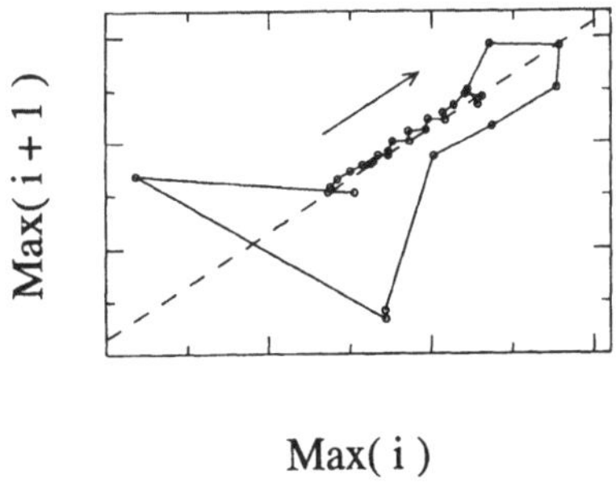

Figure 5. Projection of the first map.

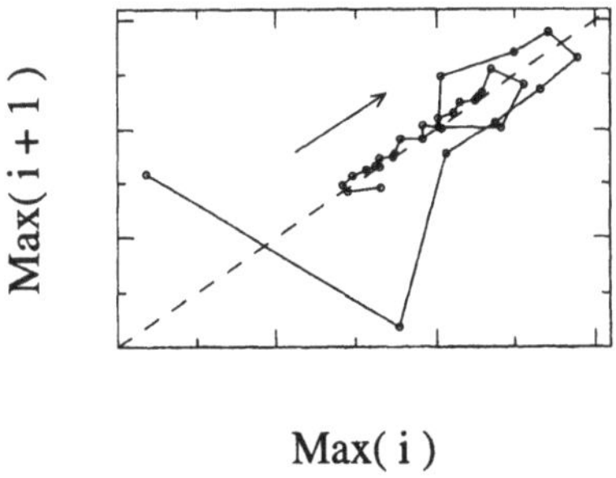

Figure 6. A second one.

Indeed, another characteristic point of type-I intermittency can be seen in figure 5, where a projection of the first return map is drawn. One can see a slow drift near the bissecting line (dashed). Another more complicated example (often seen) is drawn in figure 6. During the slow drift near the bissecting line (Fig. 5, 6), small oscillations are present with a frequency which is equal to $F_0/2$. This fact may explain the unexpected peaks in the power spectrum (Fig. 4).

Finally, the correlation dimension of the reconstructed attractor[10] is equal to : $D_2 = 2.3 \pm 0.1$. The temperature was such that $128 \leq T(t) \leq 142$ °C (average value: 134 °C).

Second example

d is still equal to 0.7 mm but I is increasing up to 320 mA. The time series (Fig. 7) shows that the laminar periods are shorter (about 12 s for the most probable length) and more complicated. This latter fact may be pointed out by a larger correlation dimension: $D_2 = 2.9 \pm 0.2$.

12 power spectra have been averaged, the result gives us a power law behaviour for the small frequencies which is not so well defined. We get $PSD(F) \propto F^a$ with $a = -1.86 \pm 0.07$ for less than one decade.

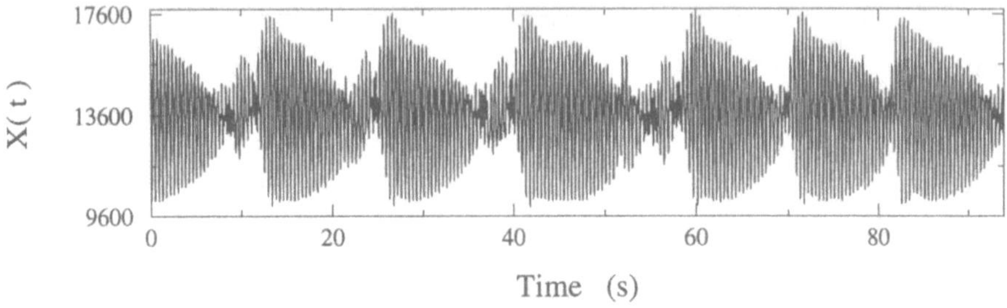

Figure 7. Time signal of the second example.

Discussion

Type-I intermittency has been observed following different procedures, for example: I is constant and there is a limit cycle, then d is increased by steps of 20 μm or, starting with d larger, d is decreased by steps of 0.1 mm. It is not easy to reproduce the same sequence because it seems to be dependent on initial conditions.

In a phenomenological point of view, the intermittent behaviour can be illustrated with two figures:

- the first one shows that oscillations are born at one edge (**A**) of the wire and travel toward the other edge (**B**).

A → → → → → → → → **B**

- the second one shows a "well" near **B**, that is to say oscillations are also born near this edge.

A → → → → → → ← ← **B**

A laminar period begins in the first situation and then, a "well" appears near **B** and moves slowly toward **A** before coming back to **B**.

The wire is adjusted horizontally by optical means when d = 0 mm and then, an angle α between the wire and the free surface can be defined: if $\alpha = 0$ the wire is horizontal. Type-I intermittency has been observed with $\alpha = 0$ and $0 < \alpha < 0.002$ rad. Then we are not able to appreciate the role of α (it is important or not) in these situations.

TYPE-II INTERMITTENCY

Very different behaviours are presented in the following. In these situations two fundamental frequencies are present. The signals have been obtained with I = 280 mA decreasing d from 0.7 mm to 0.62 mm by steps of 20 μm. For d = 0.7 mm the signal was periodic, for d= 0.68 it is no longer the case (Fig. 8).

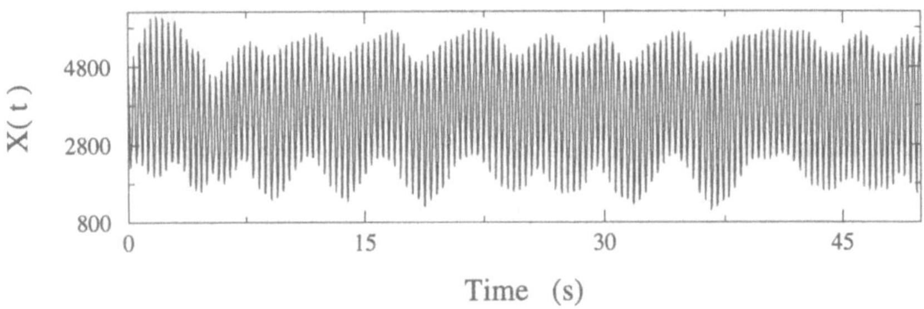

Figure 8. Time signal for d = 0.68 mm.

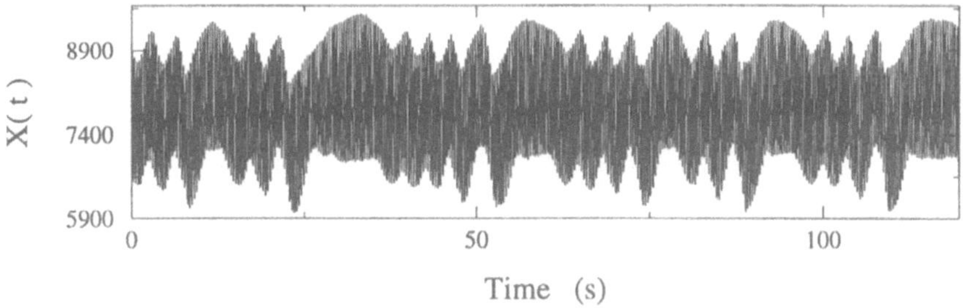

Figure 9. Time signal for d = 0.66 mm.

First example

This time series was obtained for d = 0.66 mm; the temperature of the wire is defined by: $105 \leq T(t) \leq 109$ °C.
The time signal (Fig. 9) can be divided in four laminar periods studying the maxima (Fig. 10). Two fundamental frequencies are present: a first one (near 3 Hz) and a second one (near 0.2 Hz) which appears intermittently with a growing amplitude during the laminar period. The probability distribution for the lengths is of poor interest because only fifty laminar periods have been detected.

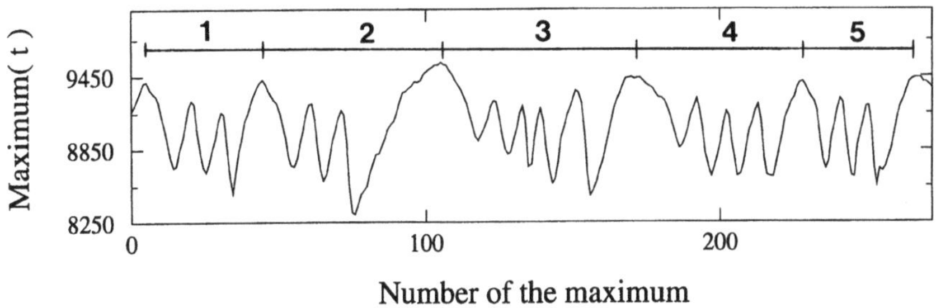

Figure 10. Time evolution of the maximum showing 5 laminar periods.

Twelve power spectra have been computed, the average is drawn in figure 11. The dashed line corresponds to a slope which is equal to -4. A power law behaviour seems to be hidden by the presence of the second frequency. The computation of the correlation dimension yields to: $D_2 = 3.3 \pm 0.3$.

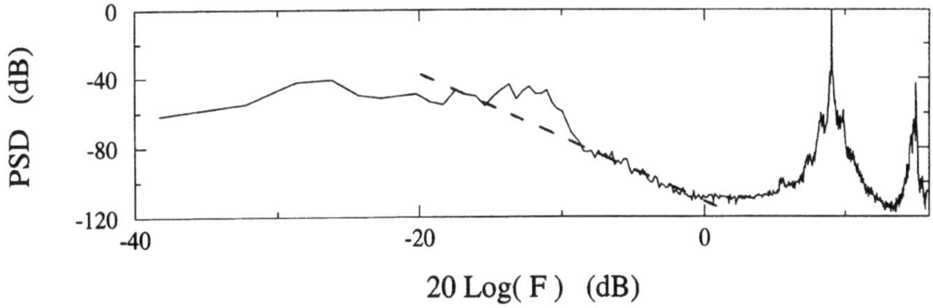

Figure 11. Averaged power spectrum for the first example (log-log scales).

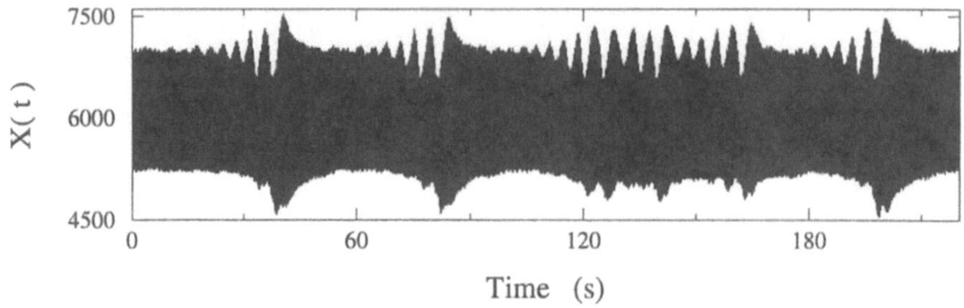

Figure 12. Long part of the time signal for the second example.

Second example

In this case, 120,000 data points were recorded for d = 0.64 mm (sampling frequency equals 40 Hz). The time signal (Fig. 12) shows a limit cycle which is unstable because of the intermittent growing influence of a second frequency. In order to detect the laminar periods, we consider the maxima of the time series. The time evolution of the maximum is illustrated in figure 13, with 8 well defined laminar periods.

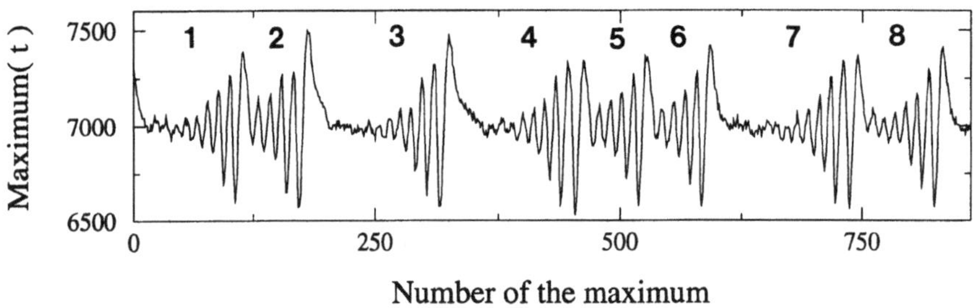

Figure 13. The evolution of the maximum shows 8 laminar periods.

The statistical features (Fig. 14, 15) are characterized by a most probable length which is about 12 s and some long lengths (up to 50 s). Richetti et al.[5] met a numerical situation where short laminar periods are present, but long ones are previleged. Because of a small slow drift and external noise, the shortest laminar periods (length $\leq$ 7 s) are sometimes difficult to define and one may make errors. However, the general aspect of figures 14, 15 may be unchanged in spite of these errors.

Fourteen power spectra have been averaged. It is not possible to point out a power law behaviour for the small frequencies (Fig. 16).

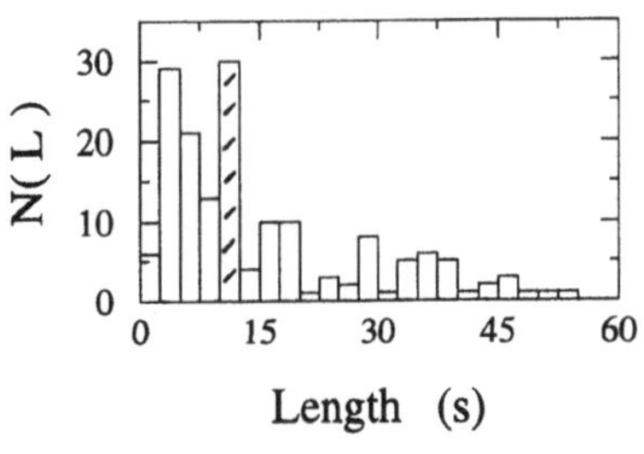

Figure 14. Probability distribution for the laminar periods.

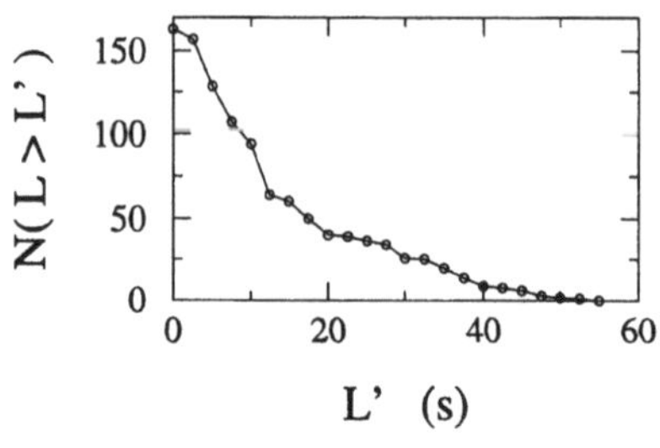

Figure 15. Number of laminar periods such that $L \geq L'$.

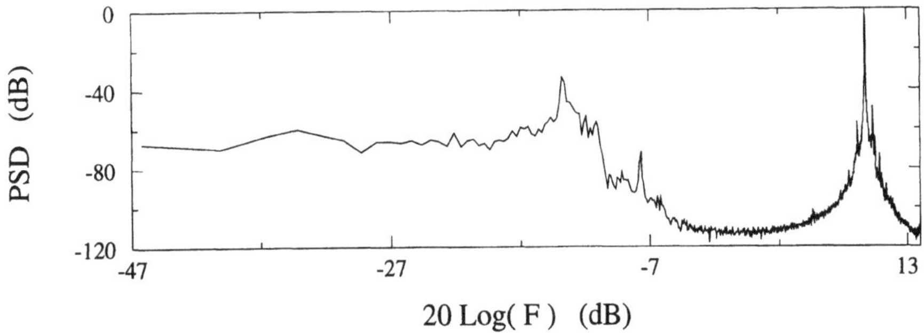

Figure 16. Averaged power spectrum (log-log scales).

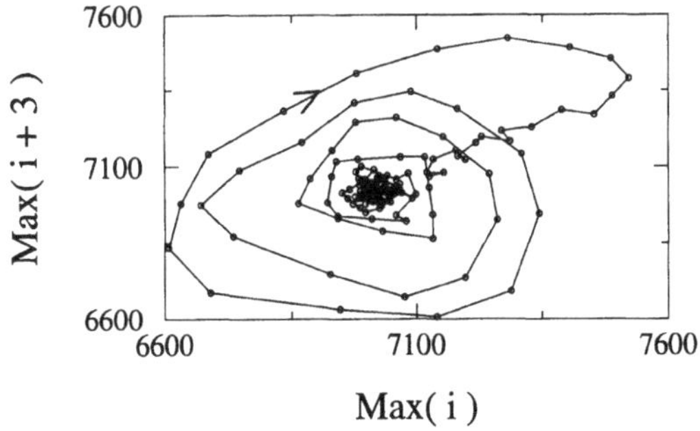

Figure 17. Projection of the third return map.

Considering the maxima of the time series, one can draw a projection of the third return map for a long laminar period. The result (Fig. 17) shows an unstable fixed point in the middle of the figure. Moreover, the correlation dimension is found to be $D_2 = 3.1 \pm 0.2$, that is to say four degrees of freedom may be involved. This result is in agreement with the projection of the map (Fig. 17) which needs three degrees of freedom. The fourth one corresponds to the fundamental frequency.

Up to now, this intermittency has not been found with $\alpha = 0$. Moreover, we do not know where does come from the second frequency.

The last time series (Fig. 18) corresponds to d = 0.62 mm. The power spectrum can be analysed as a quasi-periodic signal with two frequencies. It is confirmed by the correlation dimension: $D_2 = 2.2 \pm 0.3$.

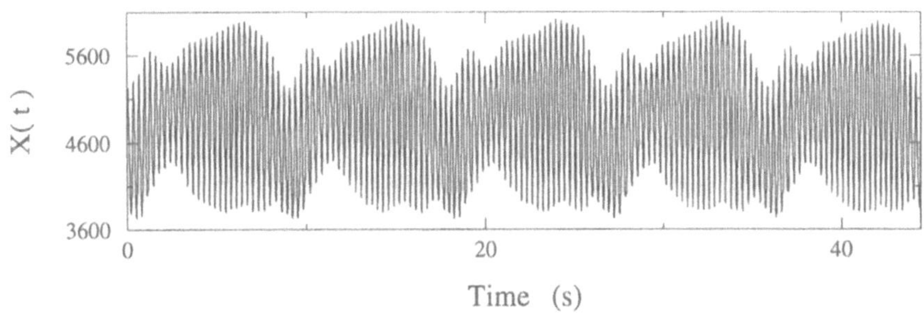

Figure 18. Time signal obtained with d = 0.62 mm.

CONCLUSION

Two transitions from a limit cycle to a chaotic state have been presented; the first one leads to type-I intermittency and the second one to type-II intermittency. Both time series are analysed by different tools (average power spectrum, map, probability distribution for the laminar periods, correlation dimension). Therefore our conclusions seem to be unambiguous.

However, the time series give us a spatially averaged information on the physical system. Then, it is not obvious to say if we are studying temporal chaos or spatio-temporal chaos even though the correlation dimensions may be larger in the case of spatio-temporal chaos.

ACKNOWLEGEMENTS

The authors thank Y. Pomeau and in particular P. Manneville for giving their opinion and some critical remarks. The authors thank P. Manneville and H. B. Stewart for pointing out references 5 and 6. They are also very grateful to M. Dubois for lighting discussions.

REFERENCES

1. Y. Pomeau and P. Manneville, "Intermittent transition to turbulence in dissipative dynamical systems", Commun. Math. Phy. 74, 189-197, 1980.
2. P. Bergé, Y. Pomeau and C. Vidal, "L'ordre dans le chaos", Hermann, Paris, 1984.
3. M. Dubois, M.A. Rubio and P. Bergé, "Experimental evidence of intermittencies associated with a subharmonic bifurcation", Phys. Rev. Letters 51, 1446, 1983.983.
4. M. A. Rubio, M. De La Torre and J. C. Antoranz, "Intermittencies and power-law low-frequency divergencies in a nonlinear oscillator", Physica D 36, 92-108,1989.989.
5. P. Richetti, F. Argoul and A. Arneodo, "Type-II intermittency in a periodically driven nonlinear oscillator", Phys. Rev. A 34, Nb 1, 1986.
6. J. Y. Huang and J. J. Kim, "Type-II intermittency in a coupled nonlinear oscillator: experimental observation", Phys. Rev. A 36 Nb 3, 1987.
7. C. Rozé, G. Gouesbet and R. Darrigo, "New investigations of oscillatory instabilities produced by temperature-controlled hot-wire heating below an interface", Proceedings of the International Conference on Multiphase Flows, September 24-27, 1991, Tsukuba, Japan.
8. Y. Pomeau, J. C. Roux,A. Rossi, S. Bachelart and C. Vidal, J. Phys. (Paris) Lett. 42, L241, 1981.
9. A. Ben-Mizrachi, I. Procaccia, N. Rosenberg, A. Schmidt and H. G. Schuster, "Real and apparent divergencies in low-frequency spectra of nonlinear dynamical systems", Phys. Rev. A 31, Nb 3, p 1830, 1985.
10. E. Ringuet, "Computation of the correlation dimension of nine different experimental attractors coming from the hot wire experiment", Internal Report ESP/ER/03/92/II.

STABILITY OF STATIONARY REGIME OF DIRECTED CRYSTALLIZATION

A.P. Gus'kov

Institute of Solid State Physics
Academy of Sciences of the USSR
Moscow District, Chernogolovka
142432, Russia

INTRODUCTION

The structure of polycrystalline materials, according to a number of publications is associated with the auto - oscillating regime of melt crystallization, as shown by Samoilivich(1979), Buevich(1988), Gus'kov(1989). Many signs of spatial periodicity, such as longtitudinal and latitudinal laminations, of distances between dendrite branches and periodical impurity distribution change, as shown by Samoilovich (1979), indicate such the regime. The probability of auto-oscillating solutions of the problem, including the equations of thermal conductivity and diffusion, as shown by Gus'kov(1989). Below is a summary of the further analysis of auto-oscillating regimes of directed crystallization. A numerical calculation of a growth decrement displayed that auto-oscillations occur at thedefinite values of governing parameters. At this various types of auto-waves of component concentration on a phase boundary can be observed.

Gus'kov(1989) presents the qualitative elucidation of a probable cause of auto-oscillation occurrences. However it cannot be satisfactory, since the numerical calculations de-

Instabilities in Multiphase Flows, Edited by G. Gouesbet
and A. Berlemont, Plenum Press, New York, 1993

monstrated the fact that the auto-oscillating regime can be manifested at such values of the governing parameters at which, according as shown by Gus'kov(1989), the regime of crystallization is stable. Herein another mechanism of auto-oscillations' occurrence is proposed. Numerical estimations of a temporal pulsation period are presented for this mechanism. The analysis of a system behavior at an asymtotic change of the governing parameters confirms validity of the mechanism of the auto-oscillating regime occurrence.

FORMULATION OF THE PROBLEM

Let's introduce the specifications: $u(y,z,\tau)$ is a temperature normalized to the temperature of a phase transition at an initial concentration of impurity;$c(y,z,\tau)$ is the impurity concentration, normalized to the initial one; y,z,τ are the dimensionless coordinates and time: $y=\alpha y_r$, $z=\alpha z_r$, $\tau=\alpha^2\chi_o\tau_r$; D- is the dimensionless diffusion coefficient in melt, $D=D_r/\chi_o$; $\kappa=\kappa_r/\chi_o$ is the coefficient of thermal conductivity, ε is the heat of a phase transition normalized to specific heat capacity and a phase transition temperature. $y_r, z_r, \tau_r, D_r, \kappa_r$ are dimension values,$\chi_o=10^{-5}м^2c^{-1}$, $\alpha_o= 10^2$ $м^{-1}$.

Let the cooling conditions be such that the plane front of crystallization moves with a constant velocity V_s in the stationary mode. The initial temperature of melt is $U_{oo}>1$,at a distance from the front the melt has a constant temperature U_o. Let's take into account thermal conductivity in the solid and liquid phases, and impurity diffusion in the liquid phase. To decrease the number of operations in equations, the x-coordinate should not be written down. The values referring to a solid phase are denoted by a dash. In the y,z,τ - coordinates, rigidly bound with the moving front, the starting equation system has the form

$$\frac{\partial u'}{\partial \tau} = \kappa' \Delta u' + V \frac{\partial u'}{\partial z}; \qquad z_0 \le z \le 0 \qquad (1)$$

$$\frac{\partial u}{\partial \tau} = \kappa\ \Delta u + V \frac{\partial u}{\partial z}; \qquad 0 \le z < \infty \qquad (2)$$

$$\frac{\partial c}{\partial \tau} = D\,\Delta c + V\frac{\partial c}{\partial z}; \qquad 0 \le z < \infty \qquad (3)$$

$$\kappa\frac{\partial u}{\partial z}\Big|_{z=+0} - \kappa'\frac{\partial u'}{\partial z}\Big|_{z=-0} = -\,\varepsilon V \qquad (4)$$

$$\frac{\partial c}{\partial z}\Big|_{z=0} = V(1-k)c\,; \qquad c(\infty)=1. \qquad (5)$$

$$U'(z_0) = U_0; \quad U(\infty) = U_{00}; \quad U(0) = U'(0); \quad c\,(\infty) = 1; \qquad (6)$$

$$V = V(u(y,0,t);\ c(y,0,t)) \qquad (7)$$

k is the equilibrium distribution coefficient.

The basic difference between (1)-(7) and problem, as shown by Gus'kov(1991), is equation (7), which generally takes into account the kinetic phenomena on the crystallization front.

We find for the stationary regime

$$\kappa'\frac{\partial^2 u'_s}{\partial z^2} + V_s\frac{\partial u'_s}{\partial z} = 0 \qquad z_0 \le z \le 0$$

$$\kappa\frac{\partial^2 u_s}{\partial z^2} + V_s\frac{\partial u_s}{\partial z} = 0 \qquad 0 \le z < \infty$$

$$D\frac{\partial^2 c_s}{\partial z^2} + V_s\frac{\partial c_s}{\partial z} = 0 \qquad 0 \le z \le \infty$$

$$\kappa\frac{\partial u_s}{\partial z}\Big|_{z=+0} - \kappa'\frac{\partial u'_s}{\partial z}\Big|_{z=-0} = -\,\varepsilon V_s$$

$$u'_s(z_0)=u_0; \quad u_s(\infty)= u_{00}; \quad u_s(0)= u'_s(0);$$

$$\frac{\partial c_s}{\partial z}\Big|_{z=0} = V_s(1-k)c_s; \qquad c_s(\infty)=1.$$

Hence

$$u_s = u_{00} + \left(\varepsilon - \frac{I_0}{V_s}\right) exp\left(-\frac{V_s}{\kappa} z\right); \quad 0 \le z < \infty;$$

$$u'_s = (u_{00} + \varepsilon) - \frac{I_0}{V_s} exp\left(-\frac{V_s}{\kappa} z\right); \quad z_0 \le z < 0;$$

where

$$I_0 = V_s(\varepsilon + u_{00} - u_0)\, exp\left(\frac{V_s}{\kappa'} z_0\right)$$

$$c_s = 1 + \frac{1-k}{k} exp\left(-\frac{V_s}{D} z\right)$$

The analysis of stability.To study the stability of stationary solutions, related to small thermal and concentration perturbations, we search for a solution in the form

$$u' = U'_s + U'_m(z) exp(w\tau + iKy); \quad (8)$$
$$u = U_s + U_m(z) exp(w\tau + iKy); \quad (9)$$
$$c = C_s + C_m(z) exp(w\tau + iKy); \quad (10)$$
$$U'_m \ll U_s; \quad U_m \ll U_s; \quad C_m \ll C_s;$$

where U_s',U_s and C_s are the solution of a stationary problem, $W=W_1+iW_2$; $K=K_1+iK_2$. Then the problem linearized over the U_m', U_m, C_m- small components has the form:

$$\kappa' \frac{\partial^2 U'_m}{\partial Z^2} + V_s \frac{\partial U'_m}{\partial Z} + \{[\kappa'(K_1^2 - K_2^2) - w_1\} - i(w_2 - 2\kappa' K_1 K_2)\} U'_m = -\frac{\partial U'_s}{\partial Z} V_m \quad (11)$$

$$\kappa \frac{\partial^2 U_m}{\partial Z^2} + V_s \frac{\partial U_m}{\partial Z} + \{[\kappa(K_1^2 - K_2^2) - w_1\} - i(w_2 - 2\kappa K_1 K_2)\} U_m = -\frac{\partial U_s}{\partial Z} V_m \quad (12)$$

$$D \frac{\partial^2 C_m}{\partial Z^2} + V_s \frac{\partial C_m}{\partial Z} + \{[D(K_1^2 - K_2^2) - w_1\} - i(w_2 - 2DK_1 K_2)\} C_m = -\frac{\partial C_s}{\partial Z} V_m \quad (13)$$

$$\kappa' \left.\frac{\partial U'_m}{\partial Z}\right|_{z=-0} - \kappa \left.\frac{\partial U_m}{\partial Z}\right|_{z=+0} = \varepsilon V_m; \quad (14)$$

$$D \frac{\partial C_m}{\partial Z} \Big|_{Z=0} = (k-1)(V_s C_m + C_s V_m) \Big|_{Z=0} \tag{15}$$

$$U'_m(-\infty)=0; \qquad U_m(\infty)=0; \qquad U_m(0)=U'_m(0) \tag{16}$$

here

$$V_m = \gamma c_{m0} + \theta u_{m0}$$

$$\gamma = \frac{\partial V}{\partial c}\Big|_{c=c_s}; \qquad \theta = \frac{\partial V}{\partial u}\Big|_{u=u_s}$$

$$c_{m0} = c_m(0); \qquad u_{m0} = u_m(0).$$

Hence, the expressions for the $C_m(Z)$, $U_m(Z)$-complex amplitudes of impurity concentration and temperature distribution are easily obtained. The derivation of a dispersion equation of an edge problem (11)-(16) is rather sophisticated. Therefore it is not presented here. Let's write it down in the general form

$$f(K,W)=0 \tag{17}$$

The following denotations are used further on:

$$Y_1 = 4D^2(K_2^2 - K_1^2)/V_s^2; \qquad Y_2 = 8D^2 K_1 K_2/V_s^2;$$

$$\Omega_1 = -4Dw_1/V_s^2; \qquad \Omega_2 = 4Dw_2/V_s^2. \tag{18}$$

$$\Phi = \frac{I_0}{\varepsilon V_s}; \qquad \mu_T = \kappa/D; \quad \mu_{T1} = \kappa'/D; \quad \mu_p = \gamma/(2\theta\varepsilon).$$

SPECTRUM OF INCREMENTS

Let's consider the dependence of a system increment on a wave number and frequency of temporal pulsations in several specific values of the governing parameters. The goal of this point is to show the fact that various types of auto-waves of a component concentration change can exist on the phase boundary partition. For the arbitrarily defined parameters the (17) equation has four independent variables: Ω_1, Ω_2, Y_1, Y_2. According to (18) the Ω_1-parameter is proportional to the actual part of a W- frequency in the (8)-(10) solution. Let's denote the value which is the increment at $W_1 \gg 0$ and decrement at $W_1 < 0$ as $\delta = -\Omega_1$. Below is the hypothesis for predominant development of a perturbation mode with the maximal increment to determine the frequency of auto-oscillations. This permits the suppositions on the character of non-stationary distribution of component concentration. Ω_2 is the frequency of temporal pulsations, Y_1 is the frequency of spatial distortions. Y_2 determines a spatial change of perturbation amplitudes. Since the surface of a phase partition in the considered problem is infinite, we assume $Y_1 = 0$ to obtain a physically observed result. Otherwise we get an infinite amplitude at $Y \Rightarrow \pm \infty$. Let's fix the values of governing parameters as:

$$\mu_p = 1; \quad \mu_T = 1,5 \cdot 10^4; \quad \mu_{T1} = 10^4; \quad k = 0,9; \quad p = -5 \cdot 10^{-3}. \tag{19}$$

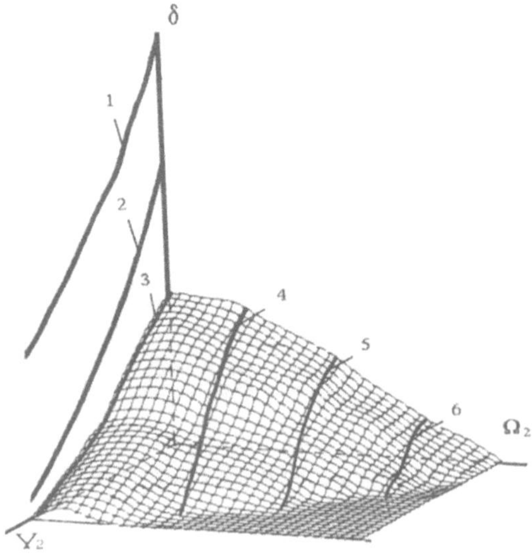

Figure 1. The dependence of the growth increment of the minory component concentration $\delta(Y_2, \Omega_2)$.

Fig.1 presents the dependence of the δ-increment of a component concentration on the frequency of temporal Ω_2- pulsations and wave Y_2-number. The 1-6 curves correspond to various values of the thermal flux Φ_1, passing through the crystallization front: $\Phi_1<\Phi_2<..<\Phi_6$. All the curves decrease monotonically, the wave number being increased. According to the hypothesis of predominant development of a perturbation mode with the maximal increment, the solidification regimes with $Y_1=0$, that is with a plane interphase boundary, correspond to the curves in Fig.1. Along the 1, 2, 3 curves the pulsation frequency is zero, it means that this is the regime with the distribution of a component concentration, being constant in time and uniform along the plane of a phase front. The uniform distribution of a component in the solid phase is also coincident with the curves. $\Omega_2=0$ on the 4 and 5 curves. At the point of maximal increment $Y_1=0$, the pulsation frequency also differs from zero. Consequently these regimes are associated with the distribution of a component concentration, pulsating in time but uniform in the plane phase front. Laminated component distribution in a solid phase corresponds to them.

Changing the values of governing parameters, let it be now:

$$\mu_p=0,2;\ \mu_T=1;\ \mu_{T1}=10^2;k=0,9;\quad p=-5\cdot10^{-3}. \qquad (20)$$

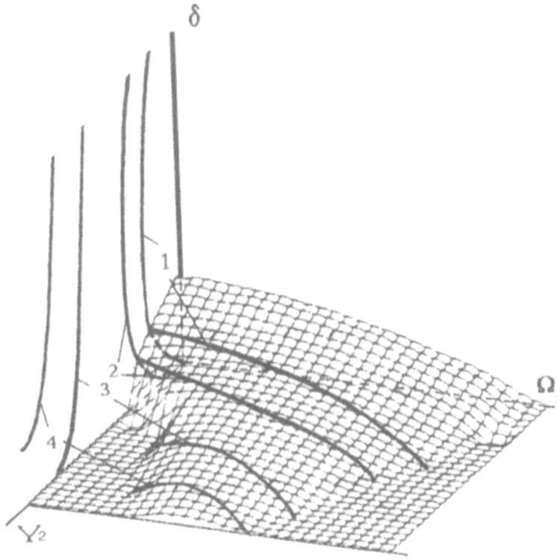

Figure 2. The dependence $\delta(Y_2,\Omega_2)$.

Fig.2 shows the example of the $\delta(\Omega,K)$ dependence, where Φ_3 and Φ_4(3 and 4 curves) are both the temporal pulsations and spatial distortions of the crystallization front. These regimes have more complicated component concentration in the solid phase. Under these conditions the minority component concentration distribution in the solid phase will be more complicated. The distribution in this case is depicted in Figure 3. Here presented an isosurface is of the impurity concentration of a two-component alloy for the case of a single-mode regime of temporary pulsations (the front is propagating in the direction of the z axis) and a two-mode regime of spatial distortions. The isosurface depicted is of a dendrite-like structure. Such formations are likely to determine the spacings between the branches of dendrites of polycrystal materials.

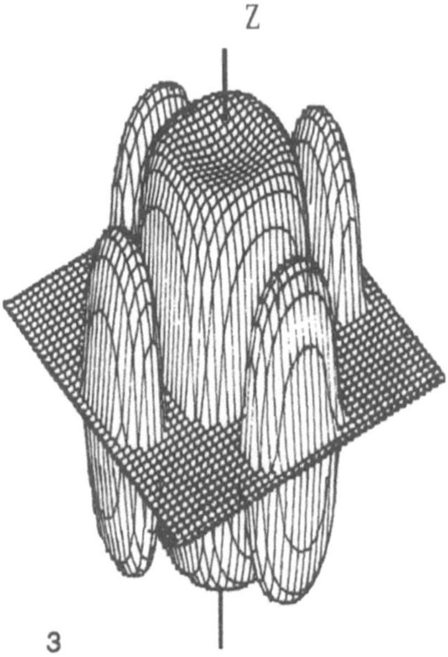

3

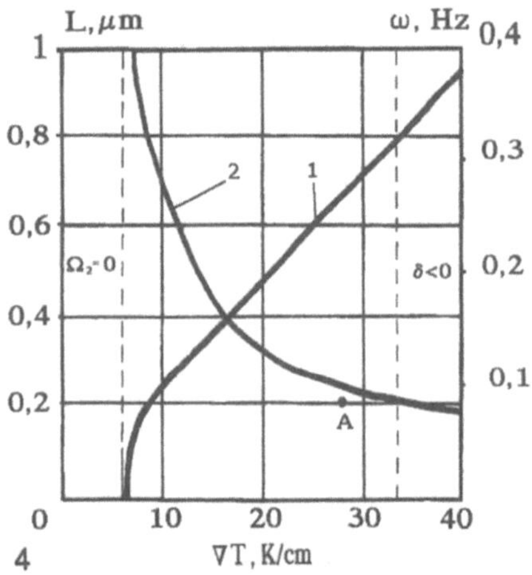

4

Figure 3. The isosurfaces of the impurity concentration of a two-component alloy.

Figure 4. Dependencies of $\omega(\nabla T)$(curve 1) and $L(\nabla T)$(curve 2).

Work as shown by Utech(1968) presents the experiment in which the Al-Cu - alloy is crystallized and the plane front is not destroyed. At this the bands of a change of a component concentration is observed along the direction of crystallization in the ingot. The experiment aims at elucidating the cause of this effect by temperature oscillations in the melt ahead the crystallization front. The oscillations are due to a temperature change in the liquid phase. However in the experiment, described in by Utech(1968) the period of the component concentration change is expected to be almost one order less than that we observe. The dependence W (ΔT) and L (ΔT) (L is the period of component distribution), displayed in Fig.4, are built at the system parameters, corresponding to experimental conditions by Utech(1968) .

The A - point is the result of the experiment. Bearing in mind the coincidence of theoretical calculations with experimental ones it may be supposed that the described in Utech (1968) component distribution is associated with the auto waves of a concentration, presented here.

THE REASONS FOR THE OCCURRENCE OF AUTO-OSCILLATIONS IN THE COMPONENT CONCENTRATION

Let's consider the two thermal fluxes, passing through the phase front and temperature gradients, associated with these fluxes. The flux $q_\varepsilon=\varepsilon u$ is generated by the energy released from a phase transition, as to the flux q_z, it is formed by external conditions. Generally such the separation is conditional and it serves for finding out qualitative regularities. However there are experiments in which one of the fluxes can be neglected. One of these extreme cases is the crystallization of strongly overcooled liquids from one center. Let the liquid be supercooled so that the heat release at a phase transition warms supercooled material up to the point of a phase transition. In the ideal case of uniform supercooling a temperature gradient is zero in a melt bulk. Only two stable stationary regimes of crystallization can be observed under these conditions: a) the regime with the static crystallization front, b) the shift of front with a velocity $V=V_\infty$, at which a thermal profile is $u=u_\infty$ at $z\gg0$ and $u=u_\infty+\varepsilon$ at $z<0$. At $0<V<V_\infty$ the thermal maximum is on the solidification front, since formation of the profile occurs due to heat conduction. Any increase of velocity leads to increase of an absolute value of the negative thermal flux to the liquid phase and the value decreases towards the solid phase. Thus the temperature on the crystallization front decreases, since the solid phase temperature behind the front increases,and consequently a velocity of the front shift becomes higher. The velocity heigh-

tening takes place until it is V∞. Decrease of velocity leads to redistribution of the fluxes, front temperature increases and a shift velocity becomes lesser. The regime is stationary at V=0, that is, any other regime is not stationary.

Another extreme case is the slow crystallization with a fixed thermal profile, determined by external conditions. The profile agrees with the regime of growing single crystalline materials. If the velocity of a phase front shift is rather low (it can be always done, since a thermal profile moves along an the ingot), then $q_\varepsilon \ll q_z$. Now the thermal profile has the form of the enhancing function. If the rate of crystallization increases in these cases then its supercooling decreases and the rate gains its previous value. Decreases of rate leads to increase of temperature on the phase front and a velocity increases. The system is stable.

The q_u/q_ε-parameter seems to determine radically the frequency of auto-waves of a component concentration on phase boundaries. In the presented examples the change of auto-oscillation frequencies by 9 orders corresponds to the change of q_u/q_ε by two orders.

In the q_u/q_ε and V a dependence of q_u/q_ε on V_z for the considered regimes is a sort of a diminishing function. The regimes of single crystal growth correspond to the values of q_u/q_ε, exceeding those of q_u/q_ε in the regime of component striped distribution.The regimes with lesser q_u/q_ε and greater V_z, than those being in the regimes of finely crystalline material production, conform with the regime of superfast quenching. However it is known that rather great values of V_z correspond to the regimes of the considered scheme, $q_u/q_\varepsilon \Rightarrow \infty$ and monotony of this dependence are violated. This is caused by the fact that the material atoms are positioned in the places of greater potential energy when the amorphous materials are produced by superfast quenching. As a result, heat release of ε- phase transition and a $q_\varepsilon \Rightarrow 0$ - flux are decreased. That is, the dependence of q_u/q_ε on V_z is increased. It should be born in mind, that at the parameters corresponding to the production of finely crystalline materials, the width of a layer, in which the $C_z(z)$ component concentration changes at phase boundaries, is comparable with interatomic spacings. Consequently these regimes are limiting for description of solidification processes by the equations of solid medium.

Gus'kov(1989) presents numerical estimations of the q_z/q_ε value for the three typical regimes of crystallization: the production of microcrystalline films, formation of microcrystalline zones at solidification of big ingots and slow crystallization with formation of striped distribution of impurity. the measurement of auto-oscillation frequencies yields 5MHz, 25Hz and 4,2 10^{-2}Hz correspondingly. The regime with a lower frequency is obviously more stable. In particular the third regime approaches the ones of production homogeneous materials. The q_z/q_ε-values for the mentioned regimes are 0,5; 5 and 60 correspondingly [1], that is a complete agreement with the mechanism of instability occurrence is achieved.

CONCLUSIONS

The interface instability at the directed crystallization is one of the reasons for the spatial inhomogeneity of a material structure.Numerical calculations of the impurity partition period are coinciding with the experimental results.

The theory explains the variety of component periodical structures.

The theory with provides the classification of solidification regimes depending on the system stability value.

REFERENCES

Buyevich,U.A., Mansurov,V.V., I.A.Nataluha I.A., (1988) Auto-oscillations, parametric modulations and the non-linear parametric resonance in the processes of volumetric crystallization, Injenrno-Fisicheski jurnal,(Russian),vol. 55, 2:285.

Gus'kov,A.P., 1989, The auto-oscillations of a component concentration ahead the front of crystallization, (Russian), Preprint of Instituteof Solid State Physics RAS.

Gus'kov,A.P., 1991, The auto-oscillations mode of binary alloy crystallization, Teplofisika Visokich Temperatur, (Russian),vol.29, 2:272.

Mullins,W.W., Sekerka,R.F., 1968, Stabilityof a plane surface of phase separation at crystallization of a diluted binary alloy, in "Problemi rosta kristallov", (Russian), Moskau.

Samoilovich Yu.A., On the possibility of melt crystallization in the mode of auto-oscillations.Teplofisika Visokich Temperatur,(Russian),vol.17, 5:992.

Ytech H.P.,Flemings M.C., 1968,The influence of a magnetic field on a thermal convection at growth of metallic crystals, in "Problemi rosta kristallov", (Russian), Moskau.

BUBBLE EVOLUTION IN LIQUID-GAS SOLUTIONS, VIEWED AS AN ELEMENTARY CATASTROPHE

J.L. Achard and E. Canot

Laboratoire des Ecoulements Géophysiques et Industriels
Institut de Mécanique de Grenoble
B.P. 53X – 38041 Grenoble Cedex
France

INTRODUCTION

The stability of gas micro-bubbles in liquid-gas solutions is of importance in a broad range of applications which include propagation and damping of sound, bubble removal during glassmaking, fermentation, initiation of boiling or cavitation. The latter application has prompted most studies on this subject : the point was to explain why real liquids cannot withstand high tensile stresses as predicted by the various homogeneous nucleation theories. It is then necessary in many cases to postulate that heterogeneous nucleation occurs via the existence of persistent nuclei. The most likely candidates for the role of nuclei are either vapour-gas bubbles trapped in crevices on the wall of a container or on the surface of suspended solid particles, or free vapour-gas micro-bubbles transported in the bulk of the fluid.

The former nuclei could preferentially act as the source of macroscopic bubble appearance if free bubbles have time to rise to the surface and thus be lost or if boiling bubbles are forced to escape from the heated solid-liquid interface. On another side, the latter nuclei, carried in the bulk of liquid, are the only one which can be envisaged and even observed in some experiments, especially in cavitation work. It is well known that a pure vapour bubble has no stable equilibrium states. Such a bubble may be stabilized if it also contains a permanent, *insoluble* gas, and if its radius is less than a critical radius which corresponds to the so-called Blake threshold pressure (Daily and Johnson, 1956).

In the case where the gas is *soluble* in the liquid the permanence of stable free micro-bubbles has been for a long time not understood. Gibbs had pointed out in 1878 that a single bubble in a large volume of liquid-gas solution can only have an unstable equilibrium state. Epstein and Plesset (1950) have shown theoretically that in liquid-gas solution of *unbounded* extent, gas bubbles grow if the solution is supersaturated and dissolve if it is undersaturated or even saturated, due to surface tension effects. This leds investigators to assume that the stabilization is due to impurities fixed at the bubble surface : gas impermeable skin of organic or surface-active material and surface electric charges.

Instabilities in Multiphase Flows, Edited by G. Gouesbet
and A. Berlemont, Plenum Press, New York, 1993

What has been shown for a single bubble in an unbounded volume of liquid-gas solution was taken, mistakenly, to be valid for a multibubble system where each micro-bubble is associated with a given volume of liquid (a "droplet" or a "spherical cell") containing a given amount of gas. Physically, the stable equilibrium state of microbubbles might result from the natural competition of the bubble free gas for the dissolved gas of the liquid phase.

So Mori et al. (1977), Cha (1981), Cha (1983), Ward et al. (1982) and Tsai and Chen (1990) addressed the stability problem by considering a bubble in a *finite* volume of a liquid-gas solution. They all showed that a stable equilibrium could exist.

There were controversies about some possible sources of errors in the previous analyses. In all of them, the nature of the equilibrium state is determined by examining whether a suitably selected thermodynamic potential is a minimum, a maximum or a point of inflection. The choice of the appropriate potential depends on the constraints which have been considered : a constant volume of the liquid-droplet for the first study, a constant volume of the total bubble-droplet system for the second one, a constant liquid pressure for the third and the fourth ones. The corresponding thermodynamic potentials are the Gibbs free energy, the Helmholtz free energy and the availability (Muller, 1985). Tsai and Chen (1990) showed moreover that the constraints adopted in the two first studies implies unphysical consequences and must be discarded.

The same conclusion concerning the existence of a stable equilibrium were obtained in a simple way by Plesset and Sadhal (1981), Ward et al. (1982) and Franklin et al.(1984). Starting from steady state balances and using physical arguments based on diffusion theory they were able to select stable solutions amongst all equilibrium solutions.

Our purpose is to develop a generalized approach which is based on the most classical method of hydrodynamics stability theory, usually called the perturbation method. The whole problem is first formulated in terms of a system of partial differential equations. Any input disturbance of infinitesimal size around a given equilibrium state is applied to the system before time $t = 0$ and is supposed to be represented through initial conditions, the boundary conditions being after $t = 0$ equal to zero. To know the long time response of the field variables to these perturbations, equations are linearized and then solved by a standard technique.

MODEL AND BALANCE EQUATIONS

Consider a spherical micro-bubble suspended in a dilute liquid-gas solution and containing the free gas and the vapour of the liquid solvent. The bubble we intend to model is typical of a system containing N equally sized micro-bubbles per unit volume of liquid. At each macroscopic point of the medium we introduce a spherical cell including such a "test" bubble of radius R at its center and a shell of surrounding liquid. The radius R^b of the cell depends on the local void fraction α and is equal to $\alpha^{-1/3}R$. There is no exchange of heat and mass at the external boundary of the cell which is allowed to expand or contract following the applied pressure pulsations. To some extent, such a model encompasses the bubble-droplet system of Mori et al.(1977) as shown in Figure 1. In their system the droplet and the surrounding liquid are assumed to be immiscible and the inert gas is assumed to be insoluble in the surrounding liquid.

In the equations that follow the subscripts i=1,2 refer the variables to the solvent (liquid constituent) and solute (gas constituent) respectively and the superscripts $k=g,l,s$ refer the variables to the gaseous phase inside the bubble, the condensed liquid phase inside the droplet and the liquid-gas interface. Moreover the subscript e refer the variables to equilibrium values of the reference state and the superscript b to the cell boundary.

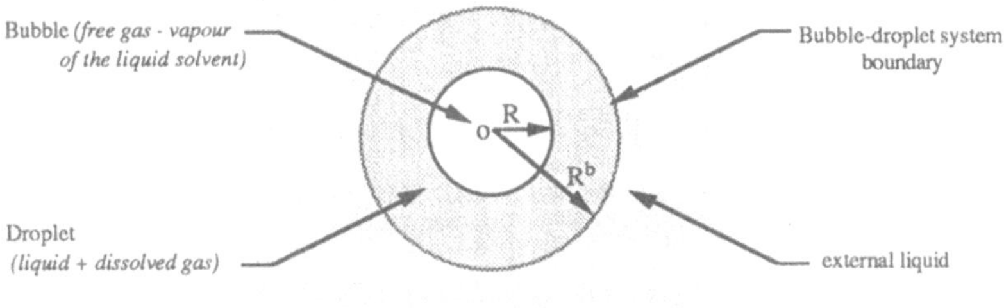

Figure 1. The physical model: a bubble droplet system.

The thermodynamic theory of mixtures (Muller,1985) introduces from the outset, besides the temperature field θ^k which is common to all participating constituents, mechanical variables for each constituent *separately*: ρ_i^k , $\mathbf{v}_i^k$ (i=1,2) their densities and velocities. For the present context it is more convenient to employ

$$\rho^k := \sum_{i=1}^{2} \rho_i^k \ , \ Y_i^k := \rho_i^k / \rho^k \ , \ \mathbf{v}^k := \sum_{i=1}^{2} Y_i^k \mathbf{v}_i^k \ , \ \mathbf{w}_i^k := \mathbf{v}_i^k - \mathbf{v}^k \tag{2.1}$$

which represent the density of the mixture, the mass fraction of constituent i, the velocity of the mixture and the diffusion velocity of constituent i relative to $\mathbf{v}^k$. The internal energy of the mixture is defined by $u^k := \sum_{i=1}^{2} Y_i^k \left[u_i^k + 1/2 \, (\mathbf{w}_i^k)^2 \right]$ where u_i^k is the internal energy of constituent i . Obviously we have the identities:

$$\sum_{i=1}^{2} Y_i^k = 1 \ , \ \sum_{i=1}^{2} Y_i^k \mathbf{w}_i^k = 0 \tag{2.2}$$

In what follows, we shall often denote for simplicity Y_2^k by Y^k and Y_1^k by $1-Y^k$. The diffusion velocities $\mathbf{w}_2^k$ and $\mathbf{w}_1^k$ must be considered as being equal to $\mathbf{j}^k/\rho^k Y^k$ and $-\mathbf{j}^k/\rho^k(1-Y^k)$ where $\mathbf{j}^k$ is the relative momentum of constituent 2 in phase *k*.

The balance equations of mass, momentum and internal energy for the mixture as a whole in each phase domain can have the same appearance as those for a single body, if the stress and the flux of internal energy of the mixture are given by $\mathbb{T}^k := \sum_{i=1}^{2} (\mathbb{T}_i^k - \rho_i^k \mathbf{w}_i^k \mathbf{w}_i^k)$ and $\mathbf{q}^k := \sum_{i=1}^{2} \left\{ \mathbf{q}_i^k - \mathbb{T}_i^k \cdot \mathbf{w}_i^k + \rho_i^k \left[u_i^k + \frac{1}{2} (\mathbf{w}_i^k)^2 \right] \mathbf{w}_i^k \right\}$, respectively (Muller, 1985).The same statement holds true for the interfacial conditions representing the balance of mass, momentum and internal energy for the mixture.

The equations for the mass fraction and the relative momentum of constituent 2 are in coordinate-free form:

$$\rho^k (\partial Y^k / \partial t + \mathbf{v}^k . \nabla Y^k) + \nabla \cdot \mathbf{j}^k = 0 \quad , \quad \mathbf{j}^k = - \rho^k D^k \nabla Y^k \qquad (k=g,l) \tag{2.3}$$

where D^l is the diffusion coefficient of the solute in the solvent and D^g the diffusion coefficient of the gas in the vapour. The second equation is the chemical constitutive Fick's law of diffusion. Thermal-diffusion or Soret effects are neglected.

The interfacial condition for the mass fraction of constituent 2 is

$$\sum_{k=g,l} \left[\rho^k (\mathbf{v}^k - \mathbf{v}^s) Y^k + \mathbf{j}^k \right] \cdot \mathbf{n}^k = 0 \tag{2.4}$$

The conditions on the cell boundary are such that there is no exchange of mass and heat with the neighboring inclusions and that the normal stress is held fixed and equal to a equilibrium pressure p_e^l:

$$(\mathbf{v}^l - \mathbf{v}^b) \cdot \mathbf{n}^l = 0 \quad , \quad \mathbf{j}^l \cdot \mathbf{n}^l = 0 \quad , \quad \mathbf{q}^l \cdot \mathbf{n}^l = 0 \quad , \quad \mathbf{n}^l \cdot \mathbb{T}^l = -\, p_e^l \, \mathbf{n}^l \qquad (2.5)$$

This last condition implies that the cell boundary may expand or contract. Finally at the center of the bubble the following conditions are assigned:

$$\mathbf{v}^g = 0 \quad , \quad \mathbf{j}^g = 0 \quad , \quad \mathbf{q}^g = 0 \qquad (2.6)$$

CONSTITUTIVE LAWS

The above problem can be considerably simplified by the following standard restrictions:

(i) The bubble remains *spherical in shape* throughout its evolution. All the required conditions (Iooss and Laure, 1989) are supposed satisfied in our case for the various fields to be spherically symmetric. Equations will be presented in the spherical Euler coordinates (r,t),v being the radial velocity velocity components.

(ii) The pressure of the system is not close to the critical pressure of the solvent. So the solution may be assumed to be *dilute*, i.e. $Y^l \ll 1$.

Now constitutive equations have to be found in order to form a complete system of equations. Some of them are common to both phases :

(i) The liquid and gas mixtures are *Newtonian fluids*. Thus the intrinsic stress is $\sum_{i=1}^{2} \mathbb{T}^k = -\, p^k \, \mathbb{I} + 2\, \mu^k \, \mathbb{D}$ $(k=g,l)$, in which p is the mixture pressure, $\mathbb{I}$ is the tensor, μ is the mixture shear viscosity coefficient and $\mathbb{D}$ is the deformation rate tensor. The bulk viscosity effects are ignored.

(ii) The liquid and gas mixtures obey the *Fourier law*. Thus the internal energy flux is $\mathbf{q}^k = -\,\lambda^k \, \nabla\, \theta^k + (\, h_1^k - h_2^k\,)\, \mathbf{j}^k + (\,1 - 2\,Y\,)\,(\,\mathbf{w}_1^k - \mathbf{w}_2^k\,)^2\, \mathbf{j}^k$ $(k=g,l)$, where λ is the thermal conductivity of the mixture and $h := u + p\,/\,\rho$ is the enthalpy. This expression is composed of a conductive flux, of contributions resulting from the interdiffusion of the two constituents and of a non intrinsic component due to relative momenta. Diffusion-thermo or Dufour effects and viscous effects in the interdiffusion term are neglected.

(iii) The liquid and gas mixture are *simple mixtures* as defined by Muller (1985). In this case the constituents behave like single fluids. Accordingly, partial internal energies u_i^k depends only on ρ_i^k, and θ^k, as for pure substances. The same statement holds for p_i^k, μ_i^k η_i^k which are partial pressures, partial chemical potential and partial entropies. There must be also a Gibbs equation for *each constituent* which implies an integrability condition which is used to simplify the internal energy equation. For the same use, another similar integrability condition can be deduced from a Gibbs equation which is valid for the *whole mixture* of whatever nature (simple or not).

Finally we need constitutive relations at the interface. Equilibration at the bubble-liquid interface are rapid so that the temperatures and the chemical potential of each constituent are continuous across the interface

$$\theta^g = \theta^l = \theta^s \quad , \quad \mu_1^g = \mu_1^l \;\text{ and }\; \mu_2^g = \mu_2^l \qquad (3.1)$$

Chemical potentials per unit of mass of each constituent in the solution can be given by

$$\mu_i^l = \mu_i^{ol}(p^l, \theta^l) + R\,\theta^l / M_i \ln(X_i^l) \quad (i=1,2) \tag{3.2}$$

where R is the mass gas constant with the value 8.314 kJ kg^{-1} K^{-1}, M_i is the molecular weight of constituent i, X_i its mole fraction. We have the identity $X_i = Y_i M / M_i$ (i=1,2) in which $M := (\sum_{i=1}^{2} Y_i / M_i)^{-1}$ is the molecular weight of the mixture. The interpretation of μ_i^o is different according to *i*. Constituent 1 behaves *ideally* in the mixture since $X_1^l \approx 1$, thus μ_1^o corresponds to the free enthalpy of the pure liquid solvent. For constituent 2 the mole fraction range is about $X_2^l \approx 0$ and μ_2^{ol} is merely a quantity independent of X_2^l which is determined experimentally.

Formula similar to (3.2) can be written down for gas mixture. It can be assumed that it is an ideal mixture for both constituents; μ_1^g and μ_2^g are therefore free enthalpies of the pure gas constituents.

Insertion of the above constitutive laws into balance equations and use of the equations (2.5) and (2.6) leads to a closed system. For the linear analysis this system must be expanded about an equilibrium state. To this end, we write for a typical variable $z = z_e + \delta z$, where δz is the perturbation field, and we retain terms through first-order (δz) perturbations. The linearized system turns out to be greatly reduced, because in the equilibrium state, all velocities vanish, temperatures and densities are uniform. Nonlinear terms in relative velocities and in gradients of temperature, densities and velocities have no contributions:

$$\frac{\partial \delta\rho^k}{\partial t} + \frac{\rho_e^k}{r^2}\frac{\partial}{\partial r}(r^2\,\delta v^k) = 0 \quad (k=g,l)$$

$$\rho_e^k \frac{\partial \delta v^k}{\partial t} + \left(\frac{\partial p^k}{\partial \rho^k}\right)_e \frac{\partial \delta\rho^k}{\partial r} + \left(\frac{\partial p^k}{\partial Y^k}\right)_e \frac{\partial \delta Y^k}{\partial r} + \left(\frac{\partial p^k}{\partial \theta^k}\right)_e \frac{\partial \delta\theta^k}{\partial r} - \frac{4}{3}\mu_e^k \frac{\partial}{\partial r}\left[\frac{1}{r^2}\frac{\partial(r^2\,\delta v^k)}{r}\right] = 0 \quad (k=g,l)$$

$$\left(\frac{\partial u^k}{\partial \theta^k}\right)_e \frac{\partial \delta\theta^k}{\partial t} - \frac{\theta_e^k}{\rho_e^{k\,2}}\left(\frac{\partial p^k}{\partial \theta^k}\right)_e \frac{\partial \delta\rho^k}{\partial t} + \frac{\theta_e^k D_e^k}{r^2}\left[\frac{1}{\rho_{2e}^k}\left(\frac{\partial p_{1e}^k}{\partial \theta^k}\right)_e - \frac{1}{\rho_{1e}^k}\left(\frac{\partial p_{2e}^k}{\partial \theta^k}\right)_e\right]\frac{\partial}{\partial r}\left(r^2\frac{\partial \delta Y^k}{\partial r}\right)$$

$$-\frac{\lambda_e^k}{r^2}\frac{\partial}{\partial r}\left(r^2\frac{\partial \delta\theta^k}{\partial r}\right) = 0 \quad (k=g,l) \tag{3.3}$$

$$\frac{\partial \delta Y^k}{\partial t} - \frac{D_e^k}{r^2}\frac{\partial}{\partial r}\left(r^2\frac{\partial \delta Y^k}{\partial r}\right) = 0 \quad (k=g,l)$$

APPROXIMATIONS

In order to make our problem as uncomplicated as possible, we shall add non-strictly necessary assumptions. The second restriction suggests setting

(i) for the *liquid phase*, that its properties are those of the pure solvent which will be assumed incompressible

$$\rho_e^l := \rho_{1\,e}^l\ ,\ \mu_e^l := \mu_{1\,e}^l\ ,\ \lambda_e^l := \lambda_{1\,e}^l\ ,\ \left(\partial u^l / \partial \theta^l\right)_e = c_1^l(\theta_e^l)\ ,\ \delta\rho^l = 0 \tag{4.1}$$

where c_1^l is the heat capacity of the liquid.

(ii) for the *gas phase*, that both constituents obey the *ideal* gas law

$$p_i^g := \rho_i^g \, R / M_i \, \theta^g \; , \; \left(\partial u_i^g / \partial \theta^g\right)_e = c_{v\,i}^g \qquad (i=1,2) \tag{4.2}$$

where $c_{v\,i}^g$ is the specific heat at constant volume of constituent i.

Several important consequences will be deduced from these simplified equations of state. By using the equilibrium conditions(3.8)$_1$ and (3.8)$_2$, we obtain (Ward et al., 1970)

$$p_1^g = (1 - X_2^l)\, p_1^*(\theta^s) \exp\left[(p^l - p_1^*)\, M_1 / (\rho_1^l \, R \, \theta^s)\right] + o\,(X_2^l) \tag{4.3}$$

This relationship where o is the small Landau symbol is a *Raoult's law* which has been modified for curvature effects. The exponential term which expresses these effects will be dropped out since it is close to one at low pressures. Similarly, by using the equilibrium conditions (3.1)$_1$ and (3.1)$_3$, we obtain

$$p_2^g / X_2^l = K_2\,(\theta^s) \tag{4.4}$$

where K_2 is the *Henry's law* constant between the solute and solvent. It will taken to be the same as the corresponding constant valid for a flat interface.

As final consequences of (4.2)$_1$ and (4.2)$_2$, we have the thermal (*Dalton's law*) and caloric equation of state for the mixture

$$p^g := p_1^g + p_2^g = \rho^g \, R \, \theta^g / M \quad , \quad (\partial u^g / \partial \theta^g) = Y^g \, c_{v\,2}^g + (1 - Y^g)\, c_{v\,1}^g \tag{4.5}$$

EQUILIBRIUM STATES AND BIFURCATING STATES

The steady state is more easily handled by using N_i^k and X_i^k ("Chemistry" formulation) as variables instead of m_i^k and Y_i^k ("Fluid Dynamics" formulation), N_i^k being the total moles number of constituent i in phase *k* and m_i^k the corresponding total mass. Not all previous linearized equations have their counterpart in defining the steady system which is much more simpler. It involves the following equations, which were already proposed by Cha (1983) and Tsai and Chen (1990) with a pure saturation condition instead of a Raoult's law:

(i) Conservation of moles

$$N_{1\,e}^g + N_{1\,e}^l = N_1 \quad , \quad N_{2\,e}^g + N_{2\,e}^l = N_2 \tag{5.1}$$

(ii) Equation of state for the gas

$$p_{1\,e}^g \frac{4}{3} \pi \, R_e^3 = N_{1\,e}^g \, \mathcal{R} \theta \quad , \quad p_{2\,e}^g \frac{4}{3} \pi \, R_e^3 = N_{2\,e}^g \, \mathcal{R} \theta \quad , \quad p_e^g = p_{1\,e}^g + p_{2\,e}^g \tag{5.2}$$

(iii) Boundary conditions

$$p_e^g - p_e^l = 2\,\sigma / R_e \quad , \quad p_{1\,e}^g = (1 - X_{2\,e}^l)\, p_1^* \quad , \quad p_{2\,e}^g = K_2 \, X_{2\,e}^l \tag{5.3}$$

(iv) Definition of $X_{2\,e}^l$

$$X_{2\,e}^l = N_{2\,e}^l / (N_{1\,e}^l + N_{2\,e}^l) \tag{5.4}$$

Where $\mathcal{R}$ is the molar gas constant with the value 8.314 $J.K^{-1}.mol^{-1}$.

In this system of *nine* equations, N_1, N_2, p_e^l, θ_e are given and so are σ, p_1^*, K_2 which depend in part on the latter variables. All these seven variables are the *control* or *external dimensional variables* of the above equilibrium problem. The corresponding nine unknowns are p_e^g, $p_{1\,e}^g$, $p_{2\,e}^g$, $N_{1\,e}^g$, $N_{1\,e}^l$, $N_{2\,e}^g$, $N_{2\,e}^l$, $X_{2\,e}^l$, R_e. They are the *state* or *internal dimensional* variables. It happens that all these equations can be condensed in a single polynomial equation for the single state variable R_e. Advantages of this straightforward operation is triple, since the resulting equation:

(i) may be more safely simplified. The point is to neglect only small terms which change the original model in a trivial way (structural stability). In our case, terms which can considered as small involve the two small dimensionless numbers $\varepsilon_1 := N_2 / N_1$ and $\varepsilon_2 := (p^l - p_1^*) / (K_2 - p^l)$.

(ii) may be more easily made nondimensional. Defining the single dimensionless state variable by $\overline{R_e} := R_e / R_0$ where $R_0 := (3 N_2 \mathcal{R}\theta / 2 \pi \sigma)^{1/2}$, we can write the simplified dimensionless equation as

$$f(U_1, U_2, \overline{R_e}) = U_2 \overline{R_e}^4 + 2 \overline{R_e}^3 + [(U_1 U_2 - 1)/2] \overline{R_e} + U_1 = 0 \qquad (5.5)$$

Here the following dimensionless control variables have been introduced:

$$U_1 := (N_1 / N_2)\left[(\sigma / R_0) / (K_2 - p^l)\right] \text{ and } U_2 := (p^l - p_1^*) / (\sigma / R_0) \qquad (5.6)$$

It can be noticed that $U_1 U_2 = \varepsilon_2 / \varepsilon_1$. To obtain (5.5) we have neglected in f the sum $-(4 \varepsilon_1 U_1 \overline{R_e}^2) + 2 \varepsilon_2 \overline{R_e}^3)$; thus, $U_1 U_2$ being fixed, this sum can be considered globally as $O(\varepsilon_1)$ or equivalently as $O(\varepsilon_2)$.

Starting with *seven* dimensional control variables, we are left with only the *two* dimensionless ones quoted above. The first one represents the force encouraging radius collapse by *dissolution*; in the opposite direction, when $U_1 = 0$, gas is insoluble in liquid. The second one represents the force encouraging radius collapse by *condensation*; on the opposite direction when $U_2 < 0$, i.e. for pressure less than saturation pressure, vapour may appear and cavitation may be observed.

(iii) may be cast into the elementary catastrophe theory or bifurcation theory. Both theories provide us with concepts and methods which are equivalent in our context. From the first theory (Majthay, 1985) we borrow notions related to singular or bifurcation equilibrium states. Equilibrium states are represented in our case by a surface M in the (U_1, U_2, $\overline{R_e}$) (Fig 2) called the *catastrophe manifold* . The *catastrophe* or *singularity set* C is the collection of points on M for which

$$\partial f / \partial \overline{R_e} = 4 U_2 \overline{R_e}^3 + 6 \overline{R_e}^2 + [(U_1 U_2 - 1)/2] = 0 \qquad (5.7)$$

C is therefore the solution set of the system composed of (5.5) and (5.7). Eliminating U_2 in these two equations yields an equivalent system of equation

$$\begin{aligned} U_1^2 + 4 U_1 \overline{R_e}^3 - 3 \overline{R_e}^4 + 4 \overline{R_e}^6 &= 0 \\ U_2 (4 \overline{R_e}^3 - U_1 / 2) &= 6 \overline{R_e}^2 + 1/2 \end{aligned} \qquad (5.8)$$

First we notice that $U_2 = 0$, $\overline{R_e} = \infty$ is solution of the second equation. From this we obtain C as a curve in the (U_1, U_2, $\overline{R_e}$) space given parametrically by

$$U_1 = 2 q^2 (\sqrt{3}/2 - q) \quad , \quad U_2 = 2 (\sqrt{3}/6 - q) / q^2 \quad , \quad \overline{R_e} = q \qquad (5.9)$$

In Figure 2, it is clear that this curve, along which the tangent plane to M is parallel to the $\overline{R_e}$ axis, is composed of two smooth pieces. According to previous similar findings of Cha (1983), the first piece laying totally in the negative pressure region is named the "cavitation limit". The second one pertaining to both negative and positive regions is named the "dissolution limit". Furthermore, there is another curve rejected to infinite radii at $U_2 = 0$. Finally, we may note that in bifurcation theory (Iooss and Joseph, 1980) the points of C are named *regular turning* points. First, they are *regular* points because for all points on M we have, U_2 being fixed

$$\partial f / \partial U_1 \neq 0 \qquad (5.10)$$

Thus conditions of the implicit function theorem hold and we can find at each point on M a unique curve $U_1 = U_1(\overline{R_e})$ through the point. Secondly, they are *turning* points since $\partial U_1 / \partial \overline{R_e}(\overline{R_e})$ changes sign. This follows from the equation valid everywhere on M

$$df\left[U_1(\overline{R_e}), \overline{R_e}\right] / d\,\overline{R_e} = \partial f / \partial \overline{R_e} + (\partial U_1 / \partial \overline{R_e})\big|_{U_2} \partial f / \partial U_1 = 0 \qquad (5.11)$$

Crossing C, $\partial f / \partial \overline{R_e}$ vanishes, so does $(\partial U_1 / \partial \overline{R_e})\big|_{U_2}$.

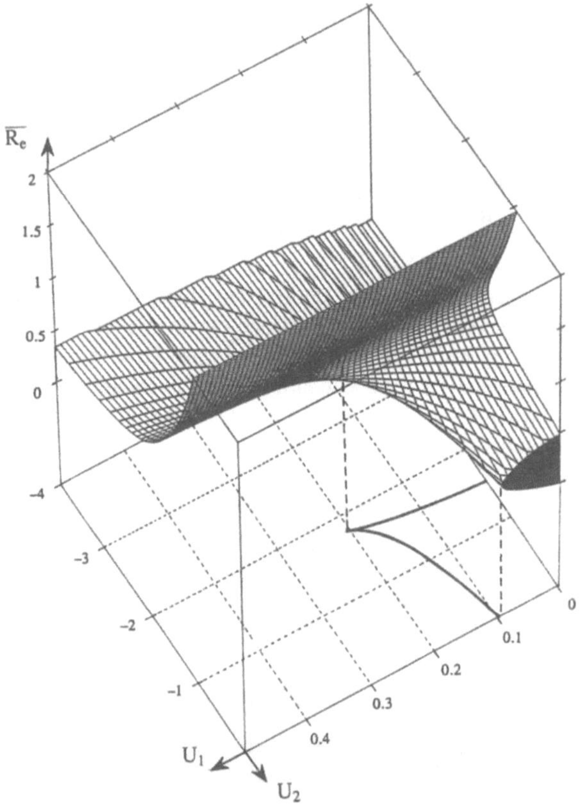

Figure 2. Top view of the equilibrium solutions surface (Catastrophe manifold M).

The *bifurcation set* B is the projection of C onto the (U_1, U_2) plane of control variables and is represented on Figure 3. It is the set, composed here of two curves, on which the number and the nature of equilibrium states change. As these changes may be triggered in the control space, this set is of the utmost importance. The first curve is defined parametrically by

$$U_1 = 2\,q^2\,(\sqrt{3}/2 - q)\ ,\ U_2 = 2\,(\sqrt{3}/6 - q)/q^2\ ,\ 0 < q < \sqrt{3}/2 \qquad (5.12)$$

and the second one by

$$U_2 = 0 \qquad (5.13)$$

Since the catastrophe manifold equation has not been derived in our case from minimizing a potential, we cannot know the stability of a given equilibrium state. The perturbation analyses in the next paragraph will show that the equilibrium is stable if and only if

$$\left(\partial U_2 / \partial \overline{R_e}\right)\Big|_{U_1} < 0 \tag{5.14}$$

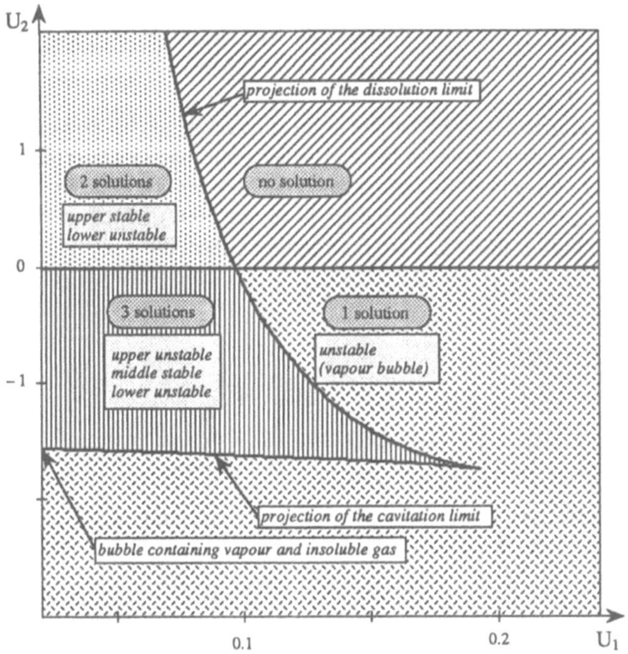

Figure 3. Bifurcation set. Number of equilibrium solutions.

Three limiting cases are of interest:

1) *Bubble containing vapour and insoluble gas*

This case corresponds to $U_1 = 0$. The catastrophe manifold reduces to a single curve. The singularity set is a point (Fig 4).

$$U_{2\,c} = -8 / 3\sqrt{3} \;,\; \overline{R_{e\,c}} = \sqrt{3}/2 \tag{5.15}$$

There are two equilibrium-radius values if $U_2 > U_{2\,c}$. If $\overline{R_e} > \overline{R_{e\,c}}$ (upper part) the equilibrium is instable and if $\overline{R_e} < \overline{R_{e\,c}}$ (lower part) the equilibrium is stable. These are well known results found by Blake.

2) *Bubble at high pressure*

This case is defined by $U_1 \to \infty$. The bifurcation set is a curve asymptotic to

$$U_2 = 1 / U_1 \tag{5.16}$$

There are two equilibrium-radius values if $U_1\, U_2 < 1$. On the upper of the catastrophe manifold the equilibrium is stable. On the lower part the equilibrium is unstable.

3) *Bubble in an infinite medium*

This case cannot only be defined in terms of control variables since the void fraction is itself a state variable as the radius. It can be shown that

$$\alpha = \varepsilon_3 \overline{R_e}^3 / (1 + \varepsilon_3 \overline{R_e}^3) + O(\varepsilon_1) \qquad (5.17)$$

where $\varepsilon_3 := 2 \rho^l R \theta_e / (M_1 K_2 U_1)$ is a third dimensionless control variable which appears only in the "Fluid Dynamics" formulation and which does not come into play in the static stability analysis. For a given $U_1 > 0$, consider a stable point on the upper part of the catastrophe manifold. By letting U_1 increase,we let α decrease down to α_c corresponding to $\varepsilon_3\overline{R_{ec}}^3 / (1 + \varepsilon_3\overline{R_{ec}}^3)$ on the singularity curve. Upon crossing this curve the bubble becomes unstable (Fig 5).

Finally it should be observed that the bifurcation set which has been exhibited (Fig 3) is a tip of a *swallowtail* after Thom. In fact, it has been mutilated for several reasons: negative parts of $\overline{R_e}$ and U_1, have been discarded since they are unphysical; one control variable is missing since it is $O(\varepsilon_2)$; control variables have not been defined in order to obtain the canonical form of the equation (5.5) but for a better physical interpretation.

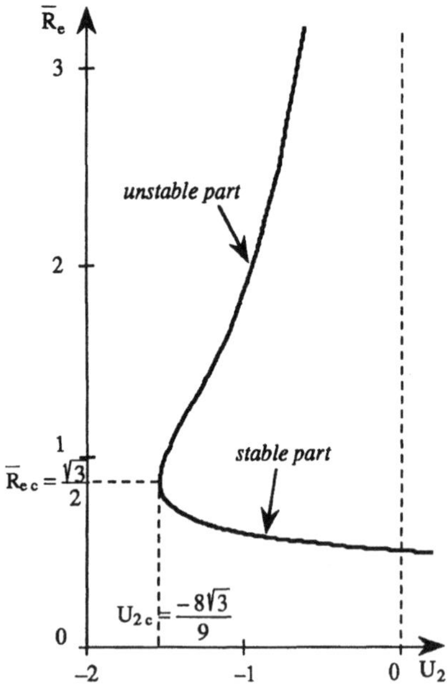

Figure 4. Equilibrium solutions curve for a bubble containing an insoluble gas and a vapour.

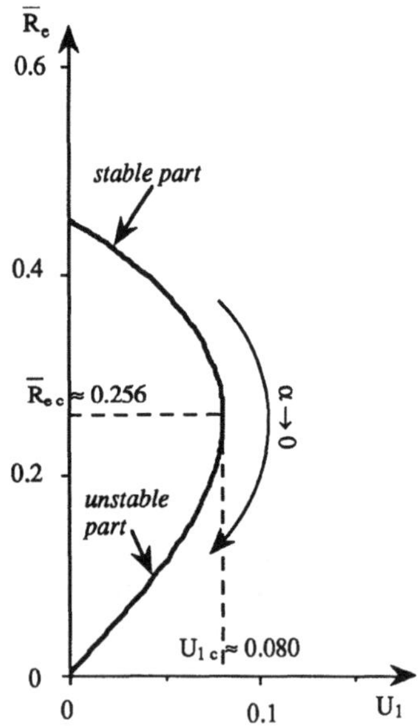

Figure 5. Void fraction effects.

LINEARIZED MODEL

The preceding static analysis is obviously incomplete and the next step is to consider the full dynamics corresponding to the whole linearized model that we have succeeded in deriving. In this paper, interest is focused on the relationship between static stability analysis and dynamic stability analysis. Therefore it is sufficient in our case to deal with a simplified linearized model which only includes the slowest transient phenomena. Stability analysis of this model will furnish three pieces of information :

(i) an expression of the catastrophe set C which has already been obtained by an other way;

(ii) the equilibrium solutions which are stable from a static point of view;

(iii) among these stable equilibrium solutions, the solutions which remain stable from a dynamical point of view.

There are many relaxation time scales in our problem. For several substances, they can by arranged in the following ascending order, where the inequality sign has been used between two successive time scales if their ratio is at least about 100.

$$R_0 / a^g << R_0^2 / v^g \approx R_0^2 / v^l << R_0^2 / D^g \approx R_0^2\, \rho_e^g\, (\partial u^g/\partial \theta^g)_e / \lambda^g << R_0^2 / \chi^l << R_0^2 / D^l$$

Here $a^g := (\partial p^g/\partial \rho^g)_\eta$ is the adiabatic sound velocity, ν is the kinematic viscosity and χ^l is the thermal diffusivity. The value of these time scales control the equalization speed of various state variables. We shall restrict ourselves to consider perturbations with the *largest* characteristic time, a liquid mass diffusion time, R_0^2 / D^l. Therefore the liquid and the gas temperatures do not deviate from the equilibrium temperature θ_e. Then, $\delta\theta^g = \delta\theta^l = 0$. The gas pressure, the gas density and the mass fraction for constituent 2 in the gas phase remain at a uniform (but non constant) value, i.e. $\delta p^g = \delta p^g(t)$, $\delta \rho^g = \delta \rho^g(t)$, $\delta X^g = \delta X^g(t)$. Solving with respect to r the gas and liquid continuity equations $(3.3)_1$ and the liquid momentum equation $(3.3)_2$ with boundary conditions $(2.6)_1$ and $(2.5)_4$ leads to the following interfacial equations:

$$\begin{aligned} &\delta v^g(R_e , t) = - \delta\dot{\rho}^g(t)\, R_e / 3\, \rho_e^g \\ &\rho_e^l \left[1 - (R_e / R_e^b) \right] R_e\, \delta\dot{v}^l (R_e , t) = \delta p^l (R_e , t) \end{aligned} \tag{6.1}$$

The interfacial momentum equation reduces to

$$\delta p^l (R_e , t) = \delta p^g (R_e , t) + (2\, \sigma / R_e^2)\, \delta R (t) \tag{6.2}$$

Henry's, Raoult's and Dalton's laws yield

$$\begin{aligned} &\delta p^g(t) = (K_2 - p_1^*)\, \delta X^l(R_e , t) \\ &\delta \rho^g(t) = \left[(K_2 M_2 - p_1^* M_1) / R\, \theta_e \right] \delta X^l(R_e , t) \end{aligned} \tag{6.3}$$

The interfacial continuity balances for the mixture and for the constituent 2 become after having introduced the small parameter $\varepsilon_4 := \rho_e^g / \rho_e^l$

$$\begin{aligned} &\delta v^l(R_e , t) - \delta\dot{R}(t) = \varepsilon_4 \left[\delta v^g(R_e , t) - \delta\dot{R}(t) \right] \\ &\partial \delta Y^l/\partial r(R_e , t) = \varepsilon_4 \left[(Y_e^l - Y_e^g) / D^l \right] \left[\delta v^g(R_e , t) - \delta\dot{R}(t) \right] \end{aligned} \tag{6.4}$$

We have to relate δX^l to δY^l

$$\delta Y^l = \delta X^l\, (M_2 / M_1) \left\{ 1 + X_e^l [(M_2 / M_1) - 1] \right\}^2 \tag{6.5}$$

The main ingredient of the linearized model is the equation of mass diffusion $(3.3)_4$ for δY^l.

Now, all equations are Laplace transformed. The equation of mass diffusion for $\delta\widehat{Y}^l(r,s)$ is first solved with respect to r, using boundary condition $(2.5)_2$ and $(6.4)_2$ in order to obtain an expression for $\delta\widehat{X}^l(R_e,s)$ in terms of ($\delta\widehat{v}^g(R_e,s) - s\, \widehat{R}(s)$). Such an expression is combined with the Laplace transformed equations (6.1), (6.2), (6.3), $(6.4)_1$ to form a closed system for $\delta\widehat{p}^g(s)$, $\delta\widehat{p}^l(s)$, $\delta\widehat{\rho}^g(s)$, $\delta\widehat{X}^l(R_e,s)$, $\delta\widehat{Y}^l(R_e,s)$, $\delta\widehat{v}^l(R_e,s)$, $\delta\widehat{v}^g(R_e,s)$, $\delta\widehat{R}(s)$. For sake of simplicity, initial conditions for all variables, except for the liquid velocity, will be taken equal to zero, then $\delta v^l(R_e,0) = \delta v_0$. Finally we can obtain a single equation for $\delta\widehat{R}(s)$, under the dimensionless form

$$\delta\widehat{\overline{R}}(\overline{s}) = \frac{\Phi(\overline{s})}{\Psi(\overline{s})}\,\delta\overline{v_0} \tag{6.6}$$

where

$$\Phi(\overline{s}) := -\varepsilon_5^2\,(1-\alpha^{1/3})\,\overline{R_e}^3\,(2\,\overline{R_e}^3 + U_1)\,(2\,\overline{R_e}^3\,A + U_1\,B)$$

$$\Psi(\overline{s}) := (4\,\overline{R_e}^6 - 3\,\overline{R_e}^4 + 2\,U_1\,\overline{R_e}^3)\,A + (2\,U_1\,\overline{R_e}^3 + U_1^2)\,B$$
$$-\varepsilon_5^2\,(1-\alpha^{1/3})\,\overline{R_e}^3\,(2\,\overline{R_e}^3 + U_1)\left[2\,\overline{R_e}^3\,A + (1-\varepsilon_4)\,U_1\,B\right]\overline{s}^2$$

$$A := \frac{1}{u}\{ch(u)\,[v\,ch(v) - sh(v)] - sh(u)\,[v\,sh(v) - ch(v)]\}$$

$$B := \frac{3}{1-\alpha}\left\{\frac{\alpha}{u^3}[u\,ch(u) - sh(u)]\,[v\,sh(v) - ch(v)] - \frac{1}{v^3}[v\,ch(v) - sh(v)]\,[u\,sh(u) - ch(u)]\right\}$$

$$\varepsilon_5 := \sqrt{\rho^l\,(D^l)^2 / 2\,\sigma\,R_0}\quad,\quad u := \overline{R_e}\,\sqrt{\overline{s}}\quad,\quad v := u/\alpha^{1/3}$$

It can be seen that the inertia term in the denomitor, which is an $O(\overline{s}^2)$ term, will not intervene in the static stability problem but only in the dynamical one. Its order of magnitude is given by ε_5^2, where ε_5 is a ratio of surface tension time scale $(\rho^l\,R_0^3/\sigma)^{1/2}$ over the liquid mass diffusion time scale R_0^2/D^l; it is usually very small.

In deriving the above transfer function we have coped with two types of inconvenience in computations. This transfer function involves essentially steady state variables of the linearized model. They have to be expressed in terms of parameters specifying a given equilibrium solution ($\overline{R_e}$ and U_1 have been selected but $\overline{R_e}$ and U_2 would have worked as well). Unfortunately most steady-state variables (ρ_e^k, Y_e^k, α...) result from a "Fluid Mechanics" formulation of the steady-state problem whereas we have developing so far a simpler "Chemistry" formulation. So we cannot avoid developing the former formulation. Secondly, to obtain in a consistent way a transfer function which is O(1) with respect to ε_1 (or equivalently to ε_2) each steady state variable must be found to its proper order, which is not necessarily O(1). For example X_e^l must be calculated to order ε_1^2.

STABILITY OF THE EQUILIBRIUM SOLUTIONS

The denominator of the transfer function (6.6) leads to the characteristic equation

$$\Psi(\overline{s}) = 0 \tag{7.1}$$

After laving neglected ε_5^2, consider surfaces in the $(U_1,\overline{R_e}, U_3)$ space (or equivalently in the $(U_2,\overline{R_e}, U_3)$ space) which are such that one root of (7.1) has a real part equal to zero. They generally separate regions (denoted D_k) having k roots with $\mathcal{Re}(\overline{s}) > 0$ and regions (denoted D_{k+1}) having k+1 roots with $\mathcal{Re}(\bar{s}) > 0$. The D-partition method is a mean of determining all these surfaces for which a root crosses the imaginary axis. For this, we set $\bar{s} = i\ \omega$ in(7.1) and we solve in terms of the frequency ω of the marginal oscillations the two equations $\mathcal{Re}[\Psi(i\ \omega)] = 0$ and $Im[\Psi(i\ \omega)] = 0$.

The *marginal stability surface* itself separates D_0 from D_1; it must be found among these surfaces by extra-considerations (a given equilibrium solution whose stability character is easy to show directly). This method has been applied in our case for equilibrium solutions which are stable from a static point of view, these solutions being sought for in the following. All these solutions have been found to be still stable from a dynamical point of view.

In order to find the static stability character of the equilibrium solutions, a simple method similar to the D-partition method has been devised. A necessary condition to apply this method is that for $\bar{s} \to 0$, $\Psi(\bar{s})$ possesses the asymptotic expansion

$$\Psi(\bar{s}) = a_0 + a_1\,\bar{s} + o(\bar{s}) \tag{7.2}$$

We restrict ourselves to follow the crossing of *real* roots. When such a root, say β, is in the neighborhood of $\bar{s} = 0$, the following relation holds

$$\Psi(\beta) = 0 \approx a_0 + a_1\,\beta \tag{7.3}$$

Thus $\beta \approx -a_0 / a_1$ and the singularity set or collection of turning points C is given by

$$a_0 = 0 \tag{7.4}$$

Moreover, knowing the signs of a_0 and a_1 provides the sign of β and which side of M about C is stable. Expanding (7.1) in a Taylor serie at $\bar{s} = 0$ gives $a_0 = a_0(U_1, \overline{R_e})$ and $a_1 = a_1(U_1, \overline{R_e}, U_3)$. It happens that $a_0 = \partial f / \partial \overline{R_e}$: this is not incidental and can be demonstrated for a large class of problems. Then it is natural to use the factorization theorem (Iooss and Joseph, 1980) which follows from equation (5.11) then we have, U_2 being fixed,

$$\beta = -\frac{a_0}{a_1} = -\left(\frac{\partial f}{\partial \overline{R_e}}\right) / a_1 = \left(\frac{\partial U_1}{\partial \overline{R_e}}\right)\bigg|_{U_2} \left(\frac{\partial f}{\partial U_1}\right) / a_1 \tag{7.5}$$

Since $(\partial f/\partial U_1)/a_1$ remains positive all along C and since $\left(\partial U_1 / \partial \overline{R_e}\right)\big|_{U_2}$ changes sign from the lower side (+) to the upper side (–), we have the following statement

$$\left.\begin{array}{l} \beta < 0 \Leftrightarrow \text{stability} \\ \beta < 0 \Leftrightarrow \left(\partial U_1 / \partial \overline{R_e}\right)\big|_{U_2} < 0 \end{array}\right\} \Rightarrow \left(\partial U_1 / \partial \overline{R_e}\right)\big|_{U_2} < 0 \Leftrightarrow \text{stability}$$

Of course, points on the lower side are unstable. It can be shown analogously that

$$\left(\partial U_2 / \partial \overline{R_e}\right)\big|_{U_1} < 0 \Leftrightarrow \text{stability}$$

Stable regions are given in figures 6 and 7.

CONCLUSION

First, a set of equations has been obtained to predict the dynamical behavior of a gas-vapour bubble in a solution. The present formulation, based on the theory of mixtures by

Muller (1985), allows a complete treatment of various phenomena which develop on a wide spectrum of time scales inside and outside the bubble. Inside the bubble, it replaces the very common assumption which states that the uniform gas pressure is given by the polytropic gas law with an arbitrary constant.

Secondly, dimensional analysis and elementary catastrophe theory have been used to capture under a *very compact form*, the whole static stability physics of a bubble in a liquid-gas solution of *bounded* extent. The catastrophe manifold (Fig 2) and its bifurcation set (Fig 3) sum up all results which have been spread out as yet in many dimensional graphs.

Thirdly, the above model is exploited to derive a linear dynamic stability analysis valid for slow transients (liquid mass diffusion controlled). Static stability results have been corroborated and extended.

Continuation of the present linear dynamic stability analysis including faster transients remains to be done and will reveal to what extent the static stability region bounded by the bifurcation (the "swallowtail tip") curve is maybe cut by new marginal dynamic stability curves.

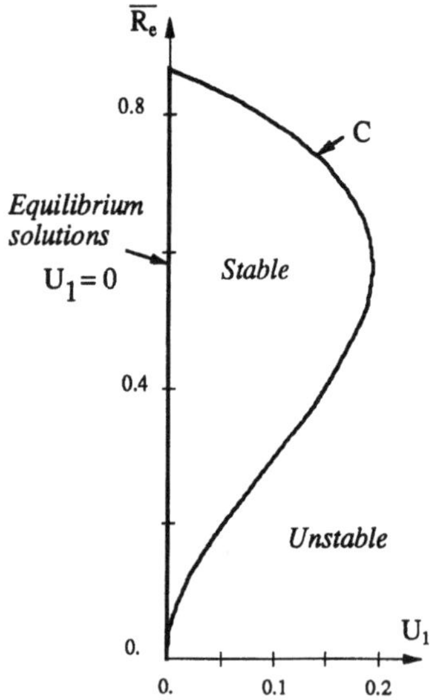

Figure 6. Projection of the singularity set C on a plane U_2 = Cte.

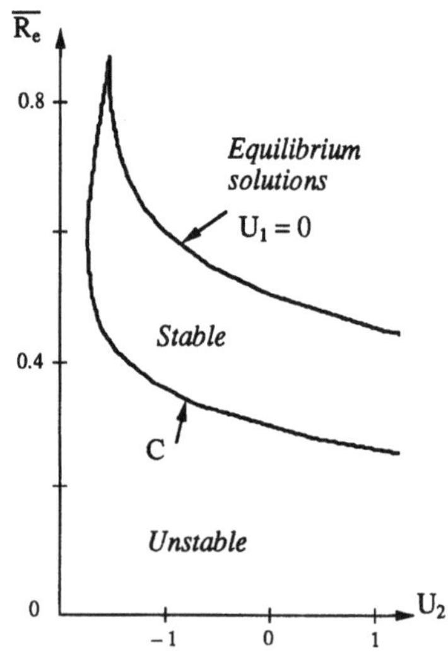

Figure 7. Projection of the singularity set C on a plane U_1 = Cte.

REFERENCES

Cha, Y.S., 1981, On the equilibrium of cavitation nuclei in liquid-gas solution, ASME *J. Fluids Engng.* 103:425.

Cha, Y.S., 1983, On bubble cavitation and dissolution, *Int. J. Heat Mass Transfer.* 26:1173.

Daily, J.W., and Johnson, V.E., Jr ,1956, Turbulence and boundary layer effects on cavitation inception from gas nuclei. *Trans* ASME. 78:1695.

Epstein, P.S., and Plesset, M.S., 1950, On the stability of gas bubble in liquid-gas solution, *J. Chem. physics.* 18:1505.

Franklin, R.E., Mc Millan, J., and Rozewicz, J., 1984, The stability and concentration of gas micro-bubbles in liquid, ASME *Int. Symp. on Cavitation inception.* New-Orleans.

Gibbs, J.W., 1961, *The Scientific Papers of J. Willard Gibbs,* Vol 1, Dover, New-York.

Iooss, G and Joseph, D.D., 1980, *Elementary Stability and Bifurcation Theory.*, Springer Verlag, New-York.

Iooss, G., Laure, P., and Rossi, M., 1989, Stability of a compressed gas bubble in a viscous liquid., *Phys. Fluids.* 1:915.

Majthay, A., 1985, *Foundations of Catastrophe Theory*, Pitman publishing, London.

Muller, I., 1985, *Thermodynamics*, Pitman publishing, London.

Mori, Y., Hijikata, K., and Nagatani, T., 1977, Fundamental study of bubble dissolution in liquid, *Int. J. Heat and mass transfer.* 20:41.

Plesset, M. S., and Sadhal, S.S., 1981, On the stability of gas bubble in liquid-gas solution, *J Proc. IUTAM Symp.* Pasadena.

Tsai, J.F., and Chen, Y.N., 1990, A generalized approach on equilibrium theory of cavitation in liquid-gas solutions, ASME *J. Fluids Engng.* 112:487.

Ward, C.A., Balakrishnan, A., and Hooper, F.C., 1970, On the thermodynamics of nucleation in weak gas-liquid solutions. *J. Basic Eng.* 92:695.

Ward, C.A., Tikuisis, P., and Venter, R.D., 1982, Stability of bubbles in a closed volume of liquid-gas solution, *J. Appl. Phys.* 53:6076.

MODELISATION OF PULVERISATION SYSTEMS :
SOME ASPECTS OF LINEAR STABILITY ANALYSIS

Ph. Camatte, I. Caré, Ch. Dumouchel, and M. Ledoux

URA CNRS 230/CORIA
University of Rouen
BP 118 - Mont Saint Aignan, Cedex, France

INTRODUCTION

Physical processes involved in pulverizing systems are classified generally as 1) pressure atomizers in which a liquid sheet is injected in a calm atmosphere or 2) air-blast atomizers : in this last case, liquid velocity is relatively low and pulverization is achieved through a high velocity air flow. In both cases destabilizing forces are due to an aerodynamic interaction between a gazeous fluid and a liquid flow. Recently pulverization processes have been classified into 1) prompt atomization Generally due to highest impaction angles between gas and liquid flows 2) classical atomization . In prompt atomization the deformation and breakup of the jet is due mainly to impacting forces (pressure diagram), analogous of what happens for a drop breakup. This situation will not be examined here. In classical pulverization, encountered in most practical cases, air and liquid flows are practically parallel (or coaxial). This last case will be considered in this paper.

A general scheme is adopted to modelize the division (for "classical" situation) of a sheet (or a cylindrical liquid jet) into drops :

- Phase 1 : An hydrodynamic instability is developping. Two cases are to be considered :

-If relative gas-liquid velocities are low (say < 30m/s) the liquid flow is perturbed as a whole and a volume breakup into ligaments is obtained. We call it pulverization.

-If relative gas-liquid velocities are high (say > 40m/s), the spatial scale (say the wavelength) or the instability is small regarding the sheet thickness. The instability is reduced to an single interface Kelvin-Helmholtz capillary instability. Small packets (or "blobs" of 10-40μm in dia) are teared off the liquid surface which is simply skimmed. This specific case will be named here atomization.

-Phase 2 : Secondary phenomena are occuring. In real (volume) pulverization, the ligaments are developping also instability leading to drop. In atomization, drops are liberated at the liquid velocity ; high gas-drop velocity is then found and secondary breakup will occur.

Adding to the complexity of this situation, it is well know that a real spray is not monosized but has to be defined through a drop size distribution.

Nevertheless some modelisation of pulverizing systems can be achieved. One can remark from the preceding scheme that the development of a primary instability is determinant for every spraying system. Although the breakup phenomena are strongly non-linear in nature, the growth of the initial perturbation is linear in its initial stage ; that leads to the idea that the length scale of the perturbation, the wave-length is representative of the future drop sizes (or initial blobs). According to this remark most of modelizing of spraying systems are based on linear theories. We want to show in this paper some recent developments of these linear methods, applied to specific geometries, as well as examples

Instabilities in Multiphase Flows, Edited by G. Gouesbet
and A. Berlemont, Plenum Press, New York, 1993

of the successful application of linear-grounded models to practical sprays calculations.

In many air craft engines, atomization is achieved by shearing a cylindrical liquid sheet between two swirled air flows. This rather complicated situation had to be modelized and a recent linear analysis is presented here. The linear instability theory has been recently applied successfully, associated to an hydrodynamical model of the internal flow in a real pressure swirl atomizer. In this case, of a cylindrical sheet in calm air, it has been shown that a non-dimensional presentation can be done of computing results. This representation is universal and can be helpful to designers.

At last, we present a case where Kelvin-Helmholtz theory, associated to simple balances of mass and energy, can give excellent results in a practical coaxial air-blast atomizer used in space rockets.

Air-blast atomization - Instability of an annular sheet

Air blast atomization is of a wide use, especially in aircraft engines where air is provided in excess by the compressor. In many aircraft atomizers an annular liquid fuel sheet is disintegrated by to coflowing air flows, one internal the other external. These air flows are generally swirling and at two different relative velocities with liquid flow.

Modelisation of such an atomizer involves two problems : 1)-instability of the annular sheet. 2)-breaking and restructuration into drops. The present paper deals with the first problem, i.e. the instability of an annular sheet surrounded by two coaxial air flows.

Few studies of this problem are found in the litterature (Crapper 1975, Martinon 1983, Guillié 1987, Meyer 1987, Dumouchel 1990, 1991) all with the same velocity on both sides of the sheet. An analysis is given here in the case of different velocities of internal and external air lows.

Linear analysis as used here is critiquable in nature since break-up processes are far from linearity. Nevertheless linear analysis has been previously used with succes in designing pressure swirl atomizers It has also to be pointed out that non linear theory (weakly non-linear or based on wortex sheet analysis) need results from linear calculations to be achieved. Furthermore interesting qualitative information can be drawn from linear analysis.

Analysis

The flow is shown in Fig.1 in stable configuration. Both liquid and gases are non-viscous (which can be accepted for low viscosity fluids as water or kerosene). A linear analysis is adopted ; for a notations are defined in Fig. 1.

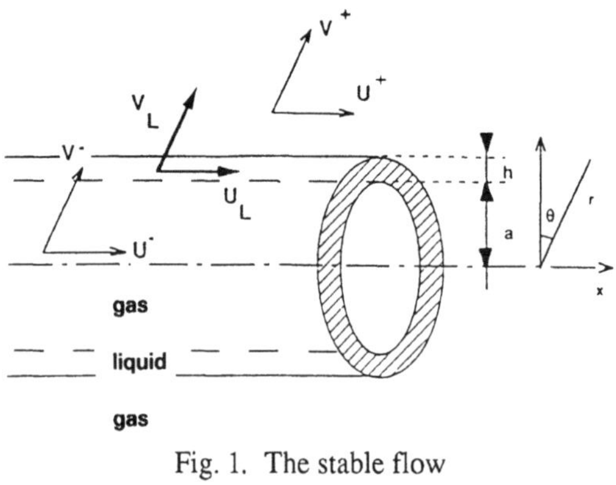

Fig. 1. The stable flow

General features

Both analysis are based on the same physical grounds. Time varying velocity field is decomposed into stable and instable part. In each flow this last part derives from a potential (φ^+, φ^- and φ_L) . η (x, t, θ) is the deviation of the surface from stable shape.

Potentials are searched under the separate from $R\,T(\theta)\,e^{iwt}$; $R = R(y)$ in case 1 (plane) ore $R = R(r)$ in case 2 (annular sheet). From following interface conditions η will be found of the form :

$$\eta = \eta_o \exp i (kx + n\theta - \omega t) \quad (1)$$

k is the axial wavenumber : $k = 2\pi/\lambda$
n is the azimuthal wavenumber
ω is a complex pulsation $\omega = \omega_r + i\omega_i$
ω_r is related to axial propagation velocity c $\omega_r = kc$
ω_i when positive leads to instability $\alpha\, e^{\omega_i t}$
η_o is the initial amplitude at each side of the sheet ; one finds then two values η_o^+ and η_o^-.

Continuity (2) and impulsion equation (3) under linearized from are written :

$$\Delta\varphi^+ = 0 \quad ; \quad \Delta\varphi^- = 0 \quad ; \quad \Delta\varphi_L = 0 \quad (2)$$

$$\frac{\partial u}{\partial t} + U\frac{\partial u}{\partial x} + \frac{V}{r}\frac{\partial u}{\partial \theta} = -\frac{1}{\rho}\frac{\partial p}{\partial x} \quad (3)$$

in each fluid (in case 1, $V \equiv 0$
Condition at the interface give :

$$\frac{\partial\varphi}{\partial r} = \frac{\partial\eta}{\partial t} + U\frac{\partial\eta}{\partial x} + \frac{V}{r}\frac{\partial\eta}{\partial\theta} \quad (4)$$

in each fluid at each interface.
Surface tension gives :

$$p_L - p_{gas} = \pm\sigma\frac{\partial^2\eta}{\partial x^2} \quad (5)$$

at each interface, under non linearized form.
Results will be shown here under non dimensionalized forms. U_M is the average velocity of the two air flows $U_M = (U^+ + U^-)/2$.

ΔU is the difference between the two air flows $\Delta U = U^+ - U^-$ ($\Delta U = 0$ leads to $U_M = U^+ = U^-$).

One introduces k_c which is the maximum value of k leading to instability in the case of symetrical plane sheet (Squire problem).

$$k_c = \frac{\rho_G U_M^2}{\sigma}$$

So

$$\tilde{h} = k_c h,\ \tilde{k} = \frac{k}{k_c},\ \tilde{\rho} = \frac{\rho_G}{\rho_L},\ \Delta\tilde{U} = \frac{\Delta\tilde{U}}{U_M},\ \tilde{\omega} = \frac{\omega}{k_c U_M}$$

By combining equations (2) to (5) a system of the following form is found for η_0^+ and η_0^-

$$F\,\eta_0^+ + G\,\eta_0^- = 0$$
$$G'\,\eta_0^+ + F\,\eta_0^- = 0 \quad (6)$$

F, G, F', G' are depending on the system studied. Detailed calculations can be obtained from the author. (6) shows that :
-The determinant FF' - GG' must be zero.
-η_0^+ and η_0^- cannot be independant. Their value can be arbitrary but their ratio η_0^+ / η_0^- can be calculated f rom - G/F or - F'/G'.
EF' - GG' gives a dispersion equations of the form :

$$H_4\tilde{\omega}^4 + H_3\tilde{\omega}^3 + H_2\tilde{\omega}^2 + H_0 = 0 \quad (7)$$

where coefficients are function of k and n. H coefficients can be found in the Annex.

Case 1 : The flat sheet

a)-U^+, U^-, U_L are parallel flows. The dispersion equation (7) has the form :

$$\left[-\omega^2 \coth 2kh-\tilde{\rho}\,(kU^+-\omega)^2+\frac{\sigma k^3}{\rho_L}\right]\left[-\omega^2 \coth 2kh-\tilde{\rho}(kU^--\omega)^2+\frac{\sigma k^3}{\rho_L}\right]-\frac{\omega^4}{sh^2 2kh} \tag{8}$$

The ratio between amplitudes on two sides is :

$$\frac{\eta_0^+}{\eta_0^-}=\frac{-\omega^2}{sh\,2kh\left[-\omega^2 \coth 2kh-\tilde{\rho}\,(kU^+-\omega)^2+\frac{\sigma k^3}{\rho_L}\right]}$$

$$=\frac{sh\,(2kh)\left[-\omega^2 \coth 2kh-\tilde{\rho}\,(kU^--\omega)^2+\frac{\sigma k^3}{\rho_L}\right]}{-\omega^2} \tag{9}$$

These equations are interesting not only as an asymptotic result ($a \to \infty$) of case 2, but also the equations allow visible simplifications in the asymptotic cas $kh \to \infty$.

b)-Asymptotic case $kh \to \infty$

This case corresponds either to great velocities U^+ and U^- (small λ, great k) or to a very thick sheets (h great).

In both cases $kh \to \infty$, $sh\,2kh \to \infty$ and term $\frac{\omega^4}{sh^2\,2kh}$ vanishes.

(8) is then factorizing in two equations of degree two which can be recognized as Kelvin-Helmholtz equations for simple interfaces. So each interface is atomized independently. η_0^+ / η_0^- has then no definite values.

c)- $U^+ = U^- = U$. This is the problem of Squire. The equation is factorized into two new equations of degree two, of the form :

$$A\,\omega^2+\tilde{\rho}\,(kU-\omega)^2-\frac{\sigma k^3}{\rho_L}=0$$

where A = tank kh or coth kh. These two values of A correspond respectively to values of η_0^+ / η_0^- equal + 1 (antisymetric perturbations) or -1 (symetric case). This coresponds to the problem analyzed by Hagerty and Shea (Camatte 1991).

A complete analysis of the results is not within the scope of this paper and is to be found in Camatte (1991 and 1992).

Annular sheet

A new parameter is now involved, the radius a. It will appear non-dimensionnally as $\tilde{a} = k_c\, a$. No place can be found here to show the analytic forms of the results.

It is to be remarked that results presented in (Camatte 1991) do not involve the hypothesis of very thin sheet (h << a) which was accepted in previous papers (Crapper, Martinon, Guillié, Mayer). It has been shown by the authors (Camatte 1991) that this hypothesis could not stand when $h/a > 10^{-5}$.

Two solutions are also found. Some fetaures can be observed (Camatte 1991 and 1992).

1)-Case $\Delta\tilde{U} = 0$

-$\tilde{\omega}_r$ has a jump around $\tilde{h}$ equal $10^{-2.5}$ $\tilde{\omega}_r$ is proportional to $\tilde{c}$ and the corresponding values of c are that of the gas (small $\tilde{h}$) or that of the liquid (high $\tilde{h}$).

-$\widetilde{\omega}_i$ is very small at lower $\widetilde{h}$. There is a very small velocity difference between the wave and the gas.

-$\widetilde{\omega}_i$ is small at higher $\widetilde{h}$. The inertia of the sheet becomes important.

-For $\widetilde{a} > 10$ solution become asymptotic to the solution relative to the flat sheet. This give a criteria to adopt then a simpler analysis (case 1).

2)-Case $\Delta\widetilde{U} = 1$

The distinction between aerodynamical (low $\widetilde{h}$.) and inertial domains (high $\widetilde{h}$.) is found again on $\widetilde{\omega}_r$ curves.

When $\widetilde{a}$ increases ($\widetilde{a} > 10$) the analysis relative to fat sheet is valid again. Notice that the analysis is also of practical use when flows are swirling.

Experimental study

In order to validate phase calculations, an experimental study was achieved. A cylindrical liquid sheet generator was especially designed and build , a complete description can be found in Camatte (1991, 1992). In addition, two coaxial air flows could be added one internal, one external in order to destabilize the sheet. Liquid sheets (water) 400μm in thickness, and respectively 4. 8, 7 and 17.8mm in internal diameter have been investigated : short time duration photographs as well as spectral analysis of the acoustic signal emitted led to a complete description of the different regimes observed for a set of values of the velocities of water flow (U_L) internal (U_i), external (U_e) air flow. The effect of air swirl was also examined. The influence of internal and sheet velocities were found determinant for $U_L \leq 0.5$m/s. The instability of the sheet is as expected. For $U_L > 1$m/s, a new phenomenon is appearing. The sheet is bubbling, i.e., a great scale deformation (bulges) is observed, the size of which is determined essentially by internal air velocity U_i. Smaller scale ripples, which can be assimilated to the "classical instability studied by linear analysis, are superimposed to the bulges.

For $U_L > 10$m/s, the deformation of the sheet is no more axisymmetric, and the disruption of the sheet is made through longitudinal threads. This effect, which is not yet explained satisfactorily at our knowledge has been denominated as "Rasta effect by the authors (Ledoux 1992).`

Pressure atomization - Practical use of linear theories

The complete modelisation of a swirl-type pressure atomizer involves three different steps : 1)-hydrodynamical model (viscous fluid) in the swirl chamber in order to determine the shape thickness and kinematics of the sheet. 2)-A model for first instability of the sheet. 3)-the modelling of breakup and droplet formation.

These three streps are rarely dealt with in papers. The authors have achieved such a work (Dumouchel 1990). Step 2n that is to say linear model of the instability of the sheet.

The first case ($U_L < 0.5$m/s) will only be considered here. A complete discussion of other situations may be found in (Camatte 1991, 1992). Photographs of plate 1 show what is observed. It can also be seen that wavelength can be clearly distinguished ; their value is then measured. These experimental values have been compared to those calculated from the previous linear theory. Comparison, shown in Figure 2 is really satisfactory.

Pressure atomization - Practical use of linear theories

The complete modelisation of a swirl-type pressure atomizer involves three different steps 1)-hydrodynamical model (viscous fluid) in the swirl chamber in order to determine the thickness and kinematics of the sheet. 2)-A model for first instability of the sheet. 3)-the modelling of breakup and droplet formation.

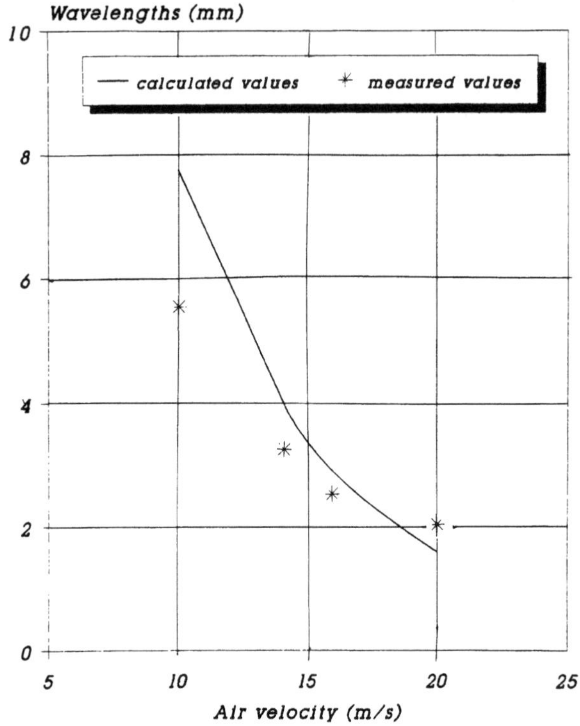

Fig. 2. Comparison between calculated and observed wavelengths

These three streps are rarely dealt with in papers. The authors have achieved such a work (Dumouchel 1990). Step 2n that is to say linear model of the instability of the sheet, will only be focused here. Although it was developed first, this model could be deduced through a simplification ($U_i = U_e = 0$) of the preceding one. The complete mathematics associated to this model may be found in (Dumouchel 1989, 1990, 1991). An emphasis will be put only on some practical aspects of the use of such a model by designers and on performances of such calculation. The following summary is based on a presentation by Dumouchel and al (1991).

The use of linear theories is based on the hypothesis made of the major effect of a dominant wavelength. Associated to a breakup model (which is possible for cylindrical or conical sheets, and more problematic for other cases) drop sizes are determined. Anyway, a single drop size is found ; Since the result of atomization in a real atomizer is polydisperse, this "global" theoretical size will be compared to different "mean" diameters deduced from the experimental drop size spectrum.

As shown previously, linear calculations can be tedious. For practical user, some simple rules, deduced from complete linear calculations, can be proposed. Two cases will be approached : the flat sheet (approximation of l inear or conical sheet of great radius), the cylindrical sheet (a succession of growing radius cylindrical sheets has been successfully replacing a conical sheet for instability calculations : see Dumouchel 1985 recall the problem is then to determine the most growing wavelength. Let k_m be the wavenumber of this wavelength, ω_m be its growth factor, h the thickness of the sheet, σ the surface tension of U_m liquid of density ρ_L, ρ_G the density of the gas, U the relative value of liquid/gas velocity and C the celerity of (growing) wave propagation.

For flate plate, we define three Weber numbers :

$$W_e = \frac{2\sigma}{\rho_L U^2 h} \qquad We_\omega = \frac{\omega_m \sigma}{\rho_G U^3} \tag{10}$$

$$W_{ek} = \frac{k_m \sigma}{\rho_G U^2} \qquad \overline{C}_{ei} = \frac{C}{U} \tag{11}$$

It was shown in (Dumouchel 1991) that for maximum growth :

$$W_e = 0.17 \qquad W_{e\omega} = 0.125$$

$$W_{ek} = 0.5 \qquad \overline{C}_{ei} = 0.75$$

For annular sheets, a new parameter, a, the radius of the sheet, induces a new number :

$$W_{ea} = \frac{\sigma}{\rho_G U^2 a} \tag{12}$$

Dominant waves are found in (Dumouchel 1991) to obey the following law :

$$W_{e\omega} = 0.8 \sqrt{W_e} \,.\, W_{ea} - 1.1\, W_e^2 + 0.11\, W_e + 0.07 \tag{13}$$

The "global" diameter d_G can be deduced, through a simple breakup model from the value of k_m :

$$d_G = 3.78 \sqrt{\frac{h}{kM}} \tag{14}$$

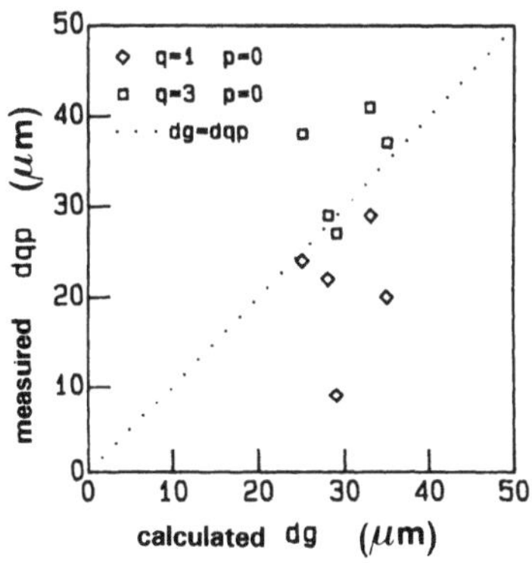

Fig. 3. Comparison between calculated and experimental mean drop sizes

Air Blast atomization - Surface skimming of the surface

When relative gas/liquid velocity becomes very important (100m/s and more) the scales of instability (λ) become very small compared to geometrical dimensions of the sheet (lets say thickness of the sheet).

In this case a five interfacial system is to be considered. An example of the application of a simple linear model to this case is given below.

Air blast atomization is widely used. In cryogenic rocket engines (like HM7 engine of european rocket Ariane) atomization of a liquid (or supposed so) jet of oxygen is achieved by a gaseous coflowing hydrogen flow. Structure of such a pulverisation is studied here through experiments and modelisation.

In such pulverisation devices great relative velocities between gas and liquid lead to a skimming of the surface : an attempt will be done here to calculate size of drops formed. The problem presents many similar features with Diesel injection. Present results should then find an application in this field.

Experiments

The atomization of a liquid cylindrical flow of initial radius a_0 by a surrounding annular air flow of lateral dimensions h has been achieved through a device described in Figure 4.

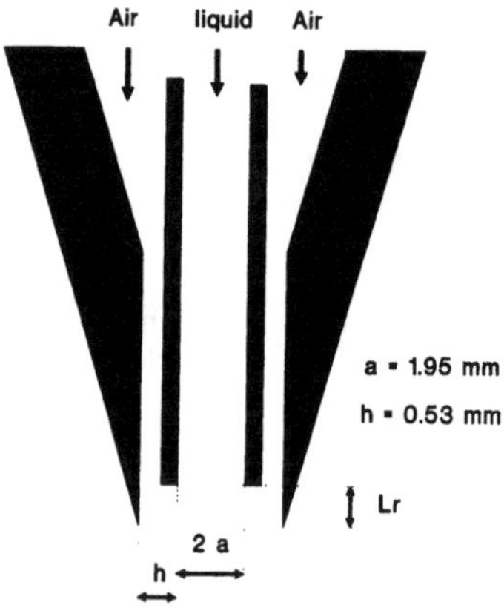

Fig. 4. The atomizer

Particular attention has been paid to axisymmetry (non swirling) of both flows. Water and kerosene have been used as substitutes to O_2. Velocity of liquids U_L was 1.4m/s. Air was used as atomizing fluid. Velocities U_G were varied from 100 to 220m/s at the nozzle exit. Photographic (short time -10μs- and long time exposure) investigations were achieved as well as LDV (for air), and Malvern spray sizing. Different procedures of enlightening of the spray may show the liquid core or the fine (scattering) spray. Dimensions of the experimental set-up are representative as well as relative velocities of those encountered on real rocket engines.

A complete structure of the spray is shown in (Caré 1990 and 1991). It will be only retained here that drops sizes could be measured, among other positions in the spray, at the interface (i.e. at the zone of their creation).

Modelisation of the atomizing zone

Drop size away from the surface (r > 5mm) can be influenced by several factors : 1)-Aerodynamical effects quoted above. 2)-Secondary atomization (breakup of drops by high velocity air stream). So calculations will be attempted here only for measurement made near the surface.

The authors developed previously a complete linear model of the instability of a (swirling) liquid jet in a (swirling) coflowing air current (Caré 1990). Due to high relative liquid-air velocities this model is in fact asymptotic to the classical Kelvin-Helmholtz theory for capillary waves at a simple interface. The dominant wavelength Λ of maximum growth Ω is then found.

$$\Lambda = \frac{3\pi\sigma}{\rho_G \, \Delta U^2} \qquad \Omega = \frac{2\rho_G \, \Delta U^3}{3\sqrt{3}\,\sigma}\left[\frac{\rho_G}{\rho_L}\right]^{1/2} \tag{15}$$

and the propagation velocity of the wave c is :

$$c = \frac{\rho_G \, U_G + \rho_L \, U_L}{\rho_L \approx U_L} \text{ according to } \frac{\rho_G}{\rho_L} = 0\,(10^{-3}) \tag{16}$$

From Λ and Ω one can deduce different parameters of the atomizing zone : SMD, (initial) angle of spray cone, skimming kinetics. Here we have to deal with two aspects of pulverisation : non linearity at breakup and polydispersion. This can be achieved in a simplified way.

Drop diameters : determination of SMD : The wave is sinusoïdal only in its initial part. Then the perturbation has the form : $\eta = \eta_0 \exp \Omega t \exp i\,(kx - \omega_r t)$ where $\omega_r = kc$. The velocity of the crest is $d\eta / dt = \Omega\eta$ where d/dt is a Lagrangian derivative. One assumes that, at breakup, the wave has the form of a cylindrical ligament of diameter $\Lambda / 2$ (non linear theory of Nayfeh (Caré 1990) gives insight of this) and amplitude $A\Lambda$ where A is an unknown constant.

In reality atomization results in a spectrum of drops of diameters d_i. N_i being the number of drops of diameter d_i formed, one can write the conservation of mass for a ligament

$$\sum N_i \, d_i^3 = 3A\,\Lambda^3 / 8 \tag{17}$$

One can also conserve surface tension energy $\sum N_i \, d_i^2 = A\Lambda^2$

SMD is defined as :

$$d_{32} = \sum N_i \, d_i^3 \, / \, \sum N_i \, d_i^2$$

Unknown constant A is then eliminated, and some effect of polydispersion is included in definition of d_{32} : $d_{32} = 3/4\Lambda$.

Mean number of drops formed by ligament is then or order A : N=(8/9) / A.

Skimming kinetics : An important parameter, not yet found at our knowledge in the litterature, is the mass flow rate of drops teared off by unit surface. This is needed to predict the evolution of the diameter of the liquid cone.

From different successive balances (summarized in Caré 1990) the skimming kinetics can be but under the form :

$$q_v = C\,\frac{\Lambda\Omega}{U_L}$$

So where q_v is the volumetric flux of liquid (m^3/m^2.s) teared off the surface, Λ and Ω respectively wavelength and growth factor of dominant (linear) wave, U_L liquid v elocity. C has to be experimentally derived. It is questionable if it is a constant . In (Caré 1991) a value of $1.8 \; 10^{-2} \pm 5.10^{-3}$ is proposed. Some further experiments from the authors show that C is in fact affected by parameters such as velocity of gas and liquid. More additional experimental information is needed. Also in (Caré 1991) some considerations on spray angle calculations may be found.

Assessment of the model

Values of SMD at the interface have been measured, through a Malvern sizer and an inversion tomographie technique. Values of these SMD are compared in Fig. 5.

With drop diameters calculed from the model described above. The agreement is remarkable.

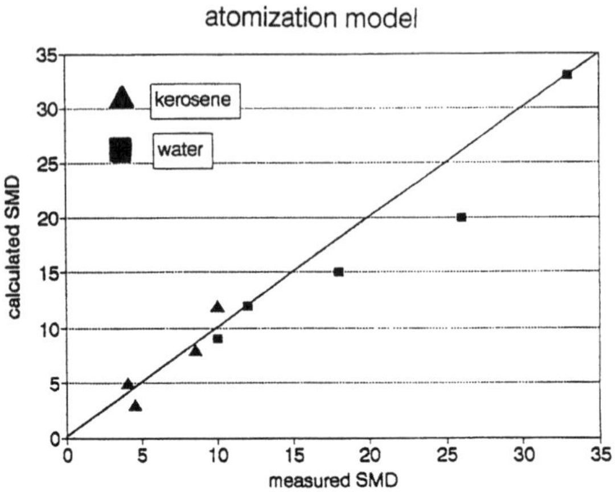

Fig. 5. Comparison between calculated and SMD measured drop sizes

CONCLUSION

The whole paper addresses one main question : Although atomization processes are mainly non-linear, may linear theories be working well?

The answer is yes for the three exemples shown here. More precisely, linear theories can provide, at least for engineering purpose, good scales of instabilities, as well as reasonably predicted "global" drop sizes.

Another question is : What can be the effects of non-linearity ? The authors are at present time working on different aspects of non-linear interactions : mode interactions, effects of an external excitation. These effects can be of practical interest : the initial (physical) perturbation leading to destabilization is not generally sinusoïdal. Following linear analysis, each sinusoïdal component of the Fourrier's decomposition of the initial perturbation is evoluting separately and the "best wins". When the non-linear (especially V.grad terms of dynamical equation) terms are taken into account this independance of sinusoïdal components is no more evident. On the opposite the authors (Internal report, Cousin 1992) showed that mode interaction can lead to non-sinusoïdal shapes : this can be interpreted as a route to polydispersion.

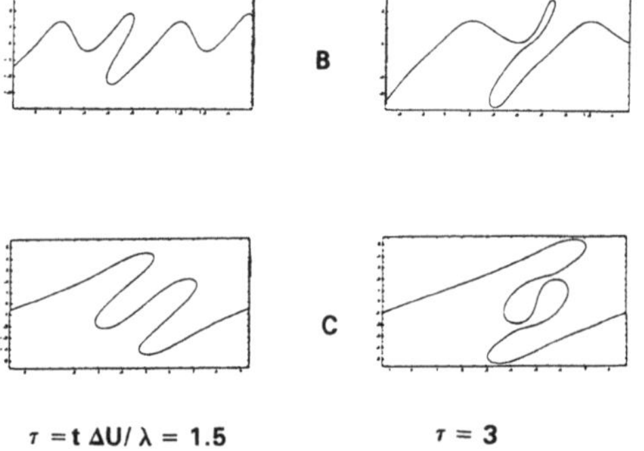

Fig. 6. Different aspects of interface due to mode interaction

Another aspect (treated in the same report) is the income of non-linearity during external excitation (acoustics, turbulence in gas or liquids...). Different techniques can be

ANNEX

This annex shows the (rather complicated) shape of the dispersion equation derived from the linear analysis of the stability of a cylindrical sheet teared off by two coaxial (swirling, internal and external) air flows.

Equations of dispersion : annular sheet

$$\omega^4 \left[-\rho_l^2 (P\,Q - R\,S) - \frac{\rho_g^2}{k^2} K\,I + \frac{\rho_g \rho_l}{k} (I\,P + K\,Q) \right]$$

$$+ 2\,\omega^3 \left[\rho_l^2 (W_0^+ + W_0^-) (P\,Q - R\,S) + 2 \frac{\rho_g^2}{k^2} (W_1 + W_2) K\,I - \frac{\rho_g \rho_l}{k} [(W_1 + W_0^-) K\,Q] \right]$$

$$[[+ (W_0^+ + W_2)\, I\,P]]$$

$$+ \omega^2 \left[-\rho_l^2 (W_0^{+2} + W_0^{-2} + 4\, W_0^- W_0^+) (P\,Q - R\,S) - \frac{\rho_g^2}{k^2} (W_1^2 + W_2^2 + 4\, W_1 W_2)\, I\,K \right]$$

$$\left[+ \frac{\rho_g \rho_l}{k} [(W_0^{+2} + W_2^2 + 4\, W_0^+ W_2)\, I\,P + (W_0^{-2} + W_1^2 + 4\, W_0^- W_1)\, K\,Q] \right]$$

$$\left[+ \rho_l\, \sigma\, (Ka^+ Q - Ka^- P) + \frac{\rho_g}{k} \sigma\, (Ka^- K - Ka^+ I) \right]$$

$$+ 2\,\omega \left[\rho_l^2 (W_0^{+2} W_0^- + W_0^{-2} W_0^+) (P\,Q - R\,S) + \frac{\rho_g^2}{k^2} (W_1^2 W_2 + W_2^2 W_1)\, K\,I \right]$$

$$+ 2\,\omega \left[\rho_l^2 (W_0^{+2} W_0^- + W_0^{-2} W_0^+) (P\,Q - R\,S) + \frac{\rho_g^2}{k^2} (W_1^2 W_2 + W_2^2 W_1)\, K\,I \right]$$

$$\left[+ \rho_l\, \sigma\, (W_0^+ Ka^- P - W_0^- Ka^+ Q) + \frac{\rho_g}{k} \sigma\, (W_2\, Ka^+ I - W_1\, Ka^- K) \right]$$

$$-\rho_l^2 W_0^{+2} W_0^{-2} (P\,Q - R\,S) - \frac{\rho_g^2}{k^2} W_1^2 W_2^2 K\,I + \frac{\rho_g \rho_l}{k} (W_0^{+2} W_2^2\, I\,P + W_1^2 W_0^{-2} K\,Q)$$

$$+ \rho_l\, \sigma\, (W_0^{-2} Ka^+ Q - W_0^{+2} Ka^- P) + \frac{\rho_g}{k} \sigma\, (W_1^2 Ka^- K - W_2^2 Ka^+ I) + \sigma^2 Ka^+ Ka^- = 0$$

$$P = \Delta\, (K'^{+}_n I^{+}_n - I'^{-}_n K^{+}_n) \qquad W_0^+ = k\, U_l + n\, \Omega_l\, (a + h)^{-2}$$

$$Q = \Delta\, (I'^{+}_n K^{-}_n - K'^{+}_n I^{-}_n) \qquad W_0^- = k\, U_l + n\, \Omega_l\, a^{-2}$$

$$R = \Delta\,(I'^{+}_{n}\,K^{+}_{n} - K'^{+}_{n}\,I^{+}_{n})$$

$$W_1 = k\,U^{+} + n\,\Omega^{+}\,(a+h)^{-2}$$

$$S = \Delta\,(K'^{-}_{n}\,I^{-}_{n} - I'^{-}_{n}\,K^{-}_{n})$$

$$W_2 = k\,U^{-} + n\,\Omega^{-}\,a^{-2}$$

$$\Delta = \left[k\,(K'^{-}_{n}\,I'^{+}_{n} - K'^{+}_{n}\,I'^{-}_{n})\right]^{-1}$$

$$Ka^{+} = \left[k^2\,(a+h)^2 + n^2 - 1\right](a+h)^{-2}$$

$$K = \frac{K^{+}_{n}}{K'^{+}_{n}} \qquad I = \frac{I^{-}_{n}}{I'^{-}_{n}}$$

$$Ka^{-} = \left[k^2\,a^2 + n^2 - 1\right]a^{-2}$$

Asymptotic expansion : case of flat plate

ka and $k(a+h) \to \infty$

$$\omega^4 \left[\frac{\rho_l^2}{k^2} + 2\,\frac{\rho_g\,\rho_l}{k^2}\coth kh + \frac{\rho_g^2}{k^2}\right]$$

$$-\,2\,\omega^3\left[2\,\frac{\rho_l^2}{k}\,U_l + \frac{\rho_g\,\rho_l}{k}\coth kh\,(2\,U_l + U^{+} + U^{-}) + \frac{\rho_g^2}{k}\,(U^{+} + U^{-})\right]$$

$$+\,\omega^2\left[6\,\rho_l^2\,U_l^2 + \rho_g\,\rho_l\coth kh\,(2\,U_l^2 + 4\,U_l\,(U^{+} + U^{-}) + U^{+2} + U^{-2})\right]$$

$$\left[+\,\rho_g^2\,(U^{+2} + U^{-2} + 4\,U^{+}\,U^{-}) - 2\,\rho_l\,k\,\sigma\coth kh - 2\,\rho_g\,k\,\sigma\right]$$

$$-\,2\,\omega\left[2\,k\,\rho_l^2\,U_l^3 + k\,\rho_g\,\rho_l\coth kh\,(U_l^2\,(U^{+} + U^{-}) + U_l\,(U^{+2} + U^{-2}))\right]$$

$$\left[+\,k\,\rho_g^2\,(U^{+}\,U^{-}\,(U^{+} + U^{-}) - 2\,k^2\,\rho_l\,\sigma\,U_l\coth kh - k^2\,\rho_g\,\sigma\,(U^{+} + U^{-})\right]$$

$$+\,k^2\,\rho_l^2\,U_l^4 + k^2\,\rho_g\,\rho_l\coth kh\,U_l^2\,(U^{+2} + U^{-2}) - 2\,k^3\,\rho_l\,\sigma\,U_l^2\coth kh$$

$$+\,\rho_g^2\,k^2\,U^{+2}\,U^{-2} - k^3\,\rho_g\,\sigma\,(U^{+2} + U^{-2}) + k^4\,\sigma^2 = 0$$

In the specific case of pressure atomization :

$U^{+} = U^{-} = 0$

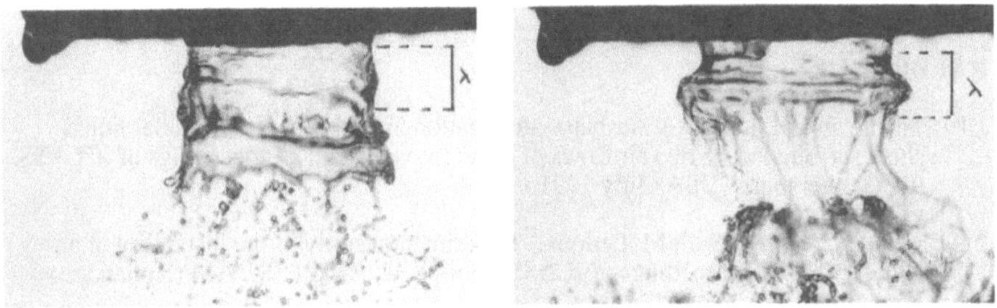

P.14. $U_l = 0.5 ms^{-1}$ $U_i = U_e = 10 ms^{-1}$ 2 cm

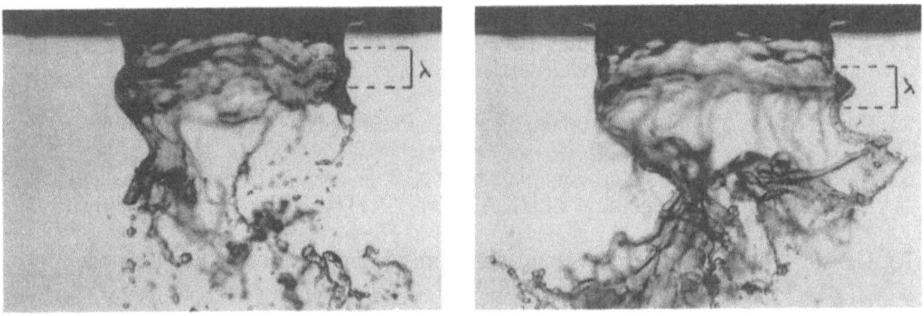

P.15. $U_l = 0.5 ms^{-1}$ $U_i = U_e = 14 ms^{-1}$ 2 cm

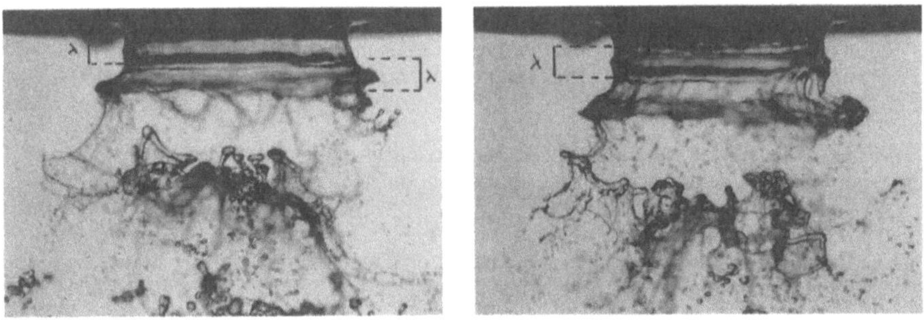

P.16. $U_l = 0.5 ms^{-1}$ $U_i = U_e = 16 ms^{-1}$ 2 cm

D=17.76mm h=0.5mm

PLATE 1

used to study the effects of non-linear terms on interface stage evolution. Authors used discretized vortex sheet calculations (see Rangel et al 1989, Camatte 1992) Volume of fluid (VOF) methods are also of interest.

As a conclusion, non-linear techniques are of a great interest in the present as a research tool, and in the future probably as a conception tool. For the moment, for the engineer analytical linear methods keeps yet many potentialities.

REFERENCES

1.P. Camatte and M. Ledoux. "Air blast atomization : instability of an annular liquid sheet surrounded by two air flows of different velocities". Proceedings of ICLASS 91, Gaithersburg (USA) July 1991.

2.P.Camatte, C. Dumouchel, M. Ledoux. "Experimental study of the instability of an annular sheet", Proceedings of ILASS, Europe 92, AMSTERDAM (Holland) september 1992.

3.P. Camatte, M. Ledoux. "Modelizaton of annular sheet instability". Brite Euram Aero 1019. Report Task 4.1.2, 1992.

4.P. Camatte. "Etude des instabilités d'une nappe liquide de géométrie annulaire en pulvérisation assistée. Modèle - Expérience. Ph.D Thesis. ROUEN 1992.

5.I. Caré. "Etude d'un injecteur coaxial assisté". Ph.D Thesis. ROUEN 1990.

6.I. Caré and M. Ledoux. "Study of an air-blast coaxial atomizer : experiments and modelization", Proceedings of ICLASS 91, GAITHERSBURG (USA) July 1991.

7.J. Cousin. "Méthode de feuillets tourbillonnaires discrétisés. Interaction de mode forçage". Internal Report (DEA). ROUEN 1992.

8.G.D. Crapper, N. Dombrowski and G.A.D. Pyott. "Kelvin-Helmholtz wave growth on cylindrical sheets". Journ. of Fluid Mech. 68, 497 (1975).

9.J.B. Dementhon, F. Vannobel, C. Dumouchel and M. Ledoux. "A new approach of the pulverisation mechanisms involved in a gasoline spray : video imaging and modelling", Proceedings of ILASS, Europe 92, AMSTERDAM (Holland) September 1992.

10.C. Dumouchel, I. Caré and M. Ledoux. "Modélisation des effets de la pression ambiante et de la viscosité sur la pulvérisation". 4èmes Journées d'études sur les Aérosols, Paris, 1-2 Déc. 1987.

11.C. Dumouchel. "Problèmes liés à la conception d'un pulvérisateur mécanique : hydrodynamique de chambre et instabilité de nappe". Ph.D Thesis. ROUEN 1989.

12.C. Dumouchel, M.I. G. Bloor, N. Dombrowski, D.B. Ingham and M. Ledoux. "The design of pressure swirl atomizers". Proceedings of 23rd Symposium (International) on Combustion (The Combustion Institute), pp. 1461-1467 (1990).

13.C. Dumouchel and M. Ledoux. "Atomization of flat and annular liquid sheets. Practical use of linear theories". Proceedings of ICLASS 91, GAITHERSBURG (USA), July 1991.

14.C. Dumouchel, M. Ledoux, M.I.G. Bloor, N. Dombrowski and D.B. Ingham. "Boundary layer characteristics in a swirl atomizer. Atomization and Sprays".

15.J. Guillé. "Etude expérimentale et théorique de l'injection aérodynamique". Ph.D Thesis, Ecole Nationale des Mines, Paris 1987.

16.W.W. Hagerty and J.F. Shea. "A study of the stability of the plane fluid sheets". Journ. Appl. Mech., 509, Dec. 1955.

17.J.M. Kendall. "Experiments on annular liquid jet instability and on the formation of liquid shells". Phys. Fluids 29, 2086 (1986).

18.M. Ledoux. 3Some remarks about modelization of sprays". Atomization and Sprays 2000 NSF. Workshop GAITHERSBURG 1991. Edited by N. Chigier.

19.C.P. Lee and T.G. Wang. "A theoretical model for the annular jet instability". Phys. Fluids 29, 2076 (1986).

20.J.G. Lee and L.D. Chen. "Linear stability analysis of gas-liquid interface". AIAA Journ. 29, 1589 (1991).

21.J. Martinon. "Instabilités d'une nappe liquide cylindrique en présence de deux écoulements gazeux". Journal de Mécanique Théorique et Appliquée 2, 629 (1983).

22.J. Meyer and D. Weihs. "Capillary instability of an annular liquid jet". Journ. Fluid Mechanics 179, 531 (1987).

23.R.D. Reitz. "Modeling atomization processes in high pressure vapozizing sprays". Atomization and Sprays Technology 3, 309 (1987).

24.H.B. Squire. "Investigation on the instability of a moving liquid film". Brit. Journ. of Appl. Phys. 4, 167 (1953).

OVERSTABILITY IN AN INFINITE LIQUID LAYER UNDER SIMULTANEOUS SURFACE TENSION, BUOYANCY AND SHEAR EFFECTS

G. Gouesbet, C. Rozé and J. Maquet

Laboratoire d'Energétique des Systèmes et Procédés
INSA de Rouen, URA CNRS 230
BP 08
76131 Mont-Saint-Aignan Cedex
France

I. INTRODUCTION

There has been a long tradition concerning the theoretical study of Rayleigh-Bénard and Bénard-Marangoni instabilities, with papers by Rayleigh (1916), Pearson (1958), Scriven and Sternling (1964) and Nield (1964) among others. The principle of exchange of stability holds for Rayleigh-Bénard problems. Conversely, for Bénard-Marangoni problems, the liquid layer may exhibit overstability. Linear analysis of overstability is much more difficult to perform than in the case of exchange of stability because eigenvalues at marginality are no more equal to 0. This is likely to be the reason why the overstability problem has been solved in the case of a pure Marangoni mechanism only about ten years ago by Takashima (1981). Thereafter, the question naturally arose to know how overstability would set in when both buoyancy and surface tension mechanisms were simultaneously acting. This problem was solved recently by

Instabilities in Multiphase Flows, Edited by G. Gouesbet and A. Berlemont, Plenum Press, New York, 1993

Gouesbet and Maquet (1989), Gouesbet et al (1990) and, independently, by Perez-Garcia and Carneiro (1991). Some specific calculations were devoted to the case of silicon oils in Gouesbet and Maquet (1989). For this liquid, it appears that the onset of oscillatory behaviour required very high driving temperature differences of about typically 500 to 5000 K, conflicting even with the very existence of the liquid state. We have then been interested in knowing whether supplementary effects like extra destabilization by basic state convection would not decrease these values to lower more reasonable ones. Pure shear being the simplest case of convection, this paper is therefore devoted to the loss of stability of an infinite horizontal liquid layer under simultaneous surface tension, buoyancy and shear effects, leading to oscillatory behaviour. The linear problem to be solved is furthermore of special mathematical interest because the introduction of shear produces a singular problem, in contrast with previous studies without any shear.

There was an original motivation in undertaking this work, namely the understanding of thermal lens oscillations as discussed in Gouesbet and Maquet (1989). The presentation of the present paper in the light of this framework is postponed to Section III.

The instability under study in this paper is investigated by means of a small disturbance (linear) analysis. The reader wanting to check and use our derivations is kindly invited to check first our computations in Gouesbet and Maquet (1989) as an exercice. This being done, we believe that the omitted material would essentially be straightforward although tedious algebra.

II.THE FORMULATION

II.1.Geometry of the Problem and θ-Equation

A homogeneous liquid layer of infinite extent is limited by a solid wall and a free surface. The depth of the layer when the surface is undeformed is d and the ambiant temperature in the gaseous medium, far enough from the surface, is T_a. We use a Cartesian coordinate system $x_i=(x,y,z)$ in which axes x and y are located on the solid wall and axis z is oriented positively towards the free surface, i.e. upwards or downwards depending on whether the free surface is above or below the solid wall (Fig 1).

As usual in Rayleigh-Bénard-Marangoni instability problems, we impose a vertical temperature gradient :

$$\frac{\partial T}{\partial x_i} = (0,\ 0,\ -k_t) \tag{1}$$

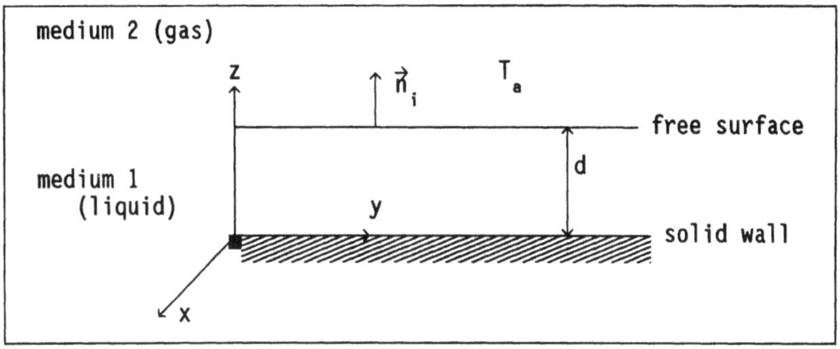

Figure 1. Geometry of the problem.

in which T is temperature and k_t is minus dT/dz. We also impose a pure shear such as the liquid velocity U_i is given by :

$$U_i = (0,\ (d-z)\ k_u,\ 0) \qquad z \in [0,d] \tag{2}$$

in which axis y is taken in the direction of the flow. Similarly as for k_t, k_u is minus dU_y/dz. Base state defined by (1)-(2) is compatible with the fluid governing equations.

We assume that the liquid complies with the Oberbeck-Boussinesq approximation (Joseph, 1976; Oertel,1982). The procedure is then similar with the one in Gouesbet and Maquet (1989). It essentially consists in writing linear perturbation equations, separating space and time dependences, invoking the normal mode decomposition and manipulating to obtain the pivot equation of the formulation concerning the $\theta(z)$ component of the temperature perturbation $f(x,y)\theta(z)$, in which $f(x,y)$ is the planform function satisfying the Helmholtz equation. This θ-equation reads:

$$B_{28}Z^2\, D^8\theta + B_{17}Z\, D^7\theta + (\, B_{06} + B_{26}Z^2 + B_{36}Z^3)\, D^6\theta$$

$$+ (\, B_{15}Z + B_{25}Z^2)\, D^5\theta + (\, B_{04} + B_{14}Z + B_{24}Z^2 + B_{34}Z^3 + B_{44}Z^4)\, D^4\theta$$

$$+ (\, B_{03} + B_{13}Z + B_{23}Z^2 + B_{33}Z^3)\, D^3\theta \tag{3}$$

$$+ (\, B_{02} + B_{12}Z + B_{22}Z^2 + B_{32}Z^3 + B_{42}Z^4\,)\, D^2\theta$$

$$+ (\, B_{01} + B_{11}Z + B_{21}Z^2 + B_{31}Z^3)\, D\theta$$

$$+ (B_{00} + B_{20}Z^2 + B_{30}Z^3 + B_{40}Z^4)\, \theta = 0.$$

in which Z=(d-z), D is the operator d/dz (coefficients B_{ij} are not given here).

II.2.Boundary Conditions

Solving (3) requires the knowledge of boundary conditions at the solid wall and at the free surface. These conditions are more complex than in the no-shear case but may be derived following the same procedure than in Gouesbet and Maquet (1989). We have five conditions at the solid wall, given below. Letters c and i indicate the so-called conductor and insulating cases respectively.

(i) first condition (basic thermal condition).

$$\theta=0, \quad z=0 \ (c) \quad \text{and} \quad D\theta=0, \quad z=0 \ (i) \tag{4-5}$$

(ii) second condition (by kinematic conditions).

$$D^2\theta=0, \quad z=0 \ (c) \quad \text{and} \quad (D^2-\alpha^2q^2)\theta = \frac{i\alpha_y k_u d}{a_T}\theta, \quad z=0 \ (i) \tag{6-7}$$

(iii) third condition (by using again kinematic conditions).

$$(D^2-\alpha^2q^2)D\theta = \frac{i\alpha_y k_u d}{a_T} D\theta, \ z=0 \ (c) \ \text{and} \ D^3\theta = -\frac{i\alpha_y k_u}{a_T}\theta, \ z=0 \ (i) \tag{8-9}$$

(iv) fourth condition (by using a temperature equation).

$$\mu a_T(D^2-\alpha^2r^2)(D^2-\alpha^2q^2)(D^2-\alpha^2)\theta = i\alpha_y k_u d \Big(\mu\, (D^2-\alpha^2r^2)(D^2-\alpha^2)\theta \tag{10}$$
$$+ \rho_0 a_T\Big[(D^2-\alpha^2q^2)D^2\theta + \frac{2i\alpha_y k_u}{a_T} D\theta\Big]\Big) - 4\mu i\alpha_y k_u\Big[D^2-\alpha^2\frac{r^2+1}{2}\Big]D\theta, z=0 \ (c)$$

$$\mu a_T(D^2-\alpha^2r^2)(D^2-\alpha^2q^2)(D^2-\alpha^2)\theta =$$
$$i\alpha_y k_u d\Big(\mu(D^2-\alpha^2r^2)(D^2-\alpha^2)\theta + \rho_0 a_T\Big[(D^2-\alpha^2q^2)D^2\theta - \frac{i\alpha_y k_u d}{a_T} D^2\theta\Big]\Big) \tag{11}$$
$$- 4\mu i\alpha_y k_u\Big[D^2-\alpha^2\frac{r^2+1}{2}\Big]D\theta - k_t\rho_0\alpha_V g\alpha^2\theta, \quad z=0 \quad (i)$$

(v) fifth condition (by using again the same temperature equation).

$$\mu a_T(D^2-\alpha^2r^2)(D^2-\alpha^2q^2)(D^2-\alpha^2)D\theta =$$
$$-i\alpha_y k_u\Big(\mu(D^2-\alpha^2r^2)(D^2-\alpha^2)\theta + \rho_0 a_T\Big[(D^2-\alpha^2q^2)D^2\theta + \frac{2i\alpha_y k_u}{a_T} D\theta\Big]\Big) \tag{12}$$
$$+i\alpha_y k_u d\Big(\mu(D^2-\alpha^2r^2)(D^2-\alpha^2)D\theta + \rho_0 a_T\Big[(D^2-\alpha^2q^2)D^3\theta - \frac{i\alpha_y k_u d}{a_T} D^3\theta\Big]\Big)$$
$$-4\mu\, i\alpha_y k_u\Big[D^2-\alpha^2\frac{r^2+1}{2}\Big]D^2\theta - k_t\rho_0\alpha_V g\alpha^2 D\theta, \quad z=0 \qquad (c)$$

$$\mu a_T(D^2-\alpha^2r^2)(D^2-\alpha^2q^2)(D^2-\alpha^2)D\theta =$$
$$-i\alpha_y k_u\Big(\mu(D^2-\alpha^2r^2)(D^2-\alpha^2)\theta + \rho_0 a_T\Big[(D^2-\alpha^2q^2)D^2\theta - \frac{i\alpha_y k_u d}{a_T} D^2\theta\Big]\Big) \tag{13}$$

$$+i\alpha_y k_u d \left[\mu(D^2-\alpha^2 r^2)(D^2-\alpha^2)D\theta + \rho_0 a_T\left[(D^2-\alpha^2 q^2)D^3\theta + \frac{i\alpha_y k_u}{a_T}(3D^2\theta - d.D^3\theta)\right]\right]$$

$$-4\mu\ i\alpha_y k_u \left[D^2-\alpha^2\frac{r^2+1}{2}\right]D^2\theta, \quad z=0\ (i)$$

The five obtained thermal conditions at the solid wall are linearly independent. We shall later observe that the determination of the θ-solution requires nine unknown elements. Since there are four boundary conditions at the free surface (see below), we conclude that there is no other independent thermal condition to be found. Also, before processing later, the reader would find convenient to re-express these boundary conditions in terms of D^n-expansions.

The four boundary conditions at the free surface read as follows.

(i) first condition (kinematic condition).

$$(D^2-\alpha^2 q^2)\theta + \bar{\omega}\ Pr\ \alpha^2\ \Delta_s = 0 \qquad (14)$$

in which we introduced the reduced time constant $\bar{\omega}$, the Prandtl number Pr and the new unknown element Δ_s :

$$\bar{\omega} = \frac{\beta\rho_0}{\mu\alpha^2}, \qquad Pr = \frac{\mu}{\rho_0 a_T}, \qquad \Delta_s = k_t\delta_s \qquad (15\text{-}17)$$

in which δ_s is the normal mode z-component of the spatial component $f(x,y)\delta_s(z)$ of the free surface perturbation location.

(ii) second condition (normal force condition).

$$Cri.\left[D^2-\alpha^2(r^2+2)\right](D^2-\alpha^2 q^2)D\theta + Cri.Sh.\frac{i}{d^3}\left[3D^2\theta - \alpha^2(r^2+2)\theta\right] +$$

$$\left[\frac{\alpha^4}{d} + Bo.\frac{\alpha^2}{d^3} + \frac{i}{d^3}.Cri.Sh.\bar{\omega}\alpha^2\right]\Delta_s = 0 \qquad (18)$$

in which we introduced the cripation number Cri, the Bond number Bo and the shear number Sh :

$$Cri = \frac{\mu a_T}{\sigma_s d}, \qquad Bo = \frac{\rho_0 g_s d^2}{\sigma_s}, \qquad Sh = \frac{d^3\alpha_y k_u}{a_T} \qquad (19\text{-}21)$$

(iii) third condition (tangential force condition).

$$(D^2+\alpha^2)(D^2-\alpha^2 q^2)\theta + Vi.d\alpha^2(D^2-\alpha^2 q^2)D\theta + \frac{2i.Sh}{d^2}\left[\frac{D\theta}{d} + \frac{1}{2}Vi.\alpha^2\theta\right]$$

$$-\ Ma.\frac{\alpha^2}{d^2}\theta + Ma.\frac{\alpha^2}{d^2}\Delta_s = 0 \qquad (22)$$

in which we introduced the viscosity number Vi and the Marangoni number Ma

$$Vi = \frac{\kappa+\epsilon}{\mu d}, \qquad Ma = -\frac{\Gamma k_t d^2}{\mu a_T} \qquad (23\text{-}24)$$

in which $(\kappa+\epsilon)$ is the sum of interfacial viscosities and Γ is $\partial\sigma_s/\partial T\ \Big|_{T=T_s}$ with σ_s the surface tension.

(iv) fourth condition (heat transfer condition).

$$d\ D\theta + Bi\ \theta - Bi\ \Delta_s = 0 \qquad (25)$$

in which Bi is the Biot number also sometimes called the Nusselt number.

II.3.The θ-Solution

We now solve the θ-equation (3). Coefficients B_{ij} form a matrix of dimension (5x9) in which the coefficients which do not appear in the equation are set to 0. (3) may then be more concisely rewritten as :

$$\sum_{j=0}^{8}\left[\sum_{i=0}^{10} E_{ij}Z^{i}\right]Z^{j}\mathcal{D}^{j}\theta = 0 \tag{26}$$

in which $\mathcal{D}$ is the operator d/dZ = -D and E_{ij} ($0\le i\le 10$, $0\le j\le 8$) is a matrix whose elements are given by :

$$E_{ij}= (-1)^{j}B_{i+j-6,j} \quad \text{if } 6\le i+j\le 10 \text{ and } =0 \text{ otherwise} \tag{27}$$

in which some E_{ij} for $6\le i+j\le 10$ are also 0 because some B_{ij}'s are 0.

The eight order equation (26) is singular, admitting a pole at Z = 0 (z = d).But the singularity is regular, according to Fuchs theorem (Ince, 1926, p 365). Then the solution may be obtained by using the Frobenius method (Ince, 1926, chapter XVI). According to this method, special solutions are free of logarithms in our case and read:

$$\theta^{\sigma}=\sum_{k=0}^{\infty} c_k^{\sigma}\, Z^{\sigma+k} \tag{28}$$

Injecting (28) in (27), we obtain :

$$\sum_{k=0}^{\infty}\sum_{j=0}^{8}\sum_{i=0}^{10} c_k^{\sigma}E_{ij}\ [\sigma+k]_j Z^{\sigma+k+i}=0 \tag{29}$$

in which :

$$[k]_j= \frac{k!}{(k-j)!},\ j>0 \quad \text{and } [k]_0=1 \tag{30}$$

Stating that the coefficient of Z^{σ} in (29), i.e. i=k=0, must be zero, we obtain :

$$E_{06}[\sigma]_6+ E_{07}[\sigma]_7+ E_{08}[\sigma]_8=0 \tag{31}$$

leading to the determination of eight σ-eigenvalues, namely $\{\sigma_i\}$ = 0, 1, 2, 3, 4, 5, 7, 8. To each σ is associated a special solution θ^{σ} (28) and the general solution of the θ-equation is a linear combination of the eight special solutions.

We now introduce :

$$f_i(\sigma+k)= \sum_{j=0}^{8} E_{ij}\ [\sigma+k]_j \tag{32}$$

and rewrite the θ-equation (26) as :

$$\sum_{k=0}^{\infty} c_k^{\sigma}\sum_{i=0}^{10} f_i(\sigma+k)\ Z^{\sigma+k+i}=0,\ \sigma\in\{\sigma_i\} \tag{33}$$

Coefficients c_k^{σ} are determined by stating that terms $Z^{k+\sigma+i}$ of the same degree must be zero, leading to the following recurrence relations :

$c_0^{\sigma}\ f_0(\sigma)=0$

$$c_1^\sigma f_0(\sigma+1)+c_0^\sigma f_1(\sigma)=0 \tag{34}$$

$$c_2^\sigma f_0(\sigma+2)+c_1^\sigma f_1(\sigma+1)+c_0^\sigma f_2(\sigma)=0$$

$$\vdots$$

$$c_{10}^\sigma f_0(\sigma+10)+c_9^\sigma f_1(\sigma+9)+\ldots\ldots+c_1^\sigma f_8(\sigma+1)+c_0^\sigma f_{10}(\sigma)=0$$

$$\vdots$$

$$c_k^\sigma f_0(\sigma+k)+c_{k-1}^\sigma f_1(\sigma+k-1)+\ldots\ldots+c_{k-9}^\sigma f_9(\sigma+k-9)+c_{k-10}^\sigma f_{10}(\sigma+k-10)=0$$

which may be rewritten as :

$$\sum_{i=0}^{\min(k,10)} c_{k-i}^\sigma \; f_i(\sigma+k-i)=0, \; k=0,1,2,\ldots \tag{35}$$

For $k \leq 9$, some of the above equations are degenerated. To accurately determine the eight arbitrary constants c_i^j associated with the eight-order θ-equation under study, a careful examination of the recurrence relations is required. This being too lengthy to be provided here, we shall be content in giving the obtained results. We find that the arbitrary constants may be taken as c_0^0, c_0^1, c_0^2, c_0^3, c_0^4, c_0^5, c_0^7 and c_0^8. For $k \leq 9$, we also obtain :

$$c_0^6 = -\frac{f_6(0)}{f_0(6)}\, c_0^0, \qquad c_5^1 = -\frac{f_5(1)}{f_0(6)}\, c_0^1, \qquad c_4^2 = -\frac{f_4(2)}{f_0(6)}\, c_0^2, \tag{36-38}$$

$$c_3^3 = -\frac{f_3(3)}{f_0(6)}\, c_0^3, \qquad c_2^4 = -\frac{f_2(4)}{f_0(6)}\, c_0^4 \tag{39-40}$$

and the following constants are zero: c_1^0, c_2^0, c_3^0, c_4^0, c_5^0, c_7^0, c_8^0, c_1^1, c_2^1, c_3^1, c_4^1, c_6^1, c_7^1, c_1^2, c_2^2, c_3^2, c_5^2, c_6^2, c_1^3, c_2^3, c_4^3, c_5^3, c_1^4, c_3^4, c_4^4, c_1^5, c_2^5, c_3^5, c_1^7.

We now introduce the notation :

$$\rho_0=8,\; \rho_1=7,\; \rho_2=5,\; \rho_3=4,\; \rho_4=3,\; \rho_5=2,\; \rho_6=1,\; \rho_7=0 \tag{41}$$

Observing that all the c_k^σ-coefficients only depend on c_0^σ, we also introduce :

$$c_k^\sigma = A_k^\sigma\, c_0^\sigma \tag{42}$$

and find that the general θ-solution may be finally written as :

$$\theta = \sum_{m=0}^{7} c_0^{\rho_m} \sum_{k=0}^{\infty} A_k^{\rho_m}\, z^{\rho_m+k} \tag{43}$$

II.4. The Characteristic Equation and the Problem Solution

The solution (43) is introduced in the boundary conditions. For z=d, we use the free surface boundary conditions (14), (18), (22) and (25) to obtain :

$$-\alpha^2 q^2 c_0^0 + 2c_0^2 + \alpha^2 \bar{\omega}\ Pr\ \Delta_s = 0 \qquad (44)$$

$$Cri.\left[\alpha^4 q^2 (r^2+2) c_0^1 - 6\alpha^2 (q^2+r^2+2) c_0^3 + 120\ c_0^5\right]$$
$$+ \frac{i}{d^3}.\ Cri.Sh.\left[\alpha^2 (r^2+2) c_0^0 - 6c_0^2\right] - \left[\frac{\alpha}{d}^4 + Bo.\frac{\alpha}{d^3}^2 + \frac{i}{d^3} Sh.Cri.\bar{\omega}\alpha^2\right]\Delta_s = 0 \qquad (45)$$

$$-\alpha^4 q^2 c_0^0 + 2\alpha^2 (1-q^2) c_0^2 + 24\ c_0^4 + Vi.d\alpha^2 (\alpha^2 q^2 c_0^1 - 6c_0^3)$$
$$\frac{2i.Sh}{d^2}\left[\frac{1}{2}\ Vi.\alpha^2 c_0^0 - \frac{c_0^1}{d}\right] - Ma.\frac{\alpha^2}{d^2} c_0^0 + Ma.\frac{\alpha^2}{d^2}\Delta_s = 0 \qquad (46)$$

$$Bi.c_0^0 - dc_0^1 - Bi.\Delta_s = 0 \qquad (47)$$

For the five boundary conditions at the solid wall $(z = 0$, (4)-(13)), they lead to :

$$\sum_{j=0}^{7} F^{\alpha}_{m\rho_j}.\ c_0^{\rho_j} = 0\ ,\ m=1,2,3,4 \text{ and } 5 \qquad (48)$$

in which superscript α is c for the conductor case and i for the insulating case. The terms $F^{\alpha}_{m\rho j}$ are displayed in Appendice II.

The set (44)-(48) is a linear homogeneous set having the form $A_{ij}V_j = 0$ in which V_j is the unknown vector $(c_0^0, c_0^1, c_0^2, c_0^3, c_0^4, c_0^5, c_0^7, c_0^8, \Delta_s)$. The existence of nontrivial solutions ($\neq$ 0) requires that the characteristic determinant Δ of the set is zero. This determinant is given by :

$$\Delta = \begin{vmatrix} F^{\alpha}_{10} & F^{\alpha}_{11} & F^{\alpha}_{12} & F^{\alpha}_{13} & F^{\alpha}_{14} & F^{\alpha}_{15} & F^{\alpha}_{17} & F^{\alpha}_{18} & 0 \\ F^{\alpha}_{20} & F^{\alpha}_{21} & F^{\alpha}_{22} & F^{\alpha}_{23} & F^{\alpha}_{24} & F^{\alpha}_{25} & F^{\alpha}_{27} & F^{\alpha}_{28} & 0 \\ F^{\alpha}_{30} & F^{\alpha}_{31} & F^{\alpha}_{32} & F^{\alpha}_{33} & F^{\alpha}_{34} & F^{\alpha}_{35} & F^{\alpha}_{37} & F^{\alpha}_{38} & 0 \\ F^{\alpha}_{40} & F^{\alpha}_{41} & F^{\alpha}_{42} & F^{\alpha}_{43} & F^{\alpha}_{44} & F^{\alpha}_{45} & F^{\alpha}_{47} & F^{\alpha}_{48} & 0 \\ F^{\alpha}_{50} & F^{\alpha}_{51} & F^{\alpha}_{52} & F^{\alpha}_{53} & F^{\alpha}_{54} & F^{\alpha}_{55} & F^{\alpha}_{57} & F^{\alpha}_{58} & 0 \\ S_{10} & 0 & 2 & 0 & 0 & 0 & 0 & 0 & S_1 \\ S_{20} & S_{21} & S_{22} & S_{23} & 0 & S_{25} & 0 & 0 & S_2 \\ S_{30}-Ma.\frac{\alpha^2}{d^2} & S_{31} & S_{32} & S_{33} & 24 & 0 & 0 & 0 & Ma.\frac{\alpha^2}{d^2} \\ Bi & -d & 0 & 0 & 0 & 0 & 0 & 0 & -Bi \end{vmatrix} \qquad (49)$$

in which α is c for the conductor case and i for the insulating case, and:

$$S_{10}= -\alpha^2 q^2, \qquad S_1= \alpha^2\bar{\omega}\, Pr, \qquad S_{20}= \frac{i}{d^3}Cri.Sh.\alpha^2(r^2+2) \tag{50-52}$$

$$S_{21}= \alpha^4 q^2(r^2+2).Cri, \qquad S_{22}= -6\,\frac{i}{d^3}Cri.Sh \tag{53-54}$$

$$S_{23}= -6\alpha^2(q^2+r^2+2).Cri, \qquad S_{25}= 5!\ Cri \tag{55-56}$$

$$S_2= -\left[\frac{\alpha^4}{d}+ Bo.\frac{\alpha^2}{d^3} + \frac{i}{d^3}\ Sh.Cri.\bar{\omega}\ \alpha^2\right] \tag{57}$$

$$S_{30}= -\alpha^4 q^2+ \frac{i}{d^2}Sh.Vi.\alpha^2, \qquad S_{31}= Vi.d.\alpha^4 q^2- \frac{2i.Sh}{d^3} \tag{58-59}$$

$$S_{32}= 2\alpha^2(1-q^2), \qquad S_{33}= -6.Vi.d\alpha^2 \tag{60-61}$$

The condition $\Delta=0$ enables us to explicitly express the critical Marangoni number as:

$$Ma = N / D \tag{62}$$

in which :

$$N= \begin{vmatrix} \text{Lines 1-7 are unchanged} & & & & & & & & \\ S_{30} & S_{31} & S_{32} & S_{33} & 24 & 0 & 0 & 0 & 0 \\ Bi & -d & 0 & 0 & 0 & 0 & 0 & 0 & -Bi \end{vmatrix} \tag{63}$$

and:

$$D = \frac{\alpha^2}{d^2} \begin{vmatrix} \text{Lines 1-7 are unchanged} & & & & & & & & \\ 1 & 0 & 0 & 0 & 0 & 0 & 0 & 0 & -1 \\ Bi & -d & 0 & 0 & 0 & 0 & 0 & 0 & -Bi \end{vmatrix} \tag{64}$$

Inputs to evaluate (62) are thermophysical and geometrical properties (μ, ρ_o, $\kappa+\epsilon$, σ_s, α_V, Γ, a_T, d), other quantities linked to the interaction between the liquid layer and the surrounding medium (Bi, g, g_s, Sh, k_t) and the characteristics of a normal mode (α, β). For these inputs, the output Ma is a pseudo-Marangoni number which is in general a complex number. We now demand that the pseudo-Marangoni number be a real number and that it is equal to its definition value (24). Therefore, it must satisfy :

$$Im\ [\ Ma(\alpha,\ \beta,\ k_t)]=0 \qquad \text{and} \qquad Re\ [\ Ma(\alpha,\ \beta,\ k_t)]= -\ \frac{\Gamma k_t d^2}{\mu a_T} \tag{65-66}$$

all inputs other than (α, β, k_t) being fixed. If the system (62), (65), (66) does not own any solution, then the system is stable with respect to overstability. We now consider the case when solutions do exist. The chosen solution procedure is then as follows.

For a given wave-number α, starting from an arbitrary value of k_t, the solution of (65) takes the form of a sequence of overstable imaginary time constants β_i. Then, using (66), we compute an associated sequence of adverse temperature gradients $k_{t,i}$. Choosing subscript i for which $|k_{t,i}|$

is the smallest, we solve again relation (65), and so on. By iterating computations on (65) and (66) until convergence, we obtain the values of critical $\beta(\alpha)$ and $k_t(\alpha)$ associated with wave-number α. Scanning over α, the α-value for which $|k_t|$ is a minimum gives us the critical wave-number α_* associated with a final critical frequency ω_* (β_*/i) and with a final critical temperature difference δT_*. We then have to scan over d to determine $\alpha_*(d)$, $\omega_*(d)$ and $\delta T_*(d)$.

II.5.Complementary Discussion

Before going to the presentation of new results, we would like to comment on the case Sh=0 discussed in Gouesbet and Maquet (1989), Gouesbet et al (1990) in order to clarify some statements against the similar work by Perez-Garcia and Carneiro (1991 and private communication). The first statement is that overstability would only appear when the rigid wall is cold. This was claimed by Takashima (1981) for a pure surface tension mechanism but we found that the statement also holds when surface tension and buoyancy simultaneously act together. This statement was meant for standard liquids in which surface tension decreases when temperature increases (Γ negative) and the coefficient of volume expansion α_v is positive. Then overstability would not appear for the case of a hot rigid wall, either a liquid layer suspended from a hot ceiling (Ra>0, Ma>0, Bo<0) or a liquid layer above a hot floor (Ra>0, Ma>0, Bo>0). However, with input parameters different than ours, Perez-Garcia and Carneiro detected cases where overstability is possible for Ra>0. For small crispation numbers, only Ma<0 could produce oscillatory motions. The analysis concerned a hot bottom floor for which standard liquids lead to Ra>0, Ma>0. Ma<0 therefore requires the use of a non-standard liquid such as the surface tension increases with temperature. This result does not contradict our claim which was devoted to standard liquids. However, for a crispation number sufficiently big (Cri>0.015), Perez-Garcia and Carneiro also detected overstability cases in the quadrant (Ra>0, Ma>0). Therefore, our claim must be softened by stating that overstability is more unlikely with hot rigid walls than with cold rigid walls, but is nevertheless possible.

The second statement concerns the inhibiting character of the Rayleigh number on overstability and addresses the issue to know how critical Ma_* changes with respect to Ra. Of course, any presentation of curves such as $Ma_*(Ra)$ implicitly assumes that both dimensionless groups are considered to be independent, i.e. the layer depth and the thermophysical properties are allowed to change along the curves. As stated by Perez-Garcia and Carneiro, for a given liquid and a given depth, Ra and Ma are just two different dimensionless forms of the temperature difference, i.e. changes of temperature difference amount to move on a staight line in Ma-Ra diagram. This is basically the reason why, to study specific situations in which both the liquid and the depth are fixed, we preferred to use a dimensional formulation in Gouesbet and Maquet (1989) and also in this paper. Conversely, in Gouesbet et al (1990), we have chosen the option of dicussing curves $Ma_*(Ra)$. We investigated the case of a liquid layer above a cold floor for a standard liquid (Ra<0, Ma<0, Bo>0) and concluded that the Rayleigh number was essentially an inhibitor of overstability although we were also able to detect a small range of input parameter values for which this statement did not hold. This might be not so surprising because, in this studied case, buoyancy is also stabilizing for the exchange of stability. However, some computations indicated that buoyancy may be stabilizing even if it is destabilizing for the exchange of stability. With positive expansion factor, this last case may be splitted in two cases namely liquid layer above a hot floor (Ra>0) and liquid layer below a cold ceiling (Ra<0). In cases where overstability is possible for Ra>0 (Ma<0, non standard liquids), Perez-Garcia and Carneiro found that buoyancy stimulates overstability. As far as our own statement is concerned, it referred to a standard liquid and to a liquid layer below a cold ceiling for which buoyancy has been found to be inhibiting (see Gouesbet and Maquet, 1989, and also dicussion of hot-wire experiments in Section III). Of course, in the framework of such numerical computations, many statements concerning non-existence must essentially own a provisional character. Definitive non-existence statements may only be reached by non-numerical mathematical analyses which must be of tremendous difficulty in the present discussed cases. Comparisons between Perez-Garcia and Carneiro results and ours are however welcome to clarify the discussed issues, this being obtained because numerical computations focussed on different and complementary cases.

III.RESULTS AND DISCUSSION

The formulation presented in the previous section provides a parameter space of big dimension. Therefore, numerical computations should only be carried out to answer specific questions. We mainly choose to complete our discussion in Gouesbet and Maquet (1989) concerning the understanding of a rather new instability (thermal lens oscillations), corresponding to our original motivation in undertaking this work

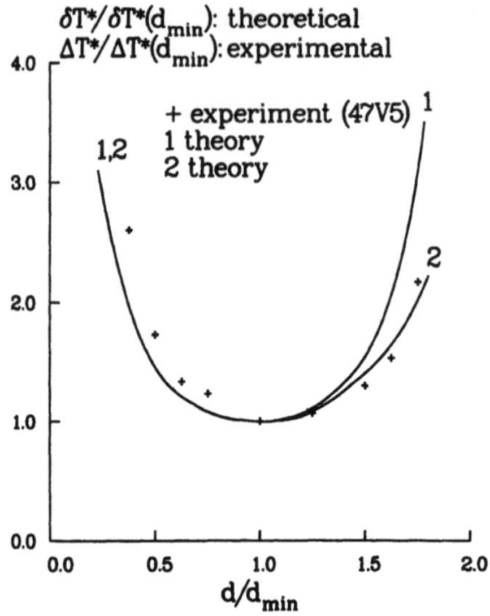

Figure 2. Silicon oil 47V5. Comparison between hot-wire experiments and theory with no shear. Reduced critical temperature differences versus reduced distances. Taken from Gouesbet and Maquet (1989).

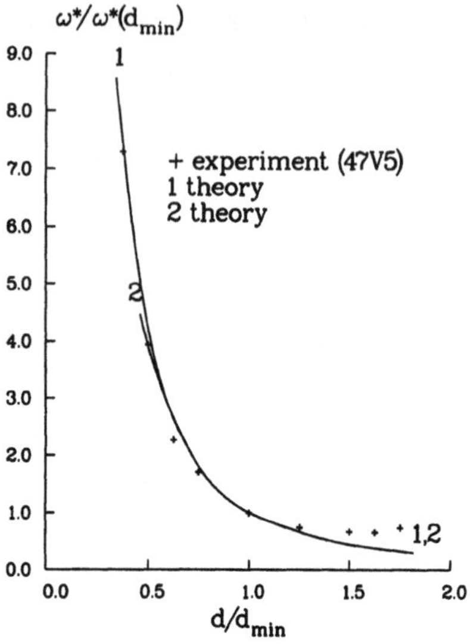

Figure 3. Silicon oil 47V5. Comparison between hot-wire experiments and theory with no shear. Reduced frequencies versus reduced distances. Taken from Gouesbet and Maquet (1989)

(Gouesbet, 1989; Anthore and al, 1982). The best up to now available understanding of this instability has been obtained by a simple model relying on concepts from nonlinear dynamical systems theory (Gouesbet 1990a and 1990b).

However, in a prior step, we wondered whether there could be any analogy between thermal lens oscillations (and associated hot wire experiments) and the case of overstability in an infinite horizontal liquid layer. To answer this question we developed the formulation in Gouesbet and Maquet (1989) for the onset of overstability when the horizontal layer is submitted to a vertical temperature gradient and may loss its stability to oscillatory behaviour through simultaneous surface tension and buoyancy mechanisms. Gouesbet et al (1990) is a by-product which does not refer any more to the analogy problem and expressed the formulation in dimensionless form. Numerical computations in Gouesbet and Maquet (1989) were carried out for a silicon oil extensively studied in the case of hot-wire heating (Weill et al 1985, Gouesbet et al 1986).

Examples of comparisons between computations and experimental data are provided in figures 2 and 3 (reproduced from Gouesbet and Maquet, 1989). Experimentally, the function $\Delta T_*(d)$, in which ΔT_* is the critical experimental temperature difference between hot-wire and gas above the liquid layer, presents a minimum at $d = d_{min}$. For a silicon oil labelled 47V5, we find $\Delta T_*(d_{min}) = 29.5$ K. Figure 2 shows the reduced critical temperature differences $\Delta T_*/\Delta T_*(d_{min})$ versus the reduced hot-wire surface distance d/d_{min}. Some theoretical results are given for curves labelled 1 and 2. They concern the case of a liquid layer suspended from a cold ceiling. In label 1 case, we essentially used the actual thermophysical and physical properties corresponding to the experiments, with also $(\kappa+\epsilon) = Bi = 0$. Label 2 is same as label 1 but with a thermal diffusivity of the liquid artificially multiplied by 17.94. Figure 3 shows reduced critical frequencies $\omega_*/\omega_*(d_{min})$ versus reduced distance d/d_{min}. Labels 1 and 2 have the same meaning as previously.

The agreement between reduced experimental and theoretical profiles in figures 2 and 3 is striking. But we did not precisely know whether this agreement supported the idea of a possible analogy or whether it was a mere coincidence, because there are also two essential differences between experiments and the case considered in the theory. The first difference is qualitative but might be essential. In numerical computations, we only detected overstability in the case of a cold solid wall, in agreement with Takashima (1981) for a pure surface tension mechanism (but see complementary discussion in II-5), while in experiments the hot-wire is hot. The second difference is quantitative and concerns the values of the temperature differences. For label 1, we find $|\delta T_*(d_{min})| \simeq 5000$ K. As expected, the artificial increase of a_T helps because we obtain $|\delta T_*(d_{min})| \simeq 1600$ K for label 2. For a further artificial increase of a_T, values of $|\delta T_*(d_{min})|$ do not decrease any more. These $|\delta T_*|$-values are in strong disagreement with the experimental value (29.5 K) and even conflict with the very existence of the liquid state.

After these results were obtained, the obvious next step was to account for the fact that the base state is convective in experiments, because in laser and hot-wire heating cases the temperature gradient is not everywhere parallel to the gravity acceleration vector, while, conversely, it was motionless in the theoretical study. The experimental convective base state structure is too complicated to be implemented in an analytical small disturbance analysis and we expect that only a rigorous numerical analysis could lead to accurate results (See Maquet et al, 1987).

However, the above discussion immediately suggests to consider a very simple convection state, namely a pure shear, and to extend the analysis of Gouesbet and Maquet (1989), Gouesbet et al (1990), to the case when surface tension, buoyancy and shear effects are simultaneously present. The aim is to examine whether the high $|\delta T_*|$-values previously mentioned

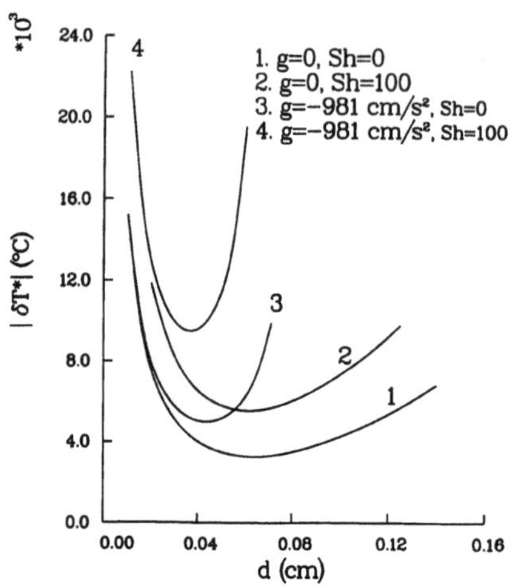

Figure 4. Silicon oil 47V5. Liquid layer suspended from a cold ceiling. Influence of shear and buoyancy on critical $|\delta T_*|$.

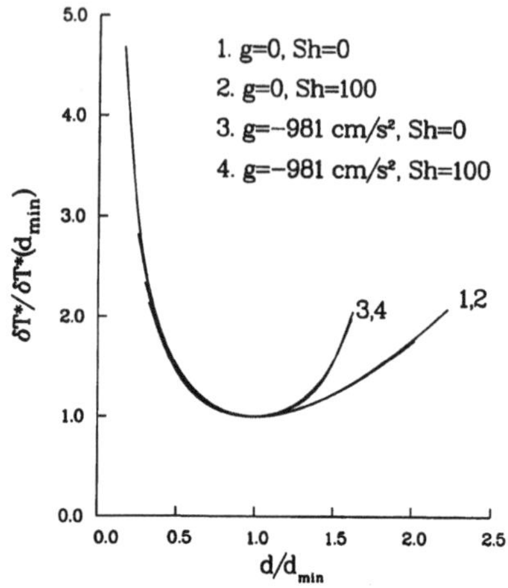

Figure 5. Silicon oil 47V5. Insensivity of reduced profiles with respect to shear number.

could be significantly decreased by shear effects and may be also possibly whether the qualitative discrepancy concerning the signs of ΔT_* and δT_* could be removed. The corresponding small disturbance analysis formulation has been presented in this paper. We now proceed to numerical results to tentatively answer the above questions.

Fig 4 presents $|\delta T_*|$-results versus d for the case of a liquid layer suspended from a cold ceiling, for a silicon oil 47V5 as in Figs 2,3. This is the conductor case. The inputs are the physical properties and $(\kappa+\epsilon)=0$, Bi=0 and the thermal diffusivity a_T is artificially multiplied by $\kappa_{eff}=2.00$. The shear number is 0 or 100, this last value having been chosen by considering orders of magnitude in hot-wire experiments. In the absence of buoyancy (g = 0), we obtain the curve labelled 1 for a pure surface tension mechanism (Sh = 0) evolving to curve 2 in the case of shear (Sh = 100). If buoyancy is added to a pure surface tension mechanism, we evolve from curve 1 to curve 3, showing that buoyancy is an overstability inhibitor although it would be destabilizing for the case of exchange of stability (but see again complementary discussion in II-5). Then, if we add a shear, we evolve from curve 3 to curve 4. Comparing curves 1, 2 and 3, 4, we observe that shear is actually stabilizing. $|\delta T_*|$-levels are becoming much higher and are still therefore unattainable in experiments.

For all curves, there still exists a $d = d_{min}$ at which $|\delta T_*(d)|$ has a

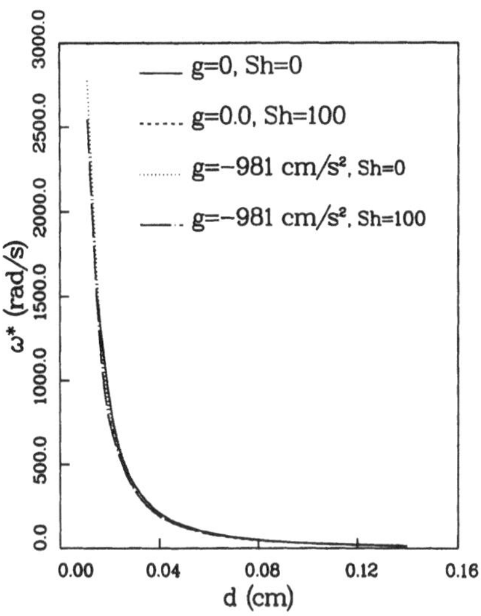

Figure 6. Silicon oil 47V5. Insensivity of shear and buoyancy on critical frequencies.

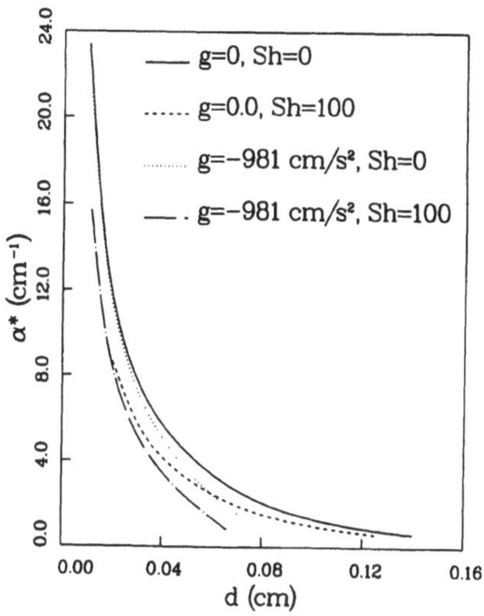

Figure 7. Silicon oil 47V5. Influence of shear and buoyancy on critical wave-numbers.

minimum. The value of d_{min} is rather insensitive to the value of the shear number. If we present the results in terms of $\delta T_*(d)/\delta T_*(d_{min})$ versus d/d_{min} (Fig 5), we observe that reduced profiles are not affected significantly by the shear. They are however modified for $d > d_{min}$ by buoyancy, as we may have expected since buoyancy becomes a predominant mechanism for thick layers.

Results for critical frequencies ω_* and critical wave-numbers α_* are provided in Figs 6 and 7 respectively. For critical frequencies, influence of buoyancy and shear is very small with respect to a pure surface tension mechanism. This influence is observable for critical wave-numbers in Fig 7 but still is not very big.

Having observed that the value of d_{min} is rather insensitive to the value of the shear number, we now choose a special value $d = d_0$ which is the d_{min} for Sh = 0. Then, Fig 8 shows $\delta T_*/\delta T_*(d_0)$ versus Sh ranging from 0 to 50. We observe that the shear number actually have a destabilizing character up to Sh $\simeq$ 40. The decrease of $\delta T_*/\delta T_*(d_0)$ is however very small (less than about 3 %).

We now consider again silicon oil 47V5 but for the case of a liquid layer above a cold floor instead of being below a cold ceiling. The other inputs are the same as for Fig 4 but for thermal diffusivity which was artificially multiplied by 17.94 (same value as for Figs 2 and 3, curve 2). $|\delta T_*|$-values versus d are given in Fig 9. In this case, the influence of shear is significant and corresponds to a destabilizing character. For

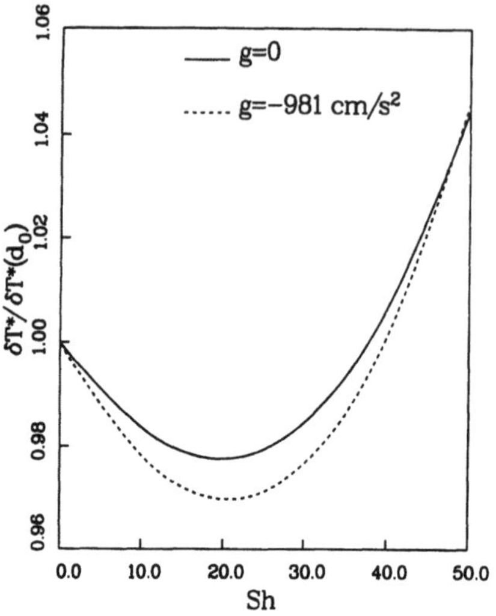

Figure 8. Silicon oil 47V5. An example showing that shear may be destabilizing.

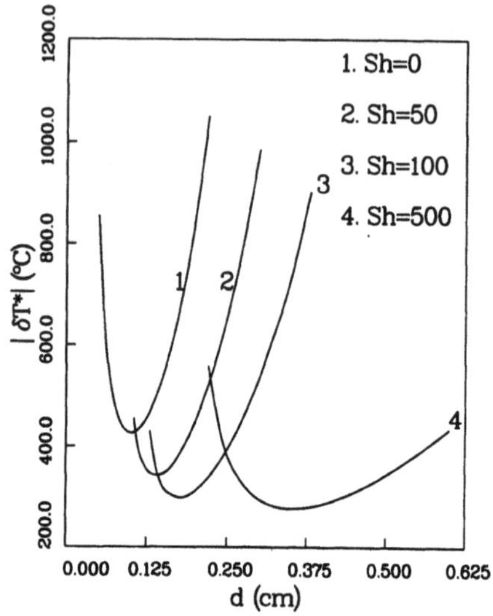

Figure 9. Silicon oil 47V5. Influence of shear on critical $|\delta T_*|$. Case of a liquid layer above a cold floor.

big Sh, the values of $|\delta T_*|$ are small enough to suggest that experimental observations of such phenomena might be feasible (notwithstanding the fact that the chosen value of thermal diffusivity was not the true one). Another result is that again in this case we never observed overstability for the case of a hot solid wall, even when the shear effect is included.

From this last result and the fact that addition of shear never produced an enough significant decrease of the critical $|\delta T_*|$ as we hoped, we may now conclude that the present work is probably not relevant to the understanding of thermal lens oscillations and associated hot-wire experiments.

Independently of our original motivation, the analysis provided in this paper suggets new experiments. This is however correct under the proviso that reasonable $|\delta T_*|$-levels are obtainable. We therefore end this paper by exhibiting results for different liquids, namely toluene and cyclohexane. Thermophysical properties of these liquids are very different from silicon oil 47V5 ones. For instance, toluene viscosity is ten times smaller, thermal diffusivity is two times bigger.

For toluene, we consider the case of a liquid layer suspended below a cold ceiling, a conductor boundary condition at the solid wall and the same values of g_s, Bi, $(\kappa+\epsilon)$ (but $\kappa_{eff}=0$) as for Fig 4. Critical temperature differences $|\delta T_*|$ are shown versus d in Fig 10. The obtained values of $|\delta T_*|$ imply that overstability should be observable. However, in

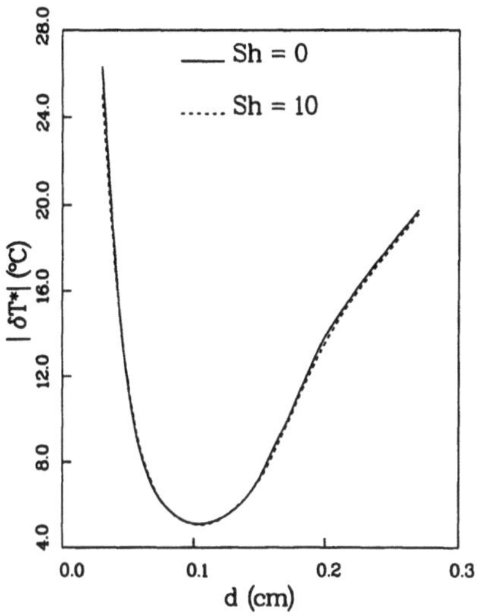

Figure 10. Toluene. Critical $|\delta T_*|$ versus d. A case showing realistic values of critical temperature differences.

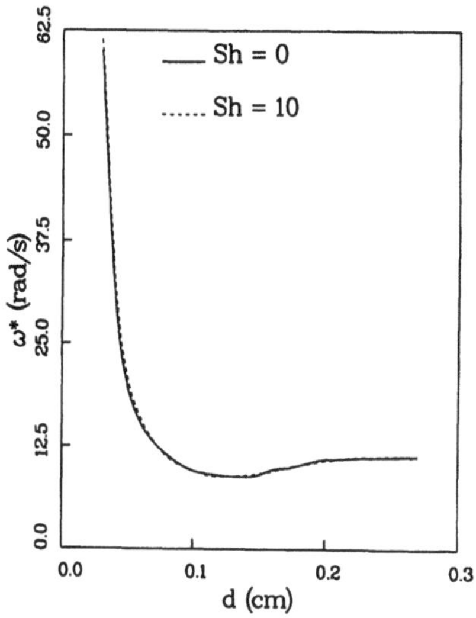

Figure 11. Toluene. Critical frequencies versus d.

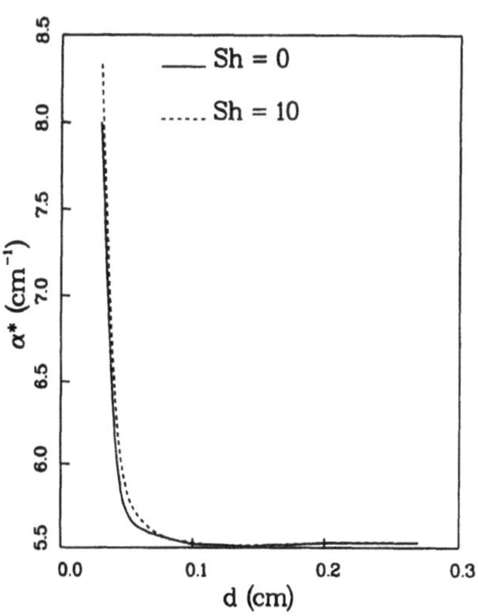

Figure 12. Toluene. Critical wave-numbers versus d.

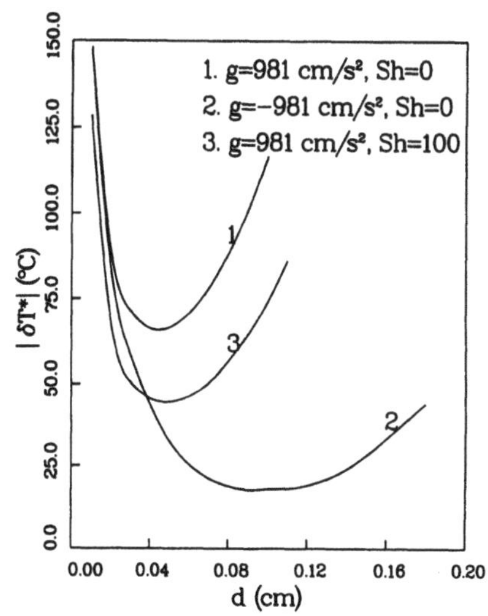

Figure 13. Cyclohexane. Critical $|\delta T_*|$ versus d.

this case, the influence of the shear number is not significant. Theoretical critical frequencies and critical wave numbers are displayed in Figs 11 and 12 respectively.

For cyclohexane, critical temperature differences $|\delta T_*|$ are shown in Fig 13. Curves 1 and 3 concern a liquid layer above a cold floor, a more

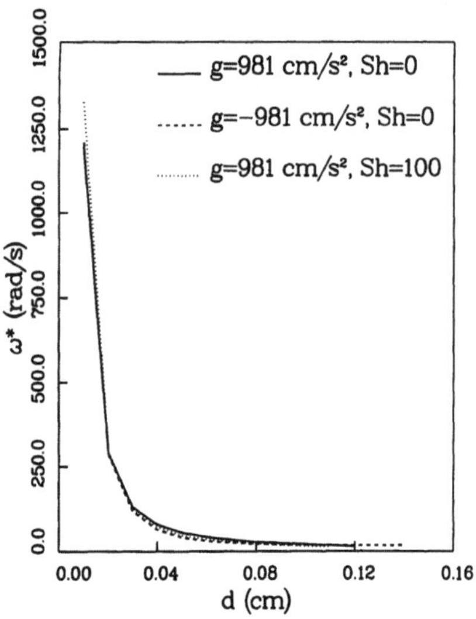

Figure 14. Cyclohexane. Critical frequencies versus d.

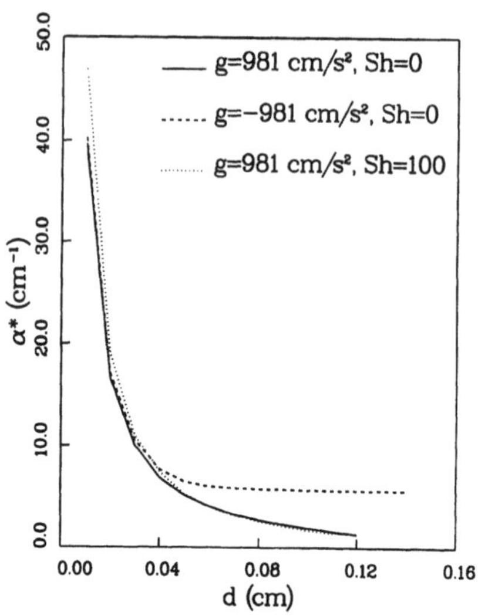

Figure 15. Cyclohexane. Critical wave-numbers versus d.

comfortable situation for experiments indeed. With a cyclohexane boiling point equal to 80.75xC, the influence of shear (which is significant) might be experimentally tested. Curve 2 corresponds to a liquid layer suspended from a cold ceiling showing that the $|\delta T^*|$-levels are much smaller. These computations have been carried out with Bi=0, $(\kappa+\epsilon)=0$ and $\kappa_{eff}=0$. Corresponding critical frequencies and wave-numbers are displayed in Figs 14 and 15 respectively. In any case, computations for a hot floor did not produced any overstability.

We end this paper with two remarks. The first one is that computations would be interpretated by separating the influence of the shear intensity k_u and of the partial wave number α_y appearing in the shear number. For a given shear prescribed by the experimental design, the most dangerous normal mode would be characterized by both α and α_y, or better by both α_x and α_y, allowing a description of the mode in term of transverse and longitudinal components. The second remark is that the large values of temperature differences to trigger overstability in Figs 10 and 13 for toluene and cyclohexane respectively certainly violate the Boussinesquian assumption. The situation was worst in previous results in this paper and previous works for silicon oils because the results were in conflict with the very existence of the liquid state. However, now that liquids for which overstability may be observed (with or without shear) do exist, it would become important to elaborate again on our formulations in Gouesbet and Maquet (1989), Gouesbet and al (1990) and in this paper, to include non-Boussinesquian effects. The conclusion of such a work might be that observations of overstability could appear much easier that can be thought at the present time.

IV.CONCLUSION

We established a small disturbance analysis of the loss of stability via overstability of an infinite horizontal liquid layer suspended from or above of a solid wall, submitted to simultaneous surface tension, buoyancy, and shear effects. This represents a generalization of a previous work in which shear effects were omitted.

We first focussed the presentation of results on the understanding of thermal lens oscillations and associated hot-wire experiments which was our original motivation in undertaking this work. We conclude that the theoretical study in this paper is probably not relevant to the understanding of these last instabilities.

Independently of this discussion concerning thermal lens oscillations, we provided examples showing that our theory is in principle amenable to experimental testing.

COMPLEMENTARY NOTATIONS

ρ_0, μ, a_T and α_V are the liquid density at the reference temperature T_0, the liquid dynamic viscosity, the liquid thermal diffusivity and the expansion factor:

$$\alpha_V = -\frac{1}{\rho_0}\frac{\partial\rho}{\partial T}\Big|_{T=T_0}$$

g is the acceleration module, with g positive when the upside of the layer is the free surface.

r^2 and q^2 are given by:

$$r^2 = 1 + \frac{\rho_0\beta}{\mu\alpha^2}, \qquad q^2 = 1 + \frac{\beta}{a_T\alpha^2}$$

β is the time constant, perturbations evolving like $\exp(\beta t)$ in which t is the time. For overstability, $\beta = i\omega$ at marginality, with ω the perturbation angular frequency. α and α_y are the normal mode wave-number and the y-partial normal mode wave-number respectively according to the planform function solution:

$$f(x,y) = \exp\,[i(\alpha_x x + \alpha_y y)], \qquad \alpha = (\alpha_x^2 + \alpha_y^2)^{1/2}$$

Note that, when there is no shear, α_y does not explicitly appear in the formulation (Gouesbet and Maquet, 1989). In the present case, it always appear in blocks $(\alpha_y k_u)$.

REFERENCES

Anthore R., Flament P., Gouesbet G., Rhazi M., Weill M.E., 1982, *Appl. Optics* 21,1,2.

Aris R.,1962, "Vectors, Tensors and the Basic Equations of the Fluid Mechanics," *Englewood Cliffs, N.J., Prentice Hall.*

Gouesbet G., Weill M.E., Lefort E., 1986, *AIAA Journal.* 24,8,1324-1330.

Gouesbet G., Maquet J., 1989, *AIAA J. of Thermophysics and Heat Transfer.* 3,1,27-32.

Gouesbet G., 1990a, *Entropie (France).* Nx153/154, 47-61.

Gouesbet G., 1990b, *Physical Review A.* 42/10, 5928-5945.

Gouesbet G., Maquet J., Roz C., Darrigo R., 1990, *Phys. of Fluids A, Vol. 2, Nx6, 903-911.*

Ince E.L., 1926, "Ordinary Differential Equations," *Dover publications, inc.*

Joseph D.D., 1976, "Stability of Fluid Motion," *Springer Verlag, Berlin, vols I and II.*

Maquet J., Gouesbet G., Berlemont A., 1987, A computer code for natural convection in an enclosed cavity with a free surface, *Paper presented at the 5th International Conference on Numerical Methods for Thermal Problems.* Montr al, Canada, June 29-30, 1987. Proceedings: *Numerical methods in thermal problems*, 5, Part 1, 472-483, edited by Lewis R.W., Morgan K., Habashi W.G., *Pineridge Press, Swansea, U.K.*

Nield D.A., 1964, *J. Fluid. Mech.* 19,341-352.

Oertel Jr. H., 1982,"Thermal Instabilities in Convective Transport and Instability Phenomena," *Zierep and Oertel, edts, G. Braun, Karlsruhe,* 3-24.

Pearson J.R.A., 1958, *J. Fluid. Mech.*, 4, 489-500.

Perez-Garcia C. and Carneiro G., 1991, *Phys. Fluids A,* 3, 2, 292-298.

Rayleigh Lord, 1916, *Phil. Mag.*, 32, 529-546.

Scriven L.E. and Sternling C.V., 1964, *J.Fluid. Mech.*, 19, 321-340.

Takashima M., 1981, *J. Phys. Soc. of Japan*, 50, nx8 , part I : 2745-2750 and part II: 2751-2756.

THE NUMERICAL INVESTIGATION OF BOILING FLOW INSTABILITY IN HEATED CHANNELS

B.I. Nigmatulin[1], V.N. Blinkov[2], O.I. Melikhov[1], and P.G. Gakhal[2]

[1]Research & Engineering Centre of LWR Safety
Electrogorsk, Moscow Region, Russia 142530
[2]Kharkov Aviation Institute, Engines Thermophysics Department
Kharkov, Ukraine 310084

To ensure the stability of boiling working fluid flow is an important problem in such applications as Boiling Water Reactors[1] and Two-Phase Thermal Control Systems for Spasecrafts[2]. The given problem was the objective of numerious investigations since 1960's. The monographs[3-5] contain the methods of linear stability analysis of boiling flow in a heated channels. The flow is assumed to be transient one-dimensional homogeneous with local losses. The system of non-linear equations is linearized. Laplace transforming is used for getting an analytical solution. The stability analysis represents the search for the boiling flow system parameters dividing the regions of stable and unstable operation. The reviews of two phase-flows stability investigations perfomed in the West by the end of 1970's were given[6-7]. Bundles of rods and tubes of nuclear reactor core or steam generator, evaporative heat-exchangers networks of two-phase thermal control system represent the systems of parallel heated channels. In such systems the oscillatory instability manifests itself as the periodic flow rate oscillations under conditions of fixed pressure drop. Parallel channels instability was studied both analytically (linear and quasilinear approaches)[8] and numerically with the use of simplified nonlinear models[9-10]. When analysing the parallel channels operation it is important to know not only the threshold of instability but also to predict the limits of parameter variations in each channel under unstable operation. Because of that and due to advances in high-speed computers and finite-difference schemes in the last years computer programs solving non-linear equations are developed even suitable for microcomputers[11].

The purposes of the presented paper are:

- to develop the mathematical model and computer program for transient boiling flow in a heated channel simulation under boundary conditions

Instabilities in Multiphase Flows, Edited by G. Gouesbet
and A. Berlemont, Plenum Press, New York, 1993

typical for parallel channels system;

- to estimate the model by means of comparisons with experiments;
- to perform the numerical calculations and to demonstrate the example of model and computer program practical application.

We applied 1P2T1W model of transient one-dimensional two-phase flow. The conservation equations can be written as:

$$\frac{\partial \rho_m}{\partial t} + \frac{1}{F}\frac{\partial}{\partial z}\left(\rho_m W F\right) = 0$$

$$\rho_m = \alpha\rho_v + \left(1 - \alpha\right)\rho_l$$

$$\frac{\partial \alpha\rho_v}{\partial t} + \frac{1}{F}\frac{\partial}{\partial z}\left(\alpha_v \rho_v W F\right) = J_{l-v}$$

$$\frac{\partial W}{\partial t} + W\frac{\partial W}{\partial z} + \frac{1}{\rho_m}\frac{\partial P}{\partial z} + \frac{1}{\rho_m} f_w - g\cos j = 0 \tag{1}$$

$$\frac{\partial}{\partial t}\left[\left(1 - \alpha\right)\rho_l U_l\right] + \frac{1}{F}\frac{\partial}{\partial z}\left[\left(1 - \alpha\right)\rho_l U_l W F\right] + \frac{1}{F} P \frac{\partial}{\partial z}$$

$$\left[\left(1 - \alpha\right) W F\right] - P\frac{\partial \alpha}{\partial t} + J_{l-v}\, i_{l\,sat} + q_{l-v} - q_{w-f}\frac{A_w}{V_w} = 0$$

$$T_v = T_{sat}(P),\ \rho_v = \rho_{v\,sat}(P),\ \rho_l = \rho_{l\,sat}\left(T_l\right)$$

The unknows in this equation set include pressure P, liquid temperature T_l, velocity W, void fraction α. The single phase flow is simulated as two-phase flow with vanishing void fraction. For solving the equations given by Eq. (1), an implicit method [12] is used having the stability criterion

$$\Delta t \le \frac{\Delta z}{W_{\max}} \tag{2}$$

which significantly increases the allowable time step compared with explicit methods and makes the use of the microcomputer feasible. Wall friction f_w is described by simple correlations for homogeneous flow as a function of Reynolds number. Interfacial heat transfer in all calculations presented in our paper was assumed to be sufficiently intensive to provide negligible thermal interphase nonequilibrium. Heat flux to coolant and wall temperature are computed on the basis of assumptions typical for the type of problem under consideration:

- the axial wall heat conduction is ignored;
- the wall is thin and radial temperature distribution is uniform.

With these assumptions the energy equation for the wall can be written in the form [13] :

$$\rho_w C_w \frac{dT_w}{dt} = q_w - \frac{q_{w-f} A_w}{V_w} \tag{3}$$

where $V_w/A_w \approx \delta$.

In the finite-difference analogs of equation (3) and of last equation in system (1) for preventing numerical instability the source term q_{w-f} should be treated in implicit manner i.e. calculated at the new time level:

$$\rho_w C_w \frac{T_{w,i}^{n+1} - T_{w,i}^{n}}{\Delta t} = q_{w,i} - \frac{q_{w-f,i}^{n+1}}{\delta_w}, \tag{4a}$$

$$\text{where } q_{w-f,i}^{n+1} = \alpha_{h,i}^{n} (T_{w,i}^{n+1} - T_{l,i}^{n+1}). \tag{4b}$$

Substituting $T_{w,i}$ from Eq.(4a) to Eq.(4b) we obtain:

$$q_{w-f,i}^{n+1} = \frac{\alpha_{h,i}^{n}}{1 + \alpha_{h,i}^{n} C_1} \left(T_{w,i}^{n} - T_{l,i}^{n+1} + C_2 q_{v,i} \right), \tag{5}$$

$$\text{where } C_1 = \frac{\Delta t}{\rho_w C_w \delta_w}, \quad C_2 = \frac{\Delta t}{\rho_w C_w}.$$

In the expression (5) heat flux at new time level is determined by the only value from new time level $T_{l,i}^{n+1}$, which is unknown variable of system (1). To calculate the heat transfer between wall and flow the correlation "critical heat flux—quality" $q_{crit} = f(x)$ [14] was used. Critical heat flux when $x < x_{critical}$ corresponds to the transition from bubble boiling to film boiling in inverted annular flow regime. Critical quality $x_{critical}$ (dryout point) determines the transition from boiling heat transfer to heat transfer at dispersed droplet regime because of dryout.

All the presented in this paper calculations were performed on IBM PC 386 according to the next algorithm:

1) The steady-state distributions of parameters along the channel under given flow rate are calculated. The inlet temperature, inlet flow rate (velocity) and outlet pressure are specified as the boundary conditions.

2) The transient response of flow on short step increase in heat power input is calculated. The inlet temperature, calculated in point 1) inlet pressure and outlet pressure are specified as the boundary conditions.

For estimation of the model the experimental data obtained on six-channel water-vapor loop of Electrogorsk LWR Research Engineering Centre[15] were used. In Fig. 1 one of the six parallel channels installed between inlet header (collector) and outlet separator is shown. In all presented calculations the single-phase adiabatic section was replaced by equivalent local throttle (pressure loss). The heated section and two-phase adiabatic section were modelled by 95 cells, with $\Delta z = 0{,}4$ m each of them (equal spaced grid). The stability boundaries, amplitudes and frequencies of oscillatory regimes were investigated in experiments. Fig.2 illustrates the applicability of the model. However the calculation of that curve required significant computational time. The ability of the model to simulate the oscillatory regimes in detail is shown in Fig.3.

At the moments $t <$ 1s the flow is steady. Then during $\Delta t =$ 1s the step perturbation in heat power input $q_w = 1.05\, q_{w0}$ occurs. The increase of q_w leads to the quality and pressure drop growth at the two-phase section and to the boiling boundary motion upstream. Since the total pressure drop remains constant the pressure drop at the economizer section decreases that causes the decrease

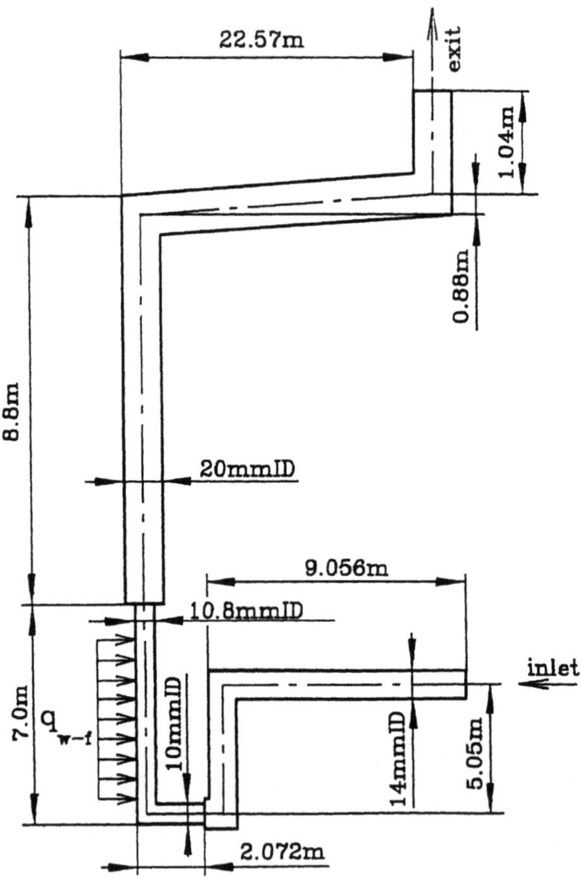

Figure 1. The geometry of one of the six identical heated channels installed on the experimental facility[15].

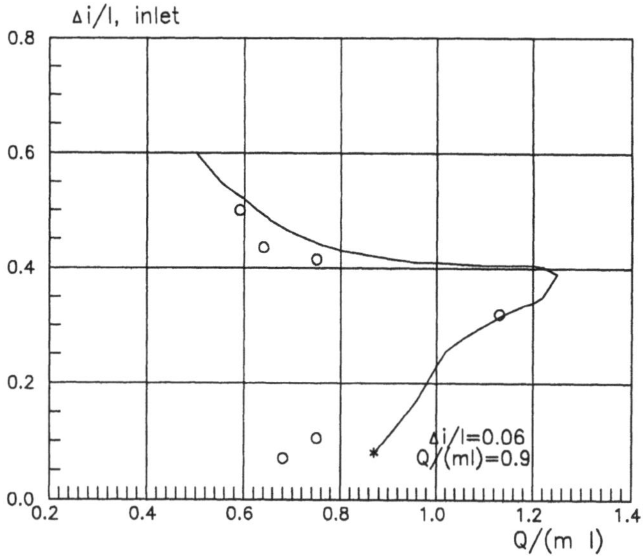

Figure 2. Comparison of experimental and calculated stability boundaries in coordinates "dimensionless initial subcooling" — "dimensionless heat load".
o — experiment[15]. P_{in} = 7.1 MPa; P_{ex} = 7.0 MPa; Q = 135.7 kW; 403K < T_{in} < 553K.

A-B of the inlet flow rate. The considered process is illustrated in Fig.4 by the links

$$\delta q_{w-f} \rightarrow \delta\Delta P_{2P} \rightarrow \delta\Delta P_{1P} \rightarrow \delta m_{in}$$
$$\delta q_{w-f} \rightarrow \delta l_{ec} \rightarrow \delta\Delta P_{1P} \rightarrow \delta m_{in}$$

The decrease of the inlet flow rate m_{in} accompanies by the increase of the exit flow rate m_{ex} (A-C, Fig.3a). The difference in the amplitudes m_{in} and m_{ex} can be explained by the fact that two-phase pressure drop depends stronger on flow rate than one-phase pressure drop. To provide the same deviation of pressure drop $\delta\,\Delta\,P_{1P} = -\,\delta\,\Delta\,P_{2P}$ it is required smaller two-phase flow rate variation than one-phase flow rate variation. After taking off the increased heat input the two-phase pressure drop decreases, one-phase pressure drop and flow rate increase (B-D, Fig.3a). The exit flow rate follows the inlet flow rate with delay. That delay can be explained as follows. The variation of the inlet flow rate causes the shift of the boiling boundary. The density wave originated near boiling boundary at its limiting left (upstream) location (Fig.3c) reaches the exit of channel when the boiling boundary reaches its limiting right (downstream) location. The velocity of the density wave is close to the homogeneous flow velocity. Thus the exit flow rate variation begins to manifest itself only after the stop of the boiling boundary. For example the decrease C-E of the exit flow rate m_{ex} is a consequence of the decrease A-B of the inlet flow rate m_{in} etc. The exit flow rate m_{ex} variation causes the two-phase pressure drop $\Delta\,P_{2P}$

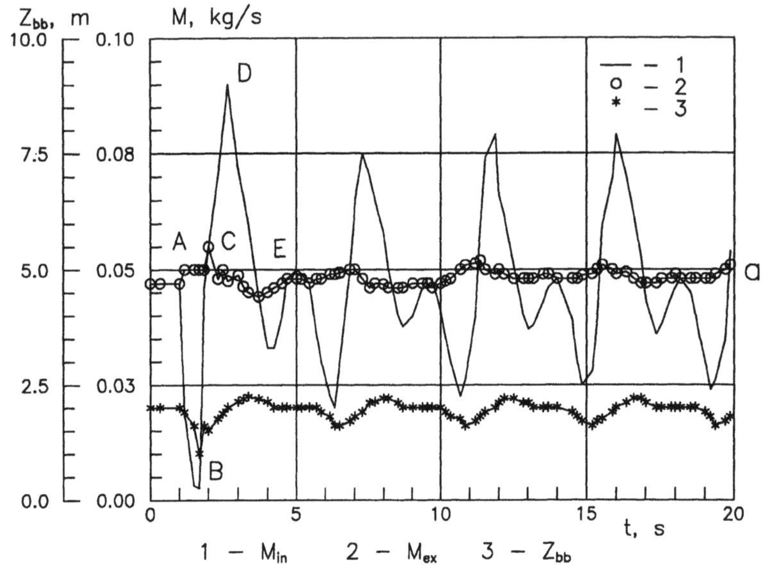
Z_bb, m
M, kg/s
10.0
7.5
5.0
2.5
0.0
0.10
0.08
0.05
0.03
0.00
A
B
C
D
E
— – 1
o – 2
* – 3
a
0
5
10
15
20
t, s
1 – M_in
2 – M_ex
3 – Z_bb

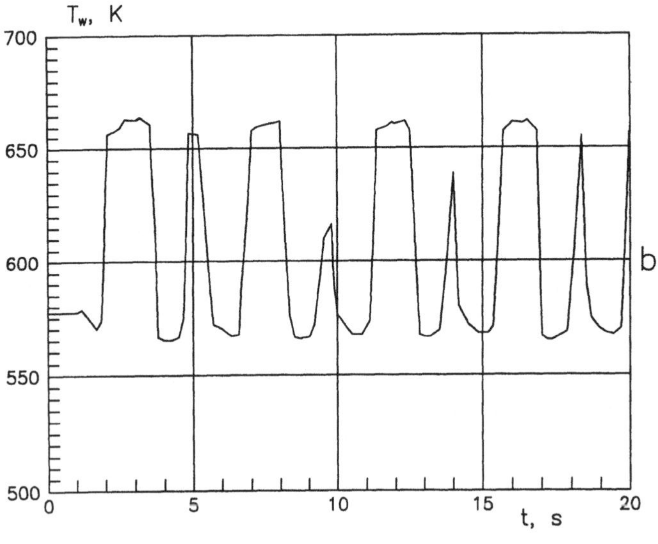
T_w, K
700
650
600
550
500
0
5
10
15
20
t, s
b

Figure 3. a, b.

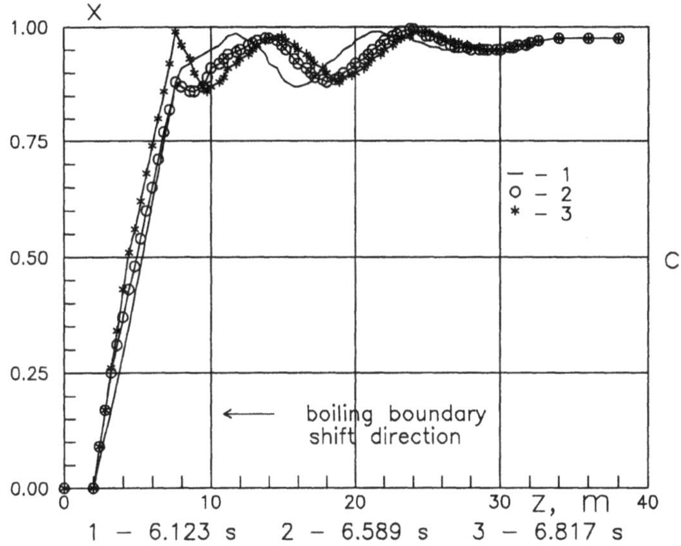

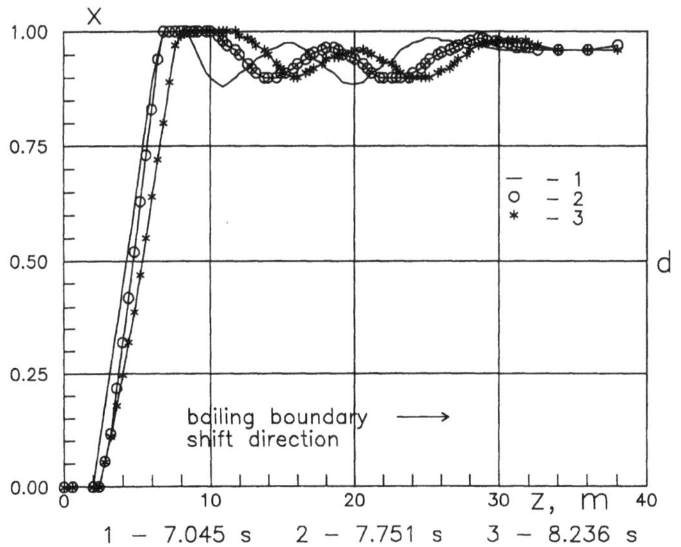

Figure 3. Modeling of oscillatory regime of heated channel with boiling water flow: P_{in} = 7.04 MPa; P_{ex} = 7.0 MPa; Q = 89.2 kW; T_{in} = 467 K.

a) M_{in}, M_{ex}, Z_{bb} ; b) T_w - wall temperature at the end of heated section (the last computing cell of heated section); c), d) - the structures of density waves.

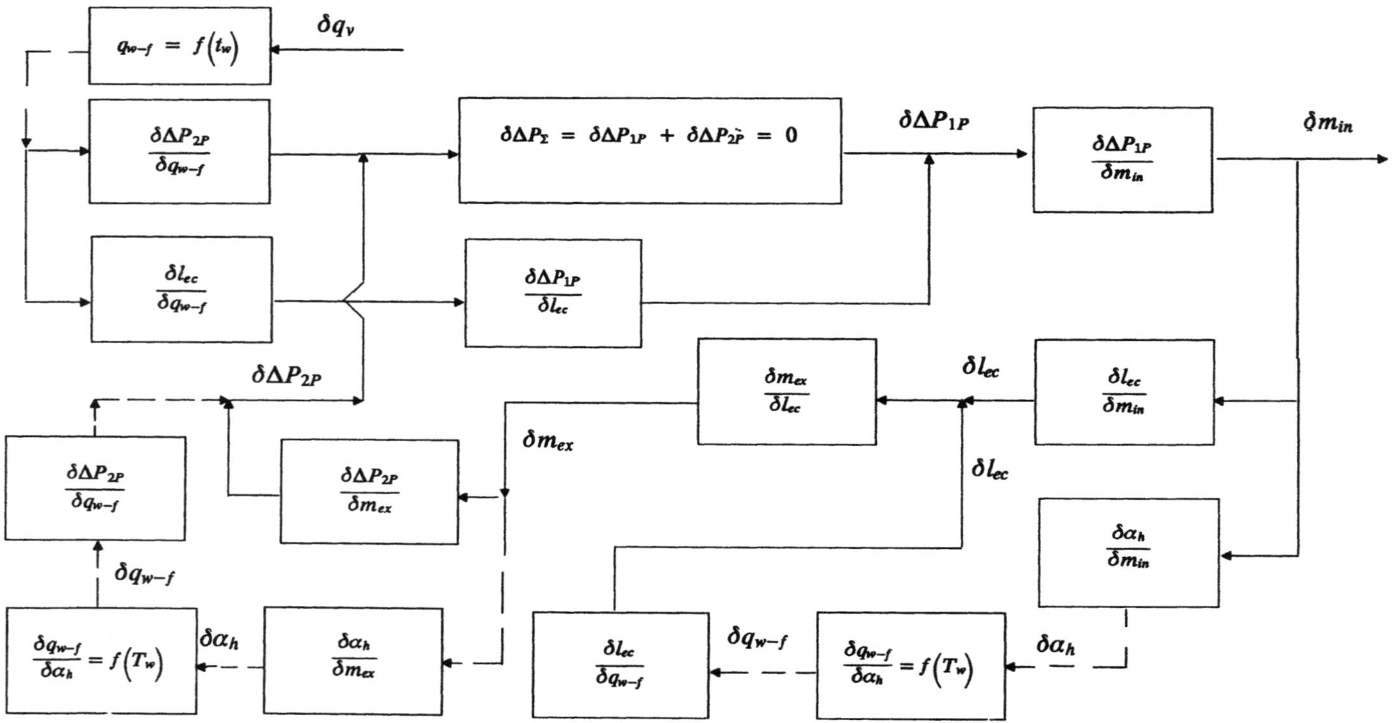

Figura 4. Schematic block diagram of density — waves oscillations.

variation. The described time-delayed feedback can be expressed as follows (Fig.4):

$$\delta\, m_{in} \longrightarrow \delta\, l_{ec} \longrightarrow \delta\, m_{ex} \longrightarrow \delta\, \Delta\, P_{2P}$$

The effect of dependence "wall heat transfer-mass flow rate" on density wave oscillations is noticeable in a single-phase convection section only because the boiling heat transfer weekly depends on mass flow rate. That effect is stabilizing and can be expressed as an aditional feedback effect:

$$\delta\, m_{in} \longrightarrow \delta\, \alpha_{h1P} \longrightarrow \delta\, q_{w-f} \longrightarrow \delta\, l_{ec} \longrightarrow \delta\, m_{ex} \longrightarrow \delta\, \Delta\, P_{2P}$$

where α_{h1P} - single-phase forced convection heat transfer coefficient.

In the considered sample (Fig.3) the period of oscillations is determined by the time required for a density wave to travel through the vertical section. It is because the deviations of hydrostatic pressure drop exceed here the deviations of friction pressure drop and in horizontal section the feedback $\delta m_{2P} \longrightarrow \delta \Delta P_{2P}$ weakens. The local extremes of oscillations (t = 4.1 s, t = 8.6 s, t = 13.1 s) arise at the moments of density waves going out from heated section that corresponds the cessation of the feedback action

$$\delta\, q_{w-f} \longrightarrow \delta\, l_{ec} \longrightarrow \delta\, m_{2P} \longrightarrow \delta\, \Delta\, P_{2P}$$

The heated wall thermal inertia causes the delay in response of heat flux q_{w-f} on variation of heat source q_v or heat transfer coefficient α_h. The duration of that delay may be estimated by value of "metal wall heating constant" $t_w = C_w \rho_w \delta / \alpha_h$ (for the case under consideration $t_w \approx 0.1 - 1$ s). The influence of the wall thermal inertia on feedbacks is shown in block diagram Fig.4 by dashed lines. It follows from calculations (see Fig.5) that wall thermal inertia, which may be characterized by the wall thickness, exerts stabilizing influence.

To conclude let us consider an example of the model practical application — the prediction and analysis of the effect of cavitating venturi to prevent the oscillations. Under the certain difference between inlet and back pressure the flow rate through venturi reaches the maximum value because of "choked flow" phenomenon occuring due to vaporization near the throat. In that regime the flow accelerates in a converging section and decelerates after vapor condensation in a shock wave in a diverging section. The condition $P_t = P_{sat}(T)$ realizes in the throat. The further pressure differential growth cannot increase the flow rate that provides the stabilizing effect. Critical flow rate is a function of inlet pressure, liquid subcooling and throat diameter. That is why, if the venturi tube would be designed on the certain steady flow rate and installed upstream from economizer section, the flow rate could never grow above the "critical" value, determined by inlet pressure and temperature. That is the difference of cavitating venturi from commonly used for damping of oscillations throttles in which pressure drop growth always causes the flow rate growth. On steady-state stable regimes of heated channel cavitating venturi works as usual hydraulic venturi tube with low pressure loss[16]. Application of cavitating venturi widenes the region of heated channel with boiling flow stable operation.

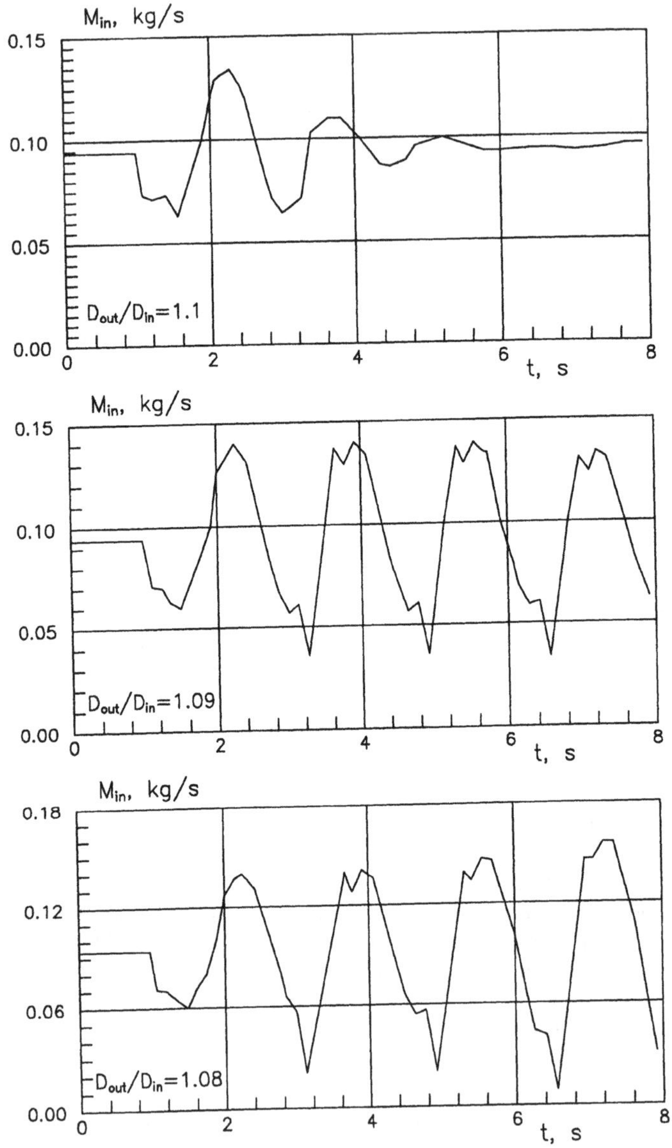

Figure 5. The effect of wall thermal inertia on stability boundary.

P_{in} = 7.178 MPa; P_{ex} = 7.0 MPa; Q = 135.72 kW; T_{in} = 523 K; D_{in} = 10 mm.

Thus the well corresponding with experimental data mathematical model and computer program are developed. It permits to calculate the stability boundaries of boiling flows in a heated channels, to analyse the steady and transient (oscillatory) regimes in detail, to predict the effect of technical steps towards providing the stability.

Acknowledgment

The authors would like to thank Prof. R.T.Lahey for direction.

References

1. R.T.Lahey and F.J.Moody. "The Thermal-Hydraulics of a Boiling Water Nuclear Reactor", American Nuclear Society, New York (1984).

2. A.A.Nikonov, G.A.Gorbenko and V.N.Blinkov. "The Two-Phase Spacecraft Thermal-Control Systems", SPA Energy, Kaliningrad (1991), (In Russian).

3. E.P.Serov and B.P.Korolkov. "The Dynamic Processes in Heat-Mass Exchangers", Energia, Moscow (1967). (In Russian).

4. I.I.Morozov and V.A.Gerliga. "The Stability of Boilers", Atomizdat, Moscow (1969). (In Russian).

5. N.S.Leleev. "The Transient Flows of Coolant in Heated Pipes of Powerful Steam-Generators", Energia, Moscow (1978). (In Russian).

6. D.Butterworth and G.F.Hewitt (ed.). "Two-Phase Flow and Heat Transfer", Oxford University Press (1977).

7. J.M.Delhaye, M.Giot and M.L.Riethmuller (ed.) "Thermohydraulics of Two-Phase Systems for Industrial Design and Nuclear Engineering", Hemisphere Publishing Corporation (1981).

8. J.Achard, D.A.Drew and R.T.Lahey, The analysis of nonlinear density-wave oscillations in boiling channels, J.Fluid Mech. 155:219 (1985).

9. H.Gurgenci, T.N.Veziroglu and S.Kakac, Simplified nonlinear descriptions of two-phase flow instabilities in vertical boiling channel, Int. J.Heat Mass Transfer. 26:671 (1983).

10. A.Clausse, D.Delmastro, R.T.Lahey, The analysis of chaotic instabilities in natural circulation boiling system, in: "Natural Circulation in Industrial Applications", EUROTHERM Seminar Nr.16 Pisa (1990).

11. V.Chatoorgoon. SPORTS - a simple non-linear thermal-hydraulic stability code, Nuclear Engineering and Design. 93:51 (1986).

12. Review and Application of the TRAC-PD2 Computer Code, EPRINP -2826 (1983).

13. R.Taleyarkhan, M.Podowski and R.T.Lahey. The effect of heater wall dynamics on density - wave oscillations in boiling channel, in: "Heat Transfer 1982", Proc. 7th Int. Conf., Munchen (1982).

14. Heat Exchanger Design Handbook. Vol.1, Heat exchanger theory, Hemisphere Publishing Corporation (1983).

15. B.I.Nigmatulin, A.P.Proshutinsky, A.G.Lobachev, E.V.Merkulov, E.I.Trubkin, N.I.Tribkin, A.U. Skonkin, V.P.Volkov, V.V.Gorshkov. Relevance of natural circulation in present and next generation reactors in the Soviet Union, in: "Natural Circulation in Industrial Applications", EUROTHERM Seminar Nr.16 Pisa (1990).

16. I.E.Idelchik. "Handbook of Hydraulic Resistance", Machinostroenie, Moscow (1975). (In Russian).

Nomenclature

A - wetted surface, m^2
C - specific heat, J/(kg K)
F - cross section area, m^2
f - channel wall friction, N/m^3
g - gravitational acceleration, m/s
i - enthalpy, J/kg
J - rate of mass exchange, $kg/(m^3 s)$
l - length, m; latent heat of vaporization, J/kg
m - mass flow rate, kg/s
P - pressure, Pa
N - power of heat source, W
q_w - volumetric heat generation rate in channel wall, W/m^3
q_{w-f} - heat flux from wall to flow, W/m^3
q_{l-v} - interfacial heat flux, W/m^3
T - temperature, K
t - time, s
U - internal energy, J/kg
V - volume, m^3
W - velocity, m/s
x - quality
z - axial coordinate, m
α - void fraction
δ - variation
ρ - density, kg/m^3
j - the angle between vertical and flow direction, radn

Subscripts

v - vapor
l - liquid
m - mixture
sat - saturation
w - wall
in - inlet
ex - exit
$1P$ - single-phase flow
$2P$ - two-phase flow
ec - economizer section
bb - boiling boundary
i - number of cell
f - flow
t - throat

Superscript

n - number of time level

A MATHEMATICAL DESCRIPTION OF THE CRITICAL HEAT FLUX AS A NON-LINEAR DYNAMIC INSTABILITY

Pablo Carrica and Alejandro Clausse

Centro Atomico Bariloche
8400 Bariloche, Argentina

ABSTRACT

The thermofluid dynamics of pool boiling heat transfer was theoretically studied. A two-phase flow integral model was formulated for the local boiling field adjacent to the heated wall. A set of three non-linear differential equations model the dynamics of the localized void fraction, bubble number density and vapor velocity. The boiling crisis at the critical heat flux is described as a dynamic transition caused by the competition of bubbles coalescence and breakup mechanisms. An analysis of a case of pool boiling in water is presented, showing the consistency of the formulation. The model has potential applications to thermal management of electronic microchips and devices for space environments.

INTRODUCTION

Boiling phenomena is one of the most complicated issues in heat transfer theory. As Lienhard (1988) has recently pointed out, to the present it was not possible to derive a complete theory from first principles to predict the heat flux vs. wall temperature relation for a given geometry. The numerous reported studies of boiling at different heat flux ranges have generally been given in the form of correlations. Kutateladze (1948) could identify the similarity groups associated with the boiling mechanisms. But furthermore no closed analytical solution exists for the intensive turbulent heat transfer with vapor bubbles generation and their departure from the heating surface.

On the other hand, the problem of the critical heat flux was always treated as a phenomenon independent from the nucleate boiling process, and almost nothing is known about their interrelation. A number of theoretical predictions of the critical heat flux in pool boiling, based on hydrodynamic instability considerations,were proposed in the past (Zuber, 1958; Kutateladze, 1948). All these theoretical equations are similar to one another and they agree equally well with the experimental data.

Instabilities in Multiphase Flows, Edited by G. Gouesbet and A. Berlemont, Plenum Press, New York, 1993

For the nucleate boiling heat transfer, it has generally been accepted that the dynamics of the liquid and gas components close to the wall is of primary importance. Therefore it is not surprising that most of the proposed theoretical models involve in the calculations hydrodynamic quantities as void fraction, bubble density, departure frequency, nucleation site density, gas velocity. However only comparisons with the heat flux - wall superheating relation and physical properties are performed to validate the theory. One reason of this is the lack of experimental data related to the hydrodynamic parameters.

The starting point for the development of a theoretical model for nucleate boiling is the identification of the different mechanisms by which the energy is transferred from the heated surface to the fluid. It has generally been accepted that the transient conduction evaporation and convection of superheated liquid are the relevant mechanisms. A discussion of the heat transfer rates associated with these mechanisms is given in Hsu y Graham (1976). Since these processes take place in a two-phase region adjacent to the wall, the macroscopic parameters related to the structure of the local two-phase flow play an important role in the heat transfer performance.

Realizing the significance of the dynamics of the localized boiling fields close to the heated wall in convective nucleate boiling, this paper presents a physical model based in differential balance equations which take into account the various interfacial transfer mechanisms. The model consists in a description of the boiling process and crisis within the frame of modern theory of systems dynamics, providing a new view of this old problem.

THEORETICAL MODEL

Let us consider a small horizontal plate heater providing a heat flux **q"** to a pool of liquid at saturated temperature (Fig. 1). A control volume (CV) is defined on top of the surface. The two-phase flow variables are averaged in the CV (Delhaye, 1968). Assuming that the CV contains n identical spherical bubbles per unit volume, the void fraction, α, the interfacial area density, a_i, and the bubbles volume, V_b, are related by:

$$\alpha = n\, V_b \tag{1a}$$

$$a_i = n\, 4\pi\, r_b^2 \tag{1b}$$

$$V_b = 4/3\, \pi\, r_b^3 \tag{1c}$$

where r_b is the bubble radius.

The objective of the analysis is to find which values will take the average two-phase flow parameters in the local CV for a certain imposed heat flux, **q"**. The starting point are the momentum, mass, and energy balances over the CV.

Vapor Momentum Balance

The balance of forces over one rising bubble is given by:

$$\rho_g V_b \frac{du}{dt} = g\,(\rho_l - \rho_g)\, V_b - \frac{1}{2} C_D\, (\pi r_b^2)\, \rho_l\, u^2 - \frac{1}{2} \rho_l V_b \frac{du}{dt} \qquad (2)$$

In the right hand side, from left to right, the buoyancy, the drag and the virtual mass (Drew et al, 1979) terms are written.

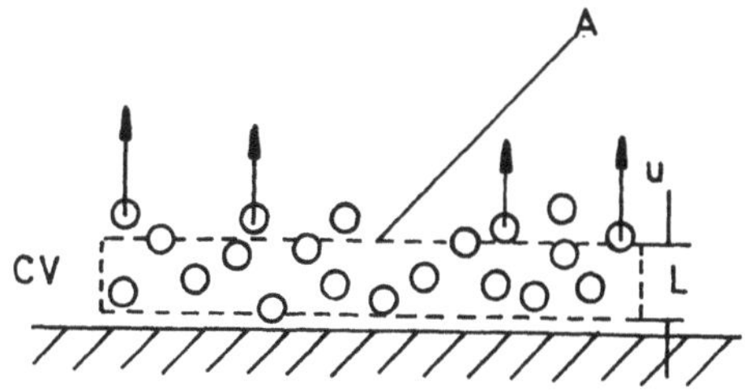

Figure 1. Diagram of the control volume (CV).

Assuming that the gas density, ρ_g, is negligible compared to the liquid density ,ρ_l, yields:

$$\frac{du}{dt} = 2g - \frac{\pi\, C_D u^2 r_b^2}{V_b} \qquad (3)$$

combining Eqs. (1) and (3) gives:

$$\frac{du}{dt} = 2g - \frac{C_D^2\, u\, a_i}{4\,\alpha} \qquad (4)$$

Vapor Mass Balance

The vapor mass balance in the control volume yields:

$$\rho_g L\, A \frac{d\alpha}{dt} + \alpha\, A\, u\, \rho_g = \Gamma \qquad (5)$$

where Γ is the evaporation rate inside the CV.

In order to evaluate Γ, let us consider the physical mechanisms of the heat transfer process. The liquid near the wall is first superheated by taking heat from the surface. Part of this superheated liquid evaporates inside the CV, whereas the rest is transported out of the CV by microconvection and Marangoni effect. The process can be thought as a collective competition between a system of "interfacial particles" and a system of superheated convective clusters (Marangoni jets, vortex turbulences, etc). Using the theory of statistical thermodynamics of interacting systems of particles, it can be shown that the equilibrium energy partition is proportional to the Boltzman factor with an appropriately chosen argument which accounts for the "particle exchanges". A reasonable choice for this argument is the interfacial area density, for it scales the drag and surface tension phenomena, which are the main mechanisms of superheated liquid transport. Therefore the fraction of the heat responsible for vapor production inside the control volume can be written as:

$$q_v = \Gamma\, h_{fg} = q''A\,(1-\alpha)\,\exp(-\lambda\, a_i) \qquad (6)$$

Where λ is a constant length parameter which characterizes the relative importance of the superheated liquid transport. The factor $(1-\alpha)$ in Eq.(6) accounts for the fraction of the CV occupied by the liquid.

Combining Eqs. (5) and (6) yields:

$$L\,\frac{d\alpha}{dt} = \frac{q''}{\rho_g\, h_{fg}}\,(1-\alpha)\,\exp(-\lambda\, a_i) - u\,\alpha \qquad (7)$$

Interfacial Area

A few empirical correlations for the interfacial area density, a_i, are available for relatively small void fractions (Kataoka and Serizawa, 1990). However, the idea of a conservation equation for a_i based on a mechanistic model that includes the dynamics of the breakup and coalescence processes for dispersed bubbles has been recently suggested (Kocamustafaogullari and Ishii, 1983; Navarro-Valenti et al, 1991). In order to consider the processes of coalescence and breakup it is convenient to write a conservation equation for the number of bubbles per unit volume.

The balance of bubbles in the control volume is given by:

$$\frac{dn}{dt} = \mathcal{B} - \mathcal{C} + \frac{1}{\mathcal{V}_b}\frac{d\alpha}{dt} \qquad (8)$$

where $\mathcal{B}$ is the rate of break up, $\mathcal{C}$ is the rate of coalescence and the last term accounts for the inflow and outflow of bubbles.

It can be shown that the rate of bubbles collision is proportional to n^2 (Navarro-Valenti, 1991). On the other hand, the probability of coalescence of two bubbles maybe taken to be proportional to the bubbles volume, $\mathcal{V}_b$. The physical meaning of this assumption is that having two bubbles in a box, the probability of coalescence is proportional to the volume they occupy. The total coalescence rate in the control volume is therefore given by:

$$\mathcal{C} = c\,\mathcal{V}_b\, n^2 \qquad (9)$$

where c is a a characteristic constant rate coefficient.

The shattering probability of a bubble is modeled in such a way that it is very low for a bubble having a volume $\mathcal{V}_b$ smaller than a critical volume $\mathcal{V}_c$, whereas for larger bubbles breakup is most likely to occur (Taylor, 1934). Following this trend, we propose:

$$\mathcal{B} = \mathbf{b}\ n \frac{\mathcal{V}_b^2}{\mathcal{V}_b^2 + \mathcal{V}_c^2} \tag{10}$$

where **b** is a constant rate coefficient.

Combining Eqs. (9) and (11) yields:

$$\frac{dn}{dt} = \mathbf{b}\ n \frac{\mathcal{V}_b^{\ 2}}{\mathcal{V}_b^{\ 2} + \mathcal{V}_c^{\ 2}} - \mathbf{c}\ \mathcal{V}_b\ n^2 + \frac{1}{\mathcal{V}_b}\frac{d\alpha}{dt} \tag{11}$$

Using Eq. (1a) to eliminate $\mathcal{V}_b$ gives:

$$\frac{dn}{dt} = \left(\frac{\mathbf{b}\ \alpha}{\alpha^2 + \mathcal{V}_c^{\ 2} n^2} - \mathbf{c} \right) \alpha\ n + \frac{n}{\alpha}\frac{d\alpha}{dt} \tag{12}$$

DIMENSIONLESS EQUATIONS

Equations (4), (7) and (12) constitute a system of three ordinary differential equations which describes the dynamics of the averaged local two-phase flow field (α, n, u) in the boiling region. The dimensionless form of these equations is:

$$\frac{d\hat{u}}{d\hat{t}} = 2 - \frac{\hat{D}\ \hat{u}^2\ \hat{a}_i}{\alpha} \tag{13a}$$

$$\hat{L}\frac{d\alpha}{d\hat{t}} = \hat{q}''\ (1-\alpha)\ \exp(-\hat{\lambda}\ a_i) - \hat{u} \tag{13b}$$

$$\frac{d\hat{n}}{d\hat{t}} = \hat{C} \left(\frac{\alpha_o\ \alpha}{\alpha^2 + \hat{n}^2} - 1 \right) \alpha\ \hat{n} + \frac{\hat{n}}{\alpha}\frac{d\alpha}{d\hat{t}} \tag{13b}$$

where the dimensionless velocity, $\hat{u}$, number density, $\hat{n}$, interfacial area density, $\hat{a}_i$, and time, $\hat{t}$, are defined by:

$$\hat{u} = \frac{u}{g^{1/2}\mathcal{V}_c^{1/6}} \tag{14a}$$

$$\hat{n} = n\ \mathcal{V}_c \tag{14b}$$

$$\hat{a}_i = a_i \mathcal{V}_c^{1/3} \tag{14c}$$

$$\hat{t} = \frac{g^{1/2}}{V_c^{1/6}} t \tag{14d}$$

In the Eqs. (13), the following dimensionless numbers are introduced:

$$\hat{D} = \frac{C_D}{4} \qquad \text{Drag number} \tag{15a}$$

$$\hat{L} = \frac{L}{V_c^{1/3}} \qquad \text{Length number} \tag{15b}$$

$$\hat{q}'' = \frac{q''}{\rho_g h_{fg} V_c^{1/6} g^{1/2}} \qquad \text{Heat flux number} \tag{15c}$$

$$\hat{\lambda} = \frac{\lambda}{V_c^{1/3}} \qquad \text{Superheat removal number} \tag{15d}$$

$$\hat{C} = c \frac{V_c^{1/6}}{g^{1/2}} \qquad \text{Coalescence number} \tag{15e}$$

$$\alpha_o = \frac{b}{c} \qquad \text{Competition number} \tag{15f}$$

STEADY STATE SOLUTIONS

The steady state solutions of Eqs. (13), ($\hat{u}_s$, $\hat{n}_s$, α_s, $\hat{a}_s$) can be found by making zero the temporal derivatives of α, $\hat{n}$ and $\hat{u}$, which gives:

$$2 \alpha_s = \hat{D} \hat{u}_s^2 \hat{a}_s \tag{16a}$$

$$\hat{q}'' = \frac{\hat{u}_s \alpha_s}{(1-\alpha_s)} \exp(\hat{\lambda} \hat{a}_s) \tag{16b}$$

$$\hat{n}_s^2 = \alpha_s (\alpha_o - \alpha_s) \tag{16c}$$

Combining Eqs. (16a) and (16b) yields:

$$\frac{\alpha_s}{(1-\alpha_s)^{2/3}} = \left(\frac{\hat{D}}{2} \right)^{1/3} \hat{a}_s^{1/3} \hat{q}''^{2/3} \exp(-2/3 \; \hat{\lambda} \; \hat{a}_s) \tag{16d}$$

Fig. 2 shows the behavior of Eqs. (16c) and (16d) in the

(α, a_i) plane for water at atmospheric pressure (see Appendix). The full line curve corresponding to Eq. (16c) represents the competition between the coalescence and breakup mechanisms. For low void fractions breakup prevails increasing the interfacial area as the amount of vapor increases, whereas at higher void fractions coalescence prevails decreasing the interfacial area as the vapor amount increases. Particularly total coalescence occurs at $\alpha = \alpha_o$. The dashed curve corresponds to Eq. (16d) for different heat fluxes. The intersections of the curves give the void fraction, α, and the interfacial area density a_i in the CV when a certain heat flux is imposed. At low heat fluxes (see the $\hat{q}'' = 2$ curve) only one solution exists, and it can be seen that the void fraction in the CV increases with **q"**. However, a_i presents a maximum as the heat flux increases, which is a consequence of the competition breakup and coalescence. For higher heat fluxes

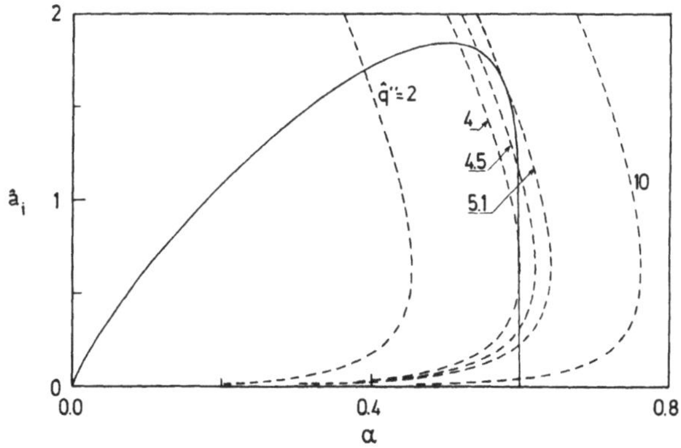

Figure 2. Graphic of Eqs. 16 showing the steady-state behavior of the void fraction and the interfacial area density.

(see $\hat{q}'' = 4.5$ curve) three possible steady state solutions coexist for a given **q"**. Furthermore, for heat fluxes higher than $\hat{q}'' = 5.1$ (corresponding to 1 Mw/m^2), there is again only one solution, which corresponds to low values of interfacial area density.

PHYSICAL INTERPRETATION OF THE SOLUTIONS

Equations (13) were solved numerically for a heat input linearly increased with time according to:

$$\hat{q}'' = 10^{-4}\ \hat{t} \tag{17}$$

Figure 3 shows the temporal evolution of the void fraction and the interfacial area density. It can be seen that during the first part of the heating both variables increase monotonically as the input power increases. This trend continues until the power reaches a critical value ($\hat{q}'' = 5.1$). A sudden transition can be observed at this particular moment: the interfacial area density drops abruptly while the void fraction develops a burst. After the transition the evolution continues smoothly in a state with high void fraction and low interfacial area density.

The physical meaning of the transition can be observed more clearly in Figure 2. The transition takes place when q" reaches the critical value for which Eq. (16d) is tangent to Eq. (16c). For higher values of q", solutions with high interfacial area density no longer exist, and the system experiences a catastrophe, jumping to the low interfacial area density state.

The transition basically is a consequence of the sudden coalescence of all the bubbles, which causes a drastic reduction of the interfacial area in the control volume. The final state

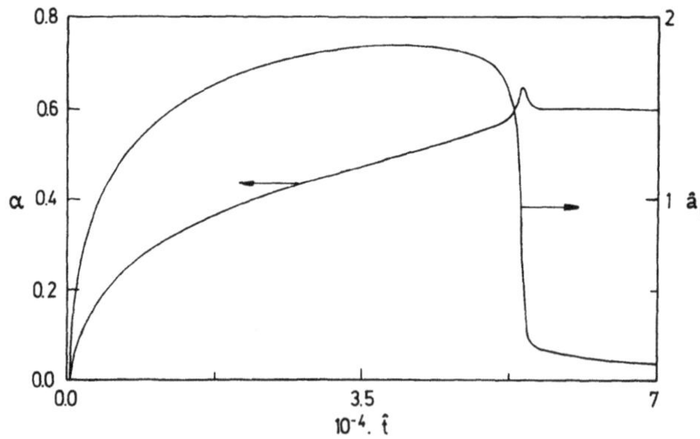

Figure 3. Temporal evolution of the void fraction and the interfacial area density during a linear power increment.

with low interfacial area can be associated to the film boiling regime. Therefore, the catastrophe can be interpreted as a mathematical dynamic description of the boiling crisis. The critical value of $\hat{q}'' = 5.1$ corresponds to the critical heat flux, which in view of this picture is related to a local sudden coalescence mechanism.

CONCLUSIONS

Several mechanisms of critical heat flux were proposed by a number of authors in the past. Hydrodynamic instabilities, thermal transients, bubbles waiting time are some of the main models. In this paper, a dynamic description of the boiling crisis was presented. The differential equations governing the local boiling field state naturally undergo a sudden catastrophe caused by the coalescence of all bubbles in the region close to the heated wall.

The original treatment of the boiling phenomena presented in this paper can be extended to more complex conditions. It is hoped that this study stimulate other researchers to further investigate boiling systems using modern dynamic theory.

APPENDIX: MODEL PARAMETERS FOR WATER AT ATMOSPHERIC PRESSURE

It is currently accepted that breakup of bubbles occurs at a critical value of the Weber number. For gravity driven flows, the Weber number can be written as:

$$We = \frac{d_b^2 \, g \, \rho_l}{\sigma}$$

The measured critical values of Weber numbers for bubbles in water are of the order of 1 (Prince and Blanch, 1990). This in turn gives a critical bubble volume of:

$$V_c = 0.01 \mathrm{cm}^3$$

Only a few experimental studies are available in the literature related to coalescence rates. Prince and Blanch (1990) presented a quite detailed experimental investigation of coalescence rates in air-water mixtures. Based on their results we estimated a value of $c = 0.8 \ \mathrm{sec}^{-1}$.

For bubbles rising in a turbulent liquid field, the drag coefficient C_D is about 8/3 (Wallis, 1969).

Liaw and Dhir (1989) have measured a void fraction between 0.6 and 0.8 (depending on the contact angle) close to the wall at critical heat flux in pool boiling in water. The theoretical point ($\hat{Q}_{CHF}$, α_{CHF}) can be used to fit the values of the remaining unknown parameters in Eqs. (13). The resulting values (considering $q''_{CHF} = 100 \ \mathrm{W/cm}^2$) are:

$$\hat{\lambda} = 0.8$$
$$\alpha_o = 0.6$$

Based also in the experiments of Liaw and Dhir (1990), we used a control volume with L = 2 mm. Up to this distance from the heater, the void fraction is uniform.

Using Eqs. (15), the set of dimensionless parameters for water are:

$$\hat{D} = 2/3$$
$$\hat{L} = 1$$

$\hat{\lambda} = 0.8$

$\hat{C} = 0.01$

$\alpha_o = 0.6$

NOMENCLATURE

A	Heated area
a_i	Interfacial area density
b	Breakup rate coefficient
$\mathcal{B}$	Rate of breakup
$\mathcal{C}$	Rate of coalescence
c	Coalescence rate coefficient
C_D	Drag coefficient
CV	Control volume
h_{fg}	Latent heat of evaporation
g	Gravity
L	Control volume high
n	Bubble number density
q"	Heat flux
q_v	Heat which produce vapor inside the CV
r	Radius
t	Time
u	Velocity
V_b	Bubble volume
V_c	Critical breakup volume

Greeks

α	Void fraction
Γ	Evaporation rate inside the CV
λ	Superheated liquid transport length
ρ	Density
σ	Surface tension

Subscripts

b	Bubble
g	Gas
l	Liquid
s	Steady state

Superscripts

^	Dimensionless

REFERENCES

Delhaye, J. M., 1968, "Equations fondamentales des ecoulements diphasiques", Partes 1 et 2, *CEA-R*-3429, Centre d'Etudes Nucleaires de Grenoble, France

Drew, D., Cheng, L. and R. T. Lahey Jr., 1979, "The analysis of virtual mass effects in two-phase flow", *Int. J. Multiphase Flow*, 5:233

Hsu, Y. Y. and Graham, R. W., 1976, "Transport Processes in Boiling and Two-Phase System", McGraw Hill, New York

Kataoka, I. and Serizawa, A., 1990, "Interfacial area concentration in bubbly flows, *Nuc. Eng. Des.*, 120:163

Kocamustafaogullari, G. and Ishii, M., 1983, "Interfacial area and nucleation site density in boiling systems", *Int. J. Heat Mass Transfer*, 26:1377

Kutateladze S., 1948, "On the transition to film boiling under natural convection", *Kotloturbostroeine*, 3:10

Liaw, S. P. and Dhir, V. K., 1989, "Void fraction measurements during saturated pool boiling of water on partially wetted vertical surfaces", *J. Heat Transfer*, 111,731.

Lienhard, J, 1988, "Things we don't know about boiling heat transfer", *Int. Comm. Heat. Mass Trans.*, 15:401 .

Navarro-Valenti, S., Clausse, A., Drew, D. and R. T. Lahey, 1991, "A contribution to the mathematical modeling of bubbly-slug flow regime transition", *Chem. Eng. Comm.*, 102:69

Prince, M and Blanch, H., 1990, "Bubble coalescence and breakup in air-sparged bubble columns, *AIChE J.*, 36:1485

Taylor, G. I., 1934, "The formation of emulsion in definable field of flow", *Proceedings of the Royal Society* (London), 146:501

Wallis,G., 1969, "One-Dimensional Two-Phase Flow", McGraw Hill

Zuber, N., 1958, "On the stability of boiling heat transfer", *Trans. ASME*, 80:711

DENSITY WAVE OSCILLATION IN A NATURAL CIRCULATION LOOP OF LIQUID NITROGEN

Mamoru Ozawa[1], Yoshihisa Asao[2] and Nobuyuki Takenaka[3]

[1]Department of Mechanical Engineering
Kansai University, Suita, Osaka 564, Japan
[2]Sumitomo Electric Industries Ltd.
Konohana, Osaka 554, Japan
[3]Department of Mechanical Engineering
Kobe University, Nada, Kobe 657, Japan

INTRODUCTION

Dynamic behavior and flow instability problems have been widely investigated in these three decades in relation with the development of nuclear power plants [Lahey & Drew 1980]. On the other hand, the dynamic behavior and/or flow instabilities in cryogenic systems have been investigated only in a limited range of experiments so far [Friedly et al. 1979, Hands 1979, Akagawa et al. 1989, Fukuda et al.1984 and so on]. Recently, increasing attention has been given to the flow instability problems in air separation systems, where thermosyphon reboiler-condensers have significant influence on the system performance. In the thermosyphon reboiler-condenser, heating of the primary fluid in the reboiler side is conducted by the condensation of the secondary fluid. Then the dynamic behavior of circulation velocities in both sides and the thermal interaction between the reboiler and the condenser play important roles in the dynamics of the total system, which makes the dynamics of the system complicated one. For better understanding such dynamics, it is necessary, at the first step, to conduct investigations in a simplified system such as without any thermal coupling.

Thus experimental investigation and numerical simulation were conducted on the boiling two-phase flow instability in a thermosyphon reboiler of liquid nitrogen (LN_2), i.e. in the natural circulation system where heating of LN_2 was carried out by Joule heating.

EXPERIMENTAL SETUP

Experiment was conducted by using an experimental setup shown in Fig. 1. Main parts of the system are LN_2 tanks, a head column, a subcooler and a test section. The head column was thermally insulated, and was filled with LN_2 supplied continuously from the LN_2 tank. The liquid level in the head column was monitorred by measuring the pressure difference and was constant throughout the experiment. The inlet section including the turbine flowmeter between the test section and the head column was submerged in the subcooler to prevent the phase change due to the pressure drop and the heat input from the ambient air. LN_2 was continuously supplied into the subcooler from the LN_2 tank. Thus the inlet subcooling was mainly controlled by the difference between the saturation temperature of LN_2 in the test section and that in the subcooler.

LN_2 supplied from the head column flowed through the inlet section and into the test section. Two-phase mixture from the test section was separated into each phase in the head column. N_2 gas in the head column was ejected to the atmosphere through the pressure control valve V_4.

The test section is a SUS 304 pipe of dimensions 5.0 mm I.D., 6.0 mm O.D. and 0.9 m in length. The test section was heated by Joule heating of A.C. electric power. The wall temperatures were measured by C-C thermocouples at every 50 mm along the test pipe.

The system pressure, i.e. the pressure at the test section exit, was P_{ex}=0.3 MPa and 0.42 MPa, the inlet subcooling ΔT_{sub}=7.0 ~ 12.0 deg. K , the heat flux q=4.2 ~ 76.2 kW/m^2 and the mass flux G=60 ~ 300 kg/m^{2}s in the present experiments. The resistance coefficient across the inlet section including the valve, the piping element and the turbine flowmeter was K_p=330 and 580, and that at the exit valve was K_v=10. The resistance coefficient K is defined by:

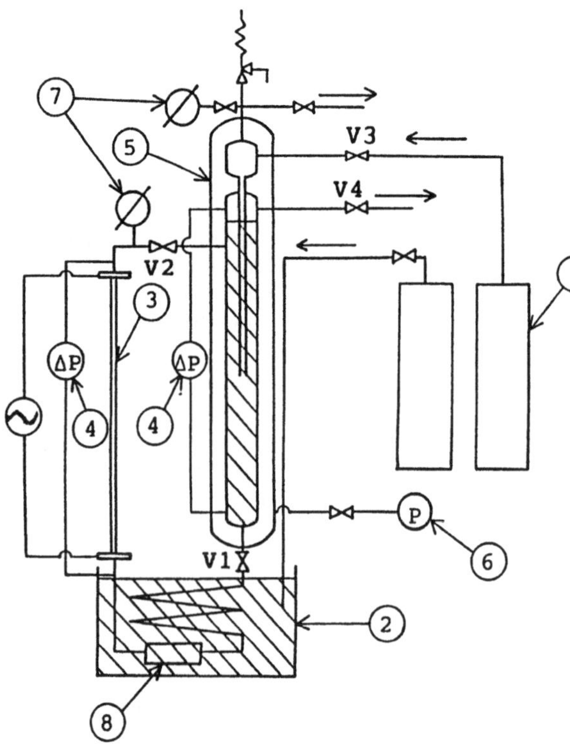

1: LN_2 tank
2: Subcooler
3: Test section
4: D.P. cell
5: Head column
6: Vacuum pump
7: Pressure gauge
8: Turbine flowmeter

Figure 1. Experimental setup.

$$\Delta P = K\frac{G^2}{2\rho_L} \quad (1).$$

CIRCULATION CHARACTERISTICS

Experimental results of the circulation velocity u are plotted against the heat flux q in Figs. 2 ~ 5. In general, the circulation velocity increases with the increase in the heat flux, and have a **maximum**. This is, of course, due to the fact that the increase in the heat flux causes the increase in the void fraction in the heated channel, which results in the increase in the driving force in the natural circulation loop. Further increase in the heat flux induces the increase in the

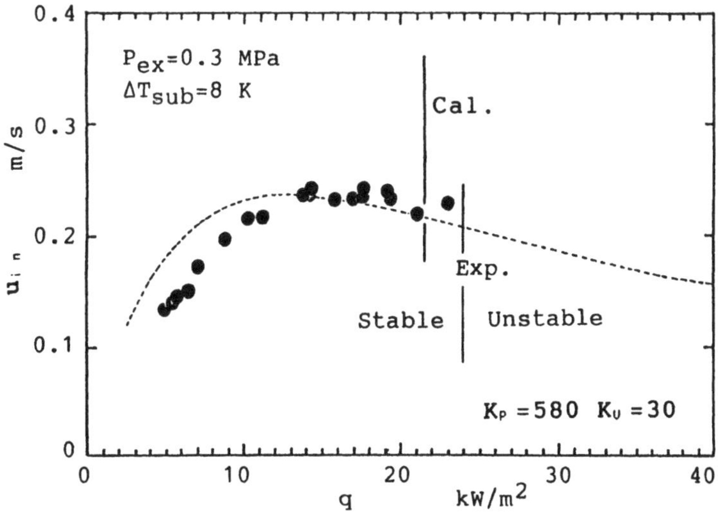

Figure 2. Circulation characteristics.

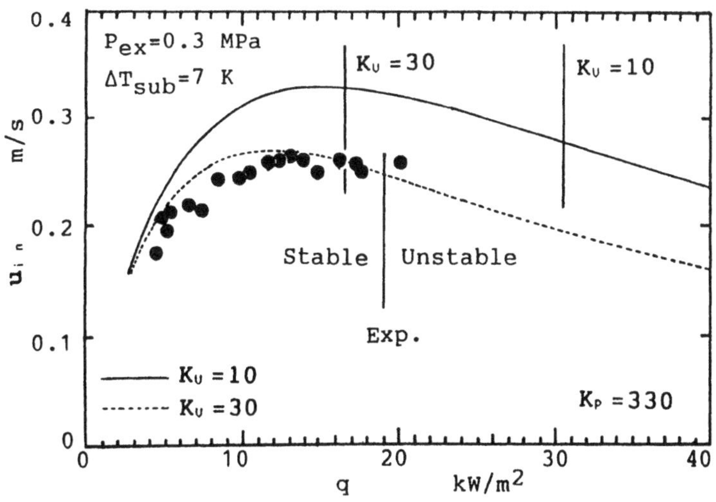

Figure 3. Circulation characteristics.

frictional pressure drop in the channel, which results in the decrease in the circulation velocity. The increase in the system pressure induces the decrease in the circulation velocity in the lower heat flux region, while the increase in the system pressure results in the increase in the circulation velocity in the higher heat flux region. The circulation velocity is a decreasing function of the system pressure. All these circulation characteristics are similar to those observed in natural circulation systems of normal fluids [Nakanishi et al. 1978].

Beyond a certain limit of heat flux, the flow instability, i.e. the density wave oscillation occurs. The threshold conditions of the density wave oscillation are shown by vertical solid lines in Figs. 2 to 5. The heat flux at the threshold condition increases with the increase in the inlet restriction and the system

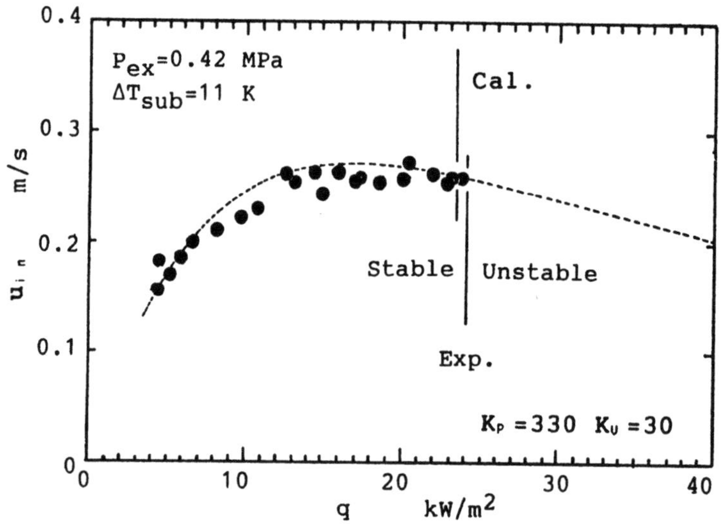

Figure 4. Circulation characteristics.

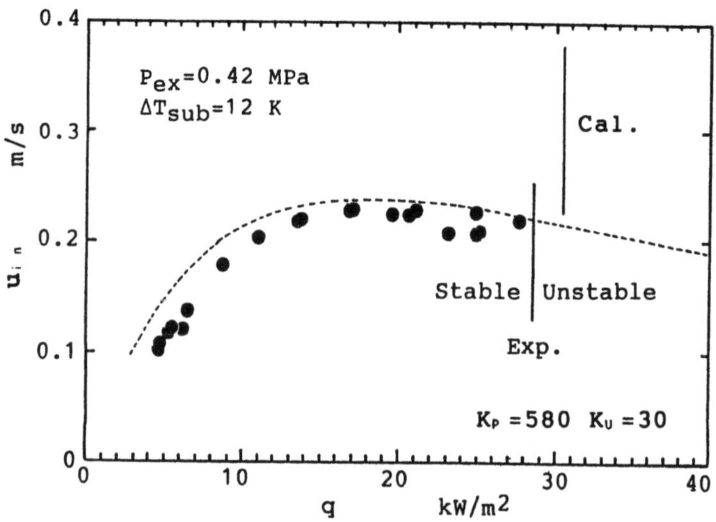

Figure 5. Circulation characteristics.

pressure. The oscillation period, which is the same order of magnitude with the residence time, is plotted against the heat flux in Fig.6. The oscillation period decreases with the increase in the heat flux, which correspond that the residence time decreases with the increase in the heat flux. The period is an increasing function of the system pressure.

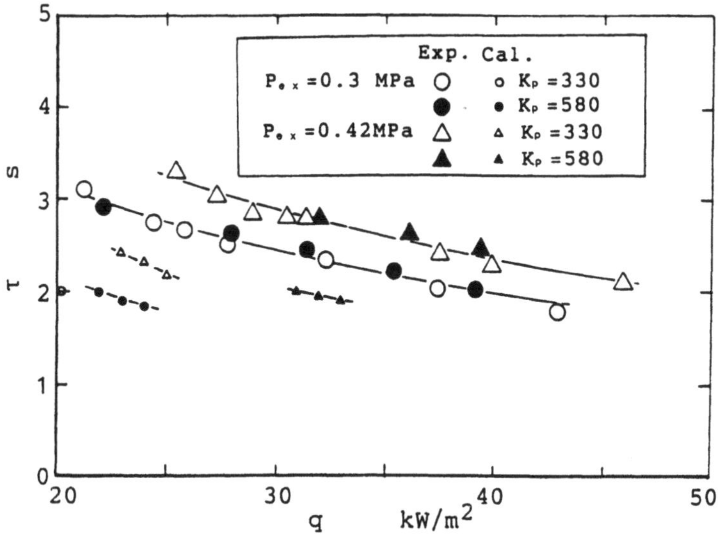

Figure 6. Oscillation period.

LUMPED-PARAMETER SIMULATION

Simulation Model

In predicting the threshold condition of the density wave oscillation in forced-flow LN_2 systems, Akagawa et al. (1988) have already verified the applicability of the linear stability analyses developed by Nakanishi et al. (1978) for normal fluids. Recently, flow instability becomes a topic of great interest with reference to non-linear dynamics, i.e. chaotic behavior of two-phase flow system. Lahey and his group are conducting extensive research using lumped-parameter model in this field [Lahey et al. 1989]. Thus in this investigation, non-linear analysis using lumped-parameter model is also adopted to simulate the density wave oscillation and to predict the threshold conditions of the oscillation.

The circulation system is simplified as shown in Fig.7. Main parts are the head column(E-A), the inlet section(A-B), the heated section(B-C) and the riser and exit section(C-D-E). The head column and the inlet section are kept in adiabatic condition, and the heat input from the surrounding is take into account in the riser and exit section. The inner diameters are represented by D_P in the inlet, riser and exit sections, D_T in the heated section, and D_H in the head column. The piping

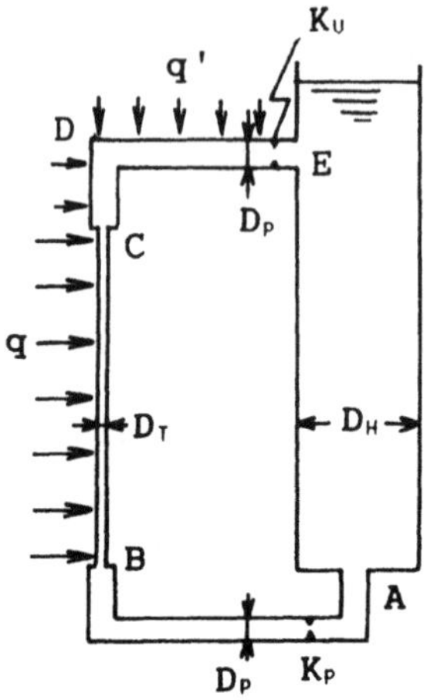

Figure 7. Simulation model.

lengths are denoted by L_{AB}, L_T, L_{CD}, L_{DE}, and L_H in the inlet, heated, riser, exit sections and the head column, respectively.

Assuming thermodynamic equilibrium, the following conservation equations are used in the analysis.

$$\frac{\partial \rho}{\partial t} + \frac{\partial}{\partial z}(\rho u) = 0 \tag{2}$$

$$\frac{\partial}{\partial t}(\rho e) + \frac{\partial}{\partial z}(\rho h u) = q_v \tag{3}$$

$$\frac{\partial}{\partial t}(\rho u) + \frac{\partial}{\partial z}(\rho u^2) = -\frac{f}{D}\frac{\rho u^2}{2} - \rho g - \frac{\partial P}{\partial z} \tag{4}$$

where q_v represents the volumetric heat flux defined by

$$q_v = \frac{4q}{D_T} \tag{5}$$

In the present simulation, the pressure is assumed to be uniform throughout the system, and the heat capacity of the pipe wall is ignored. Two-phase flow is expressed by using homogeneous flow model. The applicability of the homogeneous flow model to the heated section has been verified in the forced flow experiment under the same experimental range with the present one [Asao & Ozawa 1991]. In Eq. (3), the term $\partial(\rho e)/\partial t$ is approximated by $\partial(\rho h)/\partial t$.

Dynamic Behavior of Circulation

The heated section is divided into two regions as shown in Fig. 8, i.e. the subcooled and two-phase regions. As the heat flux is uniform along the heated section, the enthalpy along the heated section is given by

$$h = h_{in} + \frac{h_{LG}}{\lambda_G - \lambda_L} z = h_{in} + \frac{q_v}{G} z \tag{6}$$

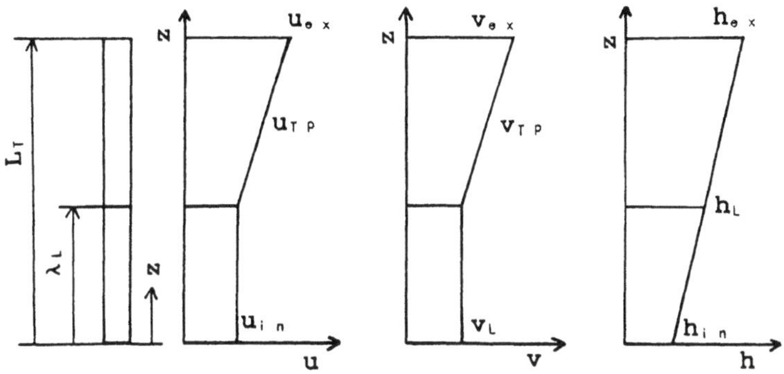

Figure 8. Distributions of velocity, specific volume and enthalpy.

where

$$\lambda_L = \frac{G}{q_v}\Delta h_{sub}, \qquad \lambda_G = \lambda_L + \frac{G}{q_v}h_{LG} \tag{7}$$

The specific volumes are expressed by

$$v = v_L \qquad (x_{eq} \leqq 0) \tag{8}$$

$$v_{TP} = v_L + \frac{v_{LG}}{\lambda_G - \lambda_L}(z-\lambda_L) = v_L + \frac{q_v v_{LG}}{Gh_{LG}}(z-\lambda_L) \qquad (x_{eq} > 0) \tag{9}$$

Using Eqs. (3) and (9), we have the velocity gradient in the two-phase region,

$$\frac{\partial u}{\partial z} = q_v \frac{v_{LG}}{h_{LG}} \equiv \Omega_{TP} \tag{10}$$

Then, the velocities in both regions are given by

$$u = u_{in} \qquad (x_{eq} \leqq 0) \tag{11}$$

$$u_{TP} = u_{in} + \Omega_{TP}(z - \lambda_L) = u_{in} + \frac{v_{LG}}{h_{LG}} q_v (z - \lambda_L) \qquad (x_{eq} > 0) \tag{12}$$

The heat fluxes in the riser and exit sections are represented by q', and then the volumetric heat flux q_v' and the characteristic parameter Ω' are given by $q_v' = 4q'/D_P$ and $\Omega' = q_v' v_{LG}/h_{LG}$, respectively. The enthalpy, specific volume and velocity can be obtained as the same manner described above.

The inertial mass in the heated section is given by

$$m_T = m_L + m_{TP} \tag{13}$$

where m_L and m_{TP} represent the inertial mass in the subcooled and two-phase regions, respectively,

$$m_L = \int_0^{\lambda_L} \rho_L dz = \rho_L \lambda_L, \quad m_{TP} = \int_{\lambda_L}^{L_T} \rho dz = \int_{\lambda_L}^{L_T} (1/v_{TP}) dz = \frac{L_{BC} - \lambda_L}{v_{ex} - v_L}\{\ln(v_{ex}/v_L)\} \tag{14}$$

The integral of the mass conservation equation across the heated section (B-C) is

$$\frac{d}{dt}\int_{B-C}\rho dz + \rho_{ex}u_{ex} - \rho_L u_{in} = 0 \quad (15).$$

Then we obtain

$$\frac{dm_T}{dt} = \rho_L u_{in} - \rho_{ex}u_{ex} \quad (16).$$

Using Eqs. (13), (14) and (16), the time derivative of the specific volume at the heated section exit is given by

$$\frac{dv_{ex}}{dt} = \frac{\rho_L u_{in} - \rho_{ex}u_{ex} - \left\{\frac{1}{v_L} - \frac{\ln(v_{ex}/v_L)}{v_{ex}-v_L}\right\}\frac{d\lambda_L}{dt}}{\frac{(L_T-\lambda_L)}{(v_{ex}-v_L)^2}\left(1 - \frac{v_L}{v_{ex}} - \ln\frac{v_{ex}}{v_L}\right)} \quad (17).$$

Just as the same manner, those at the riser exit (Point D) and the head column inlet (Point E) are obtained as

$$\frac{dv_D}{dt} = \frac{\frac{v_D-v_C}{L_{CD}}\left(\frac{u_C}{v_C} - \frac{u_D}{v_D}\right) + \left\{\frac{1}{v_C} - \frac{\ln(v_D/v_C)}{v_D-v_C}\right\}\frac{dv_C}{dt}}{\frac{1}{v_D} - \frac{\ln(v_D/v_C)}{v_D-v_C}} \quad (18),$$

$$\frac{dv_E}{dt} = \frac{\frac{v_E-v_D}{L_{DE}}\left(\frac{u_D}{v_D} - \frac{u_E}{v_E}\right) + \left\{\frac{1}{v_D} - \frac{\ln(v_E/v_D)}{v_E-v_D}\right\}\frac{dv_D}{dt}}{\frac{1}{v_E} - \frac{\ln(v_E/v_D)}{v_E-v_D}} \quad (19).$$

The energy equation (3) is integrated over the range $z=0 \sim \lambda_L$,

$$\int_0^{\lambda_L}\frac{\partial}{\partial t}(\rho_L h)dz + \int_0^{\lambda_L}\frac{\partial}{\partial z}(\rho_L h u_{in})dz = \int_0^{\lambda_L} q_v dz \quad (20).$$

Leibnitz rule is applied to Eq. (20) to give the time derivative of the boiling boundary, then we have

$$\frac{d\lambda_L}{dt} = 2u_{in} - \frac{2q_v\lambda_L}{\rho_L(h_L-h_{in})} \quad (21).$$

Using Eq. (4), the momentum balance throughout the loop A-E can be expressed by

$$\int_{A-E}\frac{\partial(\rho u)}{\partial t}dz + \int_{A-E}\frac{\partial}{\partial z}(\rho u^2)dz = -\int_{A-E}\rho g dz - \int_{A-E}\frac{f}{2D}\rho u^2 dz - \int_{A-E}\frac{\partial P}{\partial z}dz \quad (22).$$

The acceleration, friction and gravity terms are expressed by ΔP_a, ΔP_f, ΔP_g, respectively, and the momentum W_T defined by Eq. (23) is introduced,

$$W_T = \int_{A-E}\rho u dz \quad (23),$$

Then, Eq. (22) can be rewritten in the form given by

$$\frac{dW_T}{dt} = (P_A - P_E) - (\Delta P_{gT} + \Delta P_{fT} + \Delta P_{aT}) \quad (24),$$

where the frictional pressure drop ΔP_{fT} includes the flow resistance owing to expansions and contractions of the flow area and valves. The integral of the momentum equation across E-A region, i.e. in the head column is

$$\frac{dW_H}{dt} = (P_E - P_A) - (\Delta P_{gH} + \Delta P_{fH} + \Delta P_{aH}) \quad (25),$$

where

$$W_H = \int_{E-A} \rho u dz \quad (26).$$

Eliminating P_A-P_E from Eqs. (24) and (25), we have

$$\frac{d(W_T+W_H)}{dt} = -(\Delta P_{gT} + \Delta P_{fT} + \Delta P_{gH} + \Delta P_{fH}) \quad (27),$$

where the accelerational pressure drops are cancelled out.
The terms in Eq. (27) are given by

$$W_T + W_H = \rho_L \left(\frac{D_T}{D_P}\right)^2 u_{in} L_{AB} + \rho_L \left(\frac{D_T}{D_H}\right)^2 u_{in} L_{EA} + m_T u_{in}$$
$$+ \Omega_{TP} \frac{L_T - \lambda_L}{v_{ex} - v_L} (L_T - \lambda_L - v_L m_{TP}) + u_C \frac{L_{CE}}{v_E - v_C} \ln \frac{v_E}{v_C}$$
$$+ \Omega' \frac{L_{CE}}{v_E - v_C} \left(L_{CE} - v_C \frac{L_{CE}}{v_E - v_C} \ln \frac{v_E}{v_C}\right) \quad (28),$$

$$\Delta P_{gT} + \Delta P_{gH} = g m_T + g \frac{L_{CD}}{v_D - v_C} \ln \frac{v_D}{v_C} - \rho_L g (L_T + L_{CD}) \quad (29),$$

$$\Delta P_{fT} + \Delta P_{fH} = \frac{(K_A + K_P)}{2} \rho_L \left\{u_{in} \left(\frac{D_T}{D_P}\right)^2\right\}^2 + \frac{K_B}{2} \rho_L u_{in}^2$$
$$+ \frac{f_T}{2D_T} \Big[u_{in}^2 m_T + 2\Omega_{TP} u_{in} \frac{L_T - \lambda_L}{v_{ex} - v_L} (L_T - \lambda_L - v_L m_{TP})$$
$$+ \Omega_{TP}^2 \left(\frac{L_T - \lambda_L}{v_{ex} - v_L}\right)^2 \left\{\frac{(L_T - \lambda_L)(v_{ex} - 3v_L)}{2} + v_L^2 m_{TP}\right\}\Big]$$
$$+ \frac{K_C}{2} \rho_{ex} u_{ex}^2 + \frac{f_P}{2D_P} \Big[u_C^2 \frac{L_{CE}}{v_E - v_C} \ln \frac{v_E}{v_C}$$
$$+ 2\Omega' u_C \left\{\frac{L_{CE}^2}{v_E - v_C} - \left(\frac{L_{CE}}{v_E - v_C}\right)^2 v_C \ln \frac{v_E}{v_C}\right\}$$
$$+ \Omega'^2 \left\{\frac{L_{CE}^3}{2(v_E - v_C)} - \left(\frac{L_{CE}}{v_E - v_C}\right)^2 v_C L_{CE} + \left(\frac{L_{CE}}{v_E - v_C}\right)^3 v_C^2 \ln \frac{v_E}{v_C}\right\}\Big]$$
$$+ \frac{K_D}{2} \frac{u_D^2}{v_D} + \frac{(K_E + K_U)}{2} \frac{u_E^2}{v_E} + \frac{f_H L_{EA}}{2D_H} \rho L \left\{u_{in} \left(\frac{D_T}{D_H}\right)^2\right\}^2 \quad (30).$$

SIMULATION RESULTS

In the simulation, at first the steady state value of the circulation velocity was obtained by setting $d(W_T+W_H)/dt=0$ in Eq.(27), and then Eqs.(17)~(19), (21) and (27) were solved numerically by using Runge-Kutta method. The initial condition of the flow was given by the steady state value of the circulation velocity multiplied by factor 1.01.

The circulation characteristics obtained in the numerical simulation is shown by solid line in Fig.2. The characteristic curve is significantly higher than the experimental results. As is described above, the pressure drop in the heated section has been well predicted by using the homogeneous flow model. Thus the large difference between the simulation and experiment is caused by the pressure drop at the exit section, i.e. the application of homogeneous flow model to the exit restriction and/or the exit section seems to be unsuitable in the present simulation. As the circulation characteristics is a basis of the instability analysis, the resistance coefficient K_v is modified to be in good agreement between the model and the experiment. When K_v is multiplied by factor 3, the characteristic curve shown by the broken line well coincides with the experimental results. In cases of other

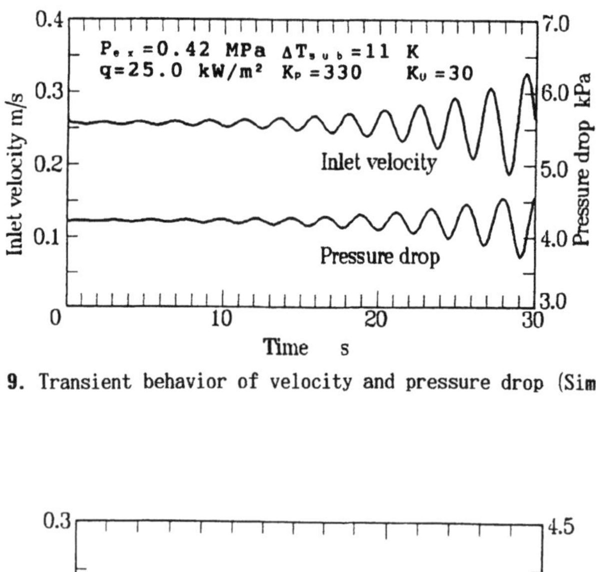

Figure 9. Transient behavior of velocity and pressure drop (Simulation).

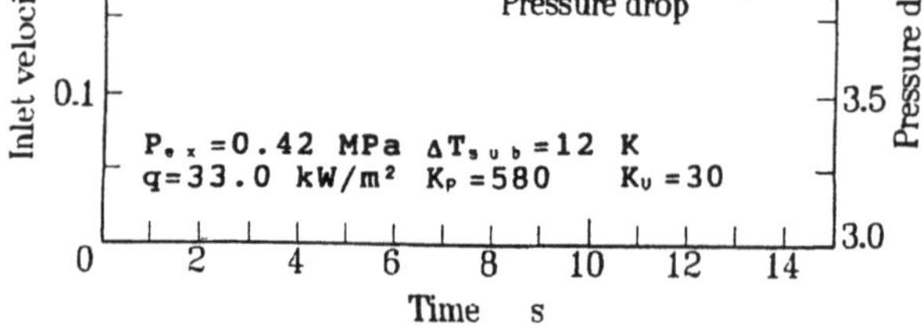

Figure 10. Transient behavior of velocity and pressure drop (Simulation).

operating conditions shown in Figs.3 to 5, this adjusting factor 3 was used throughout the present calculation. Then the numerical simulation of the density wave oscillation was conducted by using this value K_v=30.

The oscillation traces obtained in the numerical simulation are shown in Figs.9 to 11. All these figures show the feature that the small disturbance is developing with time and results in the density wave oscillation. Further calculation leads to the flow reversal condition and the present simulation model cannot be applied to such flow reversal. Further modelling is needed in this problem. Figure 11 is obtained in the neighborhood of the threshold condition, and the sustained limit cycle oscillation is observed. The oscillations of the flow rate and the pressure drop in the heated section are almost 180 deg. out of phase.

The oscillation period obtained in the simulation is shown in Fig.6. The period in the simulation is the same order of magnitude with the residence time as in the experiment, but is slightly lower than the experimental results, while the qualitative characteristics of the period well agree with the experimental results.

The threshold condition of the oscillation was obtained by such simulation. The calculated threshold conditions are also shown in Figs.2 to 5. As can be seen clearly in Fig.2, the threshold condition for K_v=10 does not agree with the experiment, while that for K_v=30 well agree with the experiment. All other figures also show the good agreement between the simulation and the experiment. This fact demonstrates that the threshold condition is well predicted by the present numerical simulation, when the circulation characteristics can be well predicted by the same model.

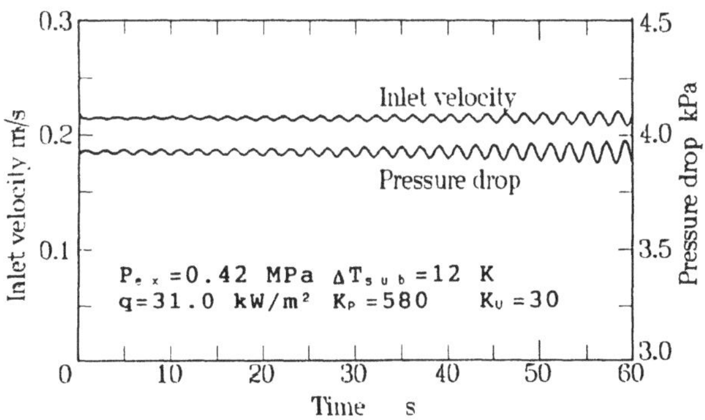

Figure 11. Transient behavior of velocity and pressure drop (Simulation).

CONCLUSION

Experimental and numerical investigations were conducted on the circulation characteristics and the density wave oscillation in the LN_2 natural circulation loop. The main feature of the circulation characteristics is similar to those for normal fluids. Moreover, the threshold condition can be well predicted by using the present numerical simulation. However, the present simulation cannot be applied to the large amplitude oscillation with flow reversal. Further development of the model is needed in this field.

NOTATIONS

D	: inner diameter	ΔT_{sub}	: subcooling (temperature)
e	: internal energy	t	: time
f	: friction factor	u	: velocity
G	: mass flux	v	: specific volume
g	: gravitational acceleration	v_{LG}	: difference in specific volumes ($=v_G-v_L$)
h	: enthalpy	W	: momentum
Δh_{sub}	: subcooling (enthalpy)	x_{eq}	: thermodynamic equilibrium quality
h_{LG}	: latent heat of vaporization	z	: axial coordinate
K	: resistance coefficient	λ_G	: characteristic length
L	: length	λ_L	: boiling boundary
m	: inertial mass	ρ	: density
P	: pressure	Ω	: characteristic parameter
ΔP	: pressure drop	q_v	: volumetric heat flux
q	: heat flux		

Subscripts

L	: liquid	ex	: exit
G	: vapor	T	: heated section
TP	: two-phase	H	: head column
in	: inlet	P	: piping element

REFERENCES

Akagawa, K., Takenaka, N., Fujii, T., Yoshida, M. and Oku, Y., 1988, Flow Instabilities in Parallel Channels with Nucleate and Film Boiling, *Proc. the Jpn.-U.S. Seminar on Two-Phase Flow Dyn.*, pp.H.1.1-14.

Asao, Y. and Ozawa, M., 1991, Boiling Heat Transfer and Flow Pattern of Liquid Nitrogen, *Trans. JSME, Ser.B*, 57-537, pp.1813-1818(1991).

Friedly, J.C., Akinjiola, P.O. and Robertson, J.M., 1979, Flow Oscillation in Boiling Channels, *AIChE Symp. Ser.*, Vol.75, No.189, pp.204-217.

Fukuda, K., Kato, A., Kondo, T. and Hasegawa, S., 1984, Two-Phase Flow Instability in a Liquid Nitrogen Heat Exchanger, *Proc. Jpn.-U.S. Seminar on Two-Phase Flow Dyn.*, F8.1-8.

Hands, B.A., 1979, The Flow Stability of a Liquid-Nitrogen Thermosiphon With 8 mm Bore Riser, *AIChE Symp. Ser.* Vol.75, No.189, pp.177-184.

Lahey, Jr., R.T. and Drew, D.A., 1980, An Assessment of the Literature to LWR Instability Modes, *NUREG/CR*-1414.

Lahey, Jr., R.T., Clause, A. and DiMarco, P., 1989, Chaos and Non-Linear Dynamics of Density-Wave Instabilities in a Boiling Channel, *AIChE Symp. Ser.*, Vol.85, No.269, pp.256-261.

Nakanishi, S., Ishigai, S., Ozawa, M., Mizuta, Y. and Tarui, H., 1978, Flow Instabilities in Boiling Channels: Density Wave Oscillation in a Single Channel Boiling System, *Theoretical and Applied Mechanics*, Vol.26, Univ. Tokyo, pp.421-430.

Nakanishi, S., Ishigai, S., Ozawa, M. and Mizuta, Y., 1978, Analytical Investigation of Density Wave Oscillation, *Technology Reports of the Osaka University*, Vol.28, No.1421, pp.243-251.

STUDY ON THE EFFECTIVE PARAMETERS FOR THE GEYSERING PERIOD

Kang Tae-il,[1] Koji Okamoto,[1] and Haruki Madarame[1]

[1] Nuclear Engineering Research Laboratory
The Faculty of Engineering
University of Tokyo
Tokai-mura, Ibaraki, 319-11, Japan

INTRODUCTION

The geysering is one of the basic two phase flow instabilities. The geysering mechanism is explained as follows (JSME, 1991): A heated vertical water column, having either a zero or very small inlet velocity, increases bubble generation after boiling, thereby decreasing the column's hydrostatic head. This reduction in head subsequently decreases saturation temperature and it allows more bubbles to be generated. Thus in certain conditions, they cause water to be expelled from the column. The column is then refilled by subcooled water from an expansion tank, with the geysering cycle being repeated. The geysering process consists of three stages, i.e., a thermal storage stage in which the subcooled water's temperature increases, and rapid boiling and subsequent refilling stages.

Although the mechanism of the geysering is simple, it is difficult to predict the geysering period correctly. Previous experimental studies (Aritomi et al., 1991; Morioka et al., 1986; Nakanishi et al., 1978) only revealed the relation between the geysering period and some macro parameters, e.g., heat flux, subcooled temperature and inlet velocity.

Nakanishi et al.(1978) experimentally investigated the geysering using R-113 as working fluid. The test section was a vertical tube with a constant low inlet velocity under atmospheric pressure. The geysering period was measured for various heat flow rates, inlet velocities, heater lengths and riser lengths, resulting in the following relations,

(1) With increasing heat flow rate, the period increased.
(2) With decreasing inlet velocity, the period increased.
(3) With increasing heater length, the period increased.
(4) With increasing riser length, the period increased.

Instabilities in Multiphase Flows, Edited by G. Gouesbet
and A. Berlemont, Plenum Press, New York, 1993

However the reasons for the above relations were not discussed.

Aritomi et al.(1991) experimentally investigated the geysering using parallel channel test section with low inlet velocity. The working fluid was water under atmospheric pressure. They evaluated the geysering period using an energy balance between the inlet subcooled temperature and the saturation temperature.

Aritomi et al.(1991) also proposed that the subcooled condition in the outlet condition was necessary for the geysering to occur. In their experiments the subcooled temperature in an expansion tank was less than 30 K, with the effects of having further subcooled temperatures not being considered.

In this study, the geysering with large subcooled temperatures in an expansion tank was experimentally investigated using a bottom closed vertical tube. The geysering was also simulated by a one dimensional numerical code, and investigated by comparing the numerical results with experimental results. The merits of the numerical code were the ability to evaluate the local physical properties, e.g., the local saturation temperature and the local void fractions. The effective parameters on the geysering period were investigated by both the experiment and the numerical simulation.

EXPERIMENT WITH LARGE SUBCOOLED TEMPERATURE

Experimental Setup and Method

Figure 1 shows the experimental test section consisting of a square vertical tube with a closed bottom, a large cylindrical expansion tank under atmospheric pressure, SUS 304 stainless steel thin plate insulated electric heaters, and thermocouples. One side of the vertical tube was made of glass in order to visualize the geysering. The expansion tank had sufficient size when compared to the vertical tube's cross section. Electric heaters were inserted into it to change the subcooled temperature in the expansion tank. Six sheathed thermocouples (C-A type) were installed in the vertical tube's heated section, i.e., at 310, 410, 510, 610, 710, and 810 mm from the tube bottom. The water was used as working fluid in the experiment, and both the subcooled temperature in an expansion tank and heat flow rate in heated section were changed as experimental parameters. In the present experiment, the maximum heat loss through the wall in the

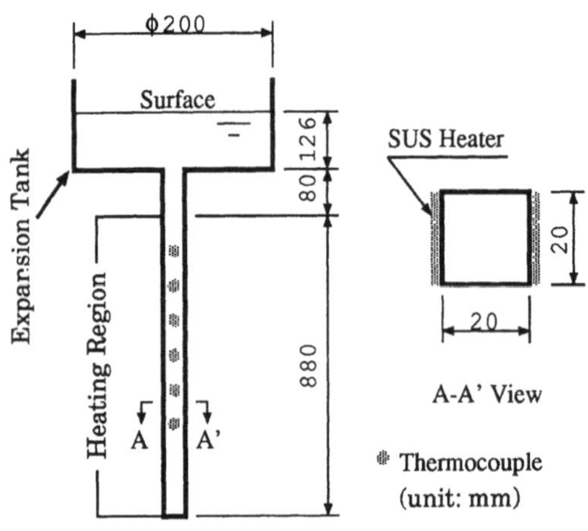

Figure 1. Experimental apparatus

heated section was evaluated as 3% and may be negligible when compared to the heat influx.

Visual Observations

Figures 2(a)-(c) show a diagram of the observed geysering process. The one cycle of the geysering was classified into three stages. As the water is heated at the thermal storage stage, the water temperature increases, and then reaches its boiling temperature. Then the water begins to boil, generating bubbles in the bottom region, followed by the bubbles coalescing into large slugs in the rapid boiling stage. The slugs subsequently travel upward and push the water out of the vertical tube. Since the water in the vertical tube's upper section has large subcooling in the refilling stage, the slugs undergo significant subcooling during mixing with the subcooled water in tube's upper section. They are condensed there rapidly and can not reach the expansion tank. Thereby, the thermal storage stage starts over again.

Geysering Temperature History

A typical geysering example with a 333 W heat flow rate and a 50 K subcooled temperature in an expansion tank is shown in fig. 3 for different thermocouple locations. Oscillating temperatures with a constant period are clearly indicated there. It should be noted that the temperatures slowly increase in the single phase thermal storage, followed by a rapid temperature increase at the rapid boiling stage, and then reaching a peak value. After this peak, a rapid decrease is observed in the refilling stage, and the thermal storage stage is repeated.

Geysering Inception

One of requirements for geysering is a temperature increase approaching the saturation temperature, such that heat influx is greater than heat efflux.

$$q_{\text{influx}} > q_{\text{efflux}} \tag{1}$$

The heat efflux contains the heat transport from the heated tube to the upper expansion tank by natural convection in single phase and subcooled boiling in two phase. Figure 5 shows geysering inception condition with respect to heat flow rate and subcooled temperature in an expansion tank. At lower heat flow rates, stable subcooled boiling was

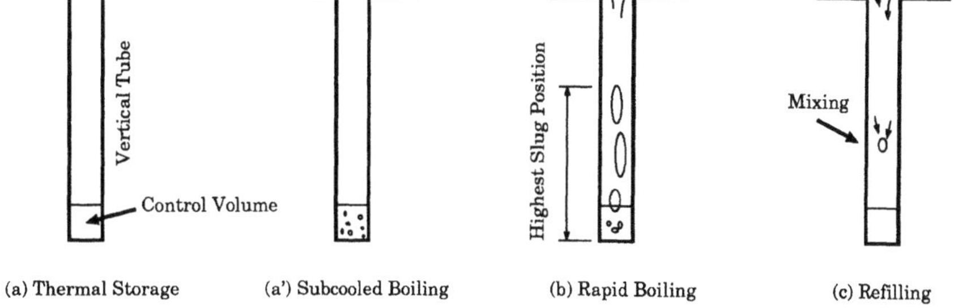

Figure 2. Schematic diagram of the geysering cycle

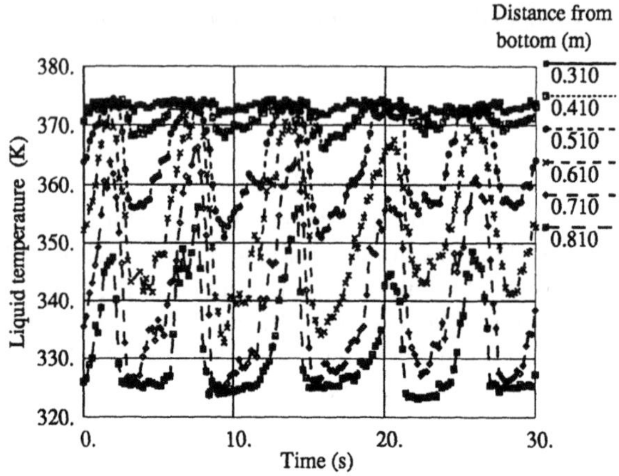

Figure 3. Experimental temperature history (Q = 333 W, $\Delta T_{\mathrm{sub}/tank}$ = 50 K)

observed. However, geysering did not occur because the heat efflux by the subcooled boiling exceeded the heat influx. In the present experiment, the heat flow rate of the geysering inception was about 280 W regardless of the subcooled temperature in an expansion tank.

Temperature Distributions

Figure 4 shows maximum and minimum water temperature distributions during a geysering cycle. Since the bottom region in heated section was nearly equal to the saturation temperature, boiling easily occurred in this region. The upper heated section had large subcooling even at the maximum temperatures, thereby condensing the slugs formed in the bottom region.

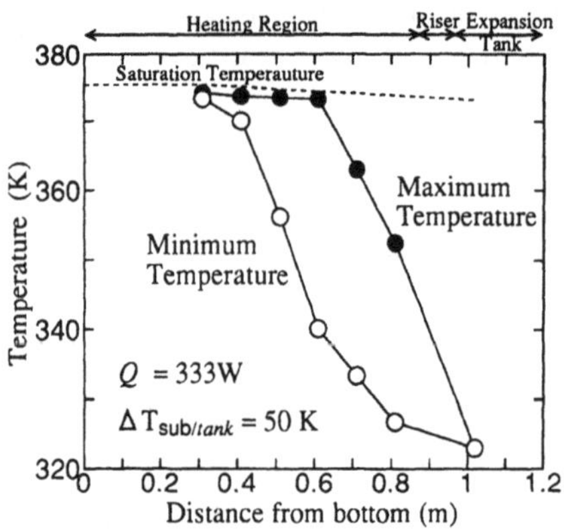

Figure 4. Experimental temperature distribution during one cycle

Experimental Geysering Periods

The geysering periods obtained from the temperature histories are shown in fig. 6. This figure shows the following relation between the geysering period and macro parameters, i.e.,subcooled temperature in an expansion tank and heat flow rate.

(1) As the heat flow rate increased, the geysering period decreased at a constant subcooled temperature in an expansion tank.

(2) As the subcooled temperature in an expansion tank increased, the geysering period decreased at a constant heat flow rate.

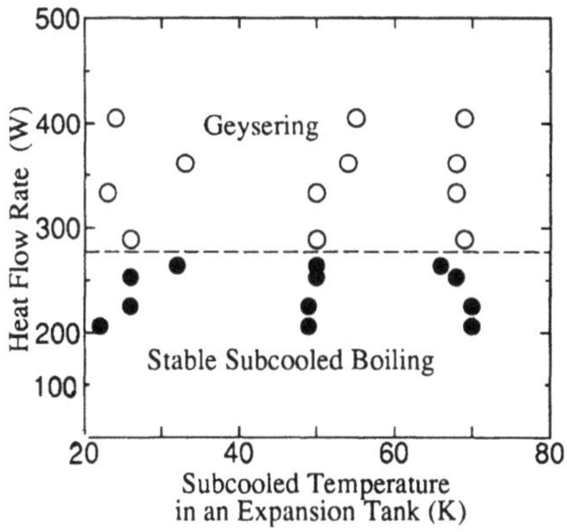

Figure 5. Experimental geysering inception condition

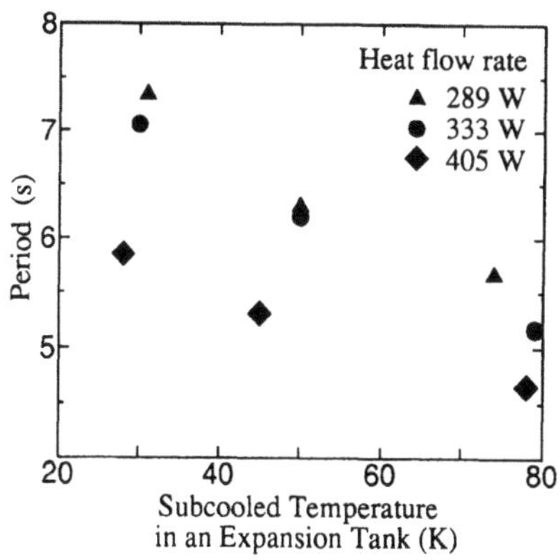

Figure 6. Experimental geysering period

NUMERICAL SIMULATION

In order to simulate the geysering, the MINI-TRAC code (Akimoto et al., 1991) was used. This code solves one dimensional mass, momentum and energy conservation equations which were based on the two fluid model, i.e., the 2V2T1P (2 Velocities, 2 Temperatures, 1 Pressure) model. As for the two fluid model constitution equations, e.g., equations for interfacial stress, interfacial heat transfer, wall stress and wall heat transfer, those in TRAC-PF1 code (Liles et al., 1986) were used.

The MINI-TRAC code contains heat conductor model, in which the two dimensional temperature distribution were calculated. The heat conductor model was coupled with the energy conservation equations for the two phase flow, consequently, the effect of the heat capacity in the heat conductor was included in the simulation.

The test section was simulated as a one dimensional vertical pipe with 49 meshes shown in fig.7. The effect of mesh length on the geysering period was relatively large. Because of the numerical diffusion and the ill-poseness of two fluid model, the simulated geysering could not be same as the experimental geysering. In this study, the constant mesh length was used in the simulation, in order to evaluate the relative tendency of the geysering period.

As boundary conditions for the numerical simulation, a zero inlet velocity in the bottom and constant atmospheric pressure in the top were set as shown in fig.7. For the initial conditions, a uniform 300 K fluid temperature and a constant heat flow rate were applied. After several minutes, the geysering initiated periodic boiling and mixing. An example of the simulated geysering at the same condition of fig.3 is shown in fig.8. The experimental and simulated temperature oscillations were found to be very similar.

The numerically simulated and experimental geysering periods are compared in fig. 9. The simulated relations between the period and the macro parameters agreed well with the experimental relations. Namely, as the heat flow rate increased, the geysering period decreased. And as the subcooled temperature in an expansion tank increased, the geysering period decreased.

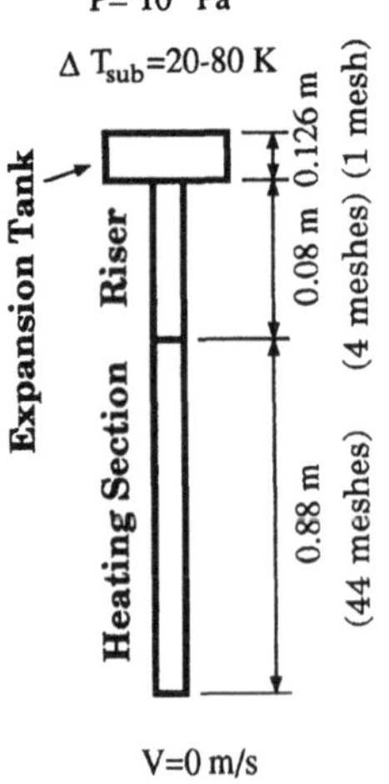

Figure 7. Simulation model

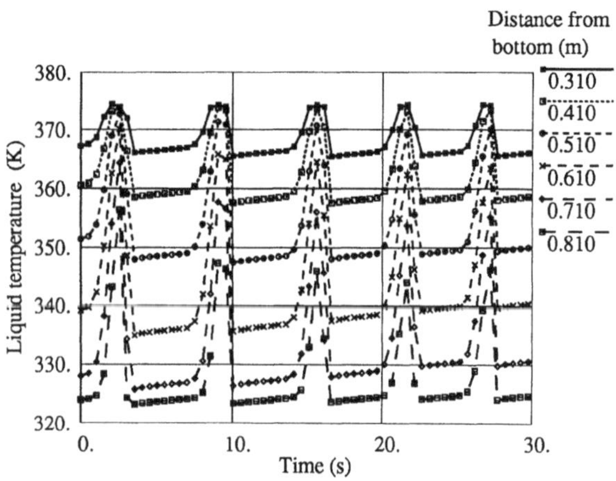

Figure 8. Simulated temperature oscillation ($Q = 330$ W, $\Delta T_{\text{sub}/tank} = 50$ K)

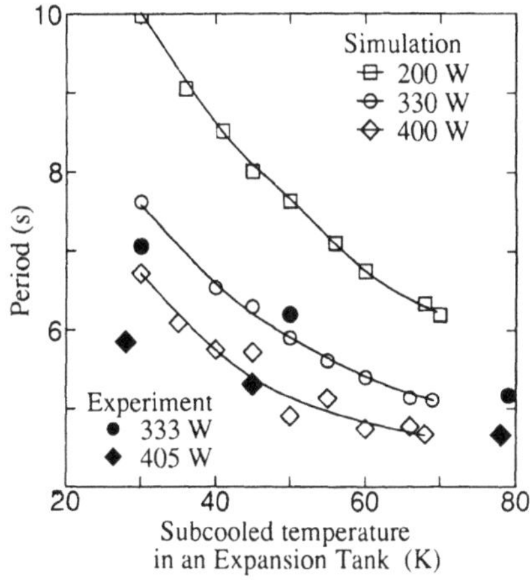

Figure 9. Geysering period

Although geysering did not occur in the experiment with a 200 W heat flow rate, it did in the numerical code. Because some of the heat transports, e.g., the subcooled boiling and the natural convection were not contained in the MINI-TRAC code, the heat efflux simulated in the code was less than that in the experiment. Therefore the geysering occurred in the simulation with low heat flow rate than indicated in equation (1). Because some of heat transports were not contained in the code, the temperature distribution in the code was also deferent from that in the experiment.

DISCUSSION

Geysering Period

The influence of the macro parameters on the geysering period could be investigated by evaluating the numerical results and the experimental results. As mentioned in section 1, the geysering consists of the three stages, i.e., rapid boiling, refilling of the subcooled water from the upper tank and thermal storage in single phase. Usually, the time of the thermal storage stage is longer than that of the other two stages. The time of the thermal storage stage was represented by τ in the following equation which was obtained considering the energy balance during the thermal storage stage.

$$\tau = \rho\ Cp\frac{(T_{sat} - T_l)}{q'''} \qquad (2)$$

where T_{sat} is saturation temperature at the boiling point, and T_l is fluid temperature at the boiling point just after the refilling stage. q''' is heat flow rate per unit volume.

In the experiment, the boiling was initiated at the bottom of the test section, regardless of the test conditions. Therefore the static pressure and/or the saturation temperature at the boiling point were considered to be constant in any test conditions. ($T_{sat} \approx$ constant).

By evaluating the numerical results and the experimental results locally, the relations between the geysering period and the macro parameters were explained using equation (2) as follows.

Subcooled temperature in an expansion tank. Both the experiment and numerical simulation showed that as the subcooled temperature in the expansion tank decreased, the geysering period increased at a constant heat flow rate. To further investigate the above relation, the geysering process was more closely examined.

When the subcooled temperature in the expansion tank was high, subcooled temperature in the upper tube section was also high, hence in the rapid boiling stage the slugs were condensed more rapidly in the upper tube section. Therefore the highest slug position, i.e., the existing region of the slugs, became shorter than case of the smaller subcooling in the expansion tank. Figure 10 shows the highest slug positions in the simulation and the experiment, where as the subcooled temperature in an expansion tank increased, the highest slug position decreased. The mixing rate in refilling stage was related to the void fraction, i.e., the highest slug position. Therefore, with shortening the slug region, the void fraction in the heating tube also decreased, resulting in the smaller mixing rate in refilling stage. When the mixing rate in the refilling stage decreased, the difference between the temperature before and after mixing decreased, thereby the subcooled temperature, i.e., $(T_{sat} - T_l)$, in the heated tube decreased.

Figure 11 shows the relation between the subcooled temperature in the expansion tank and that in the bottom region after mixing, $(T_{sat} - T_l)$. The lines denote the results from the numerical simulation. The marks denote the experimental subcooled temperature at 0.31m from the bottom. Because of the heat transport by the subcooled boiling, the temperature in the bottom region is almost the same with that at the 0.31m in the experiment as shown in fig.4. Figure 11 shows the inverse relation between the subcooled temperature in the expansion tank and that in the bottom region.

Thus, as the subcooled temperature in an expansion tank decreases, the geysering period at a constant heat flow rate increases.

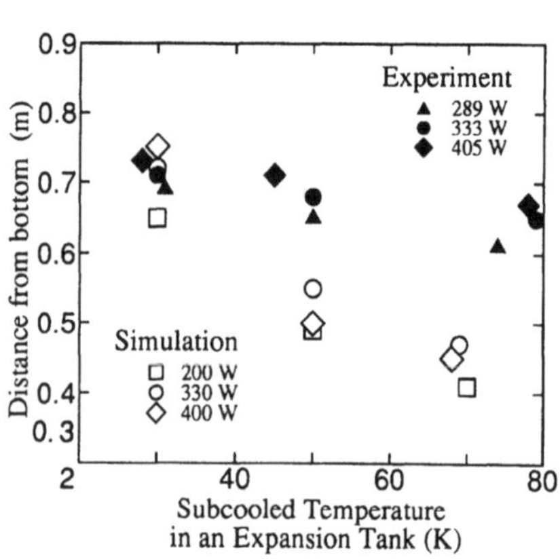

Figure 10. Highest slug position

Heat flow rate In both the experimental and simulation results, as the heat flow rate increased, the geysering period decreased at a constant subcooled temperature in the expansion tank. With increasing heat flow rate, the mixing rate of refilling stage also increased, causing $(T_{sat} - T_l)$ to decrease. From equation (2), a larger heat flow rate will reduce the geysering period.

The above discussion shows that equation (2) could explain the relation between the geysering period and the macro parameters. Therefore the local subcooled temperature at the boiling point and/or the mixing condition in the refilling stage were found to be the effective parameters on the geysering period.

Experiment of Nakanishi et al.(1978)

In order to confirm the above results, the experiment of Nakanishi et al.(1978) were simulated by the MINI-TRAC code. As mentioned in section 1, their test section was a vertical tube with the constant low inlet velocity in atmospheric pressure. The working fluid was R-113 with the constant inlet subcooled temperature and almost zero subcooled temperature in the expansion tank.

In the simulation, water was used as working fluid with 36 meshes. As boundary conditions for the numerical simulation, constant inlet velocity in the bottom and constant atmospheric pressure in the top were set as shown in fig.12. In order to avoid the numerical instability, the 3K subcooled temperature in the expansion tank was used.

Although the working fluid was not the same with the experiment, the relative tendency of the geysering period could be discussed. The simulated relations between the period and the macro parameters mentioned in section 1 agreed well with the experimental relations. For example, in the simulation, as increasing the heat flow rate and/or as decreasing the inlet velocity, the period increased as shown in fig.13.

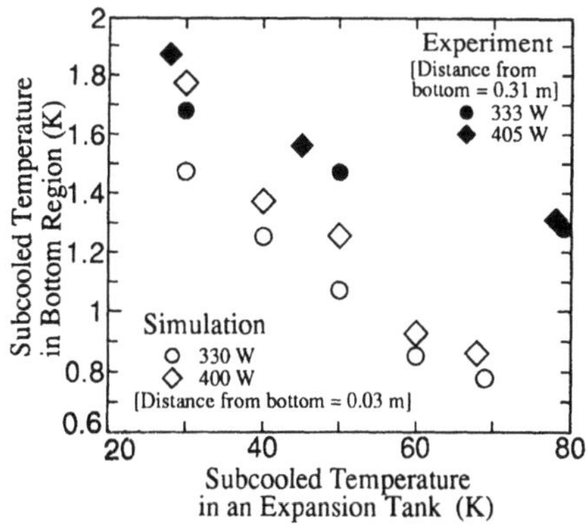

Figure 11. Relation between subcooled temperature

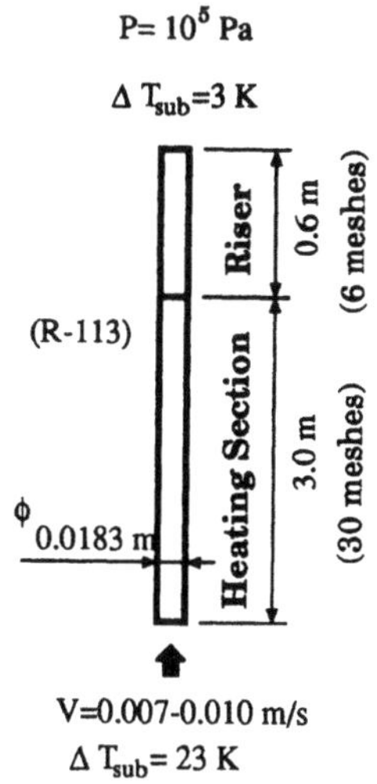

Figure 12. Simulation model for Nakanishi's experiment

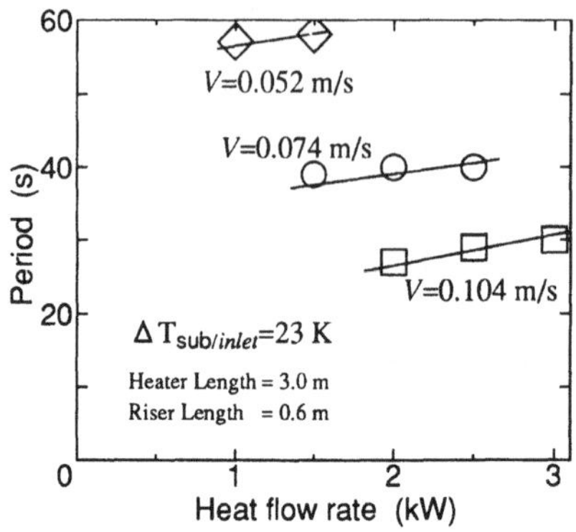

Figure 13. Simulated geysering period for Nakanishi's experiment

Because of the inlet flow, the position of the boiling point changed with test conditions, resulting in the change of T_{sat}. While, the upper tank temperature was nearly equal to the saturation temperature at atmospheric pressure, the mixing condition at the refilling stage was similar in any conditions, causing T_l to be nearly equal constant. ($T_l \approx constant$). By evaluating the numerical results locally, the relations between the geysering period and the macro parameters were explained using the equation (2) as follows.

Inlet velocity In both the experiment and the simulation, as the inlet velocity decreased, the geysering period increased. With decreasing inlet velocity, the boiling point moved upstream, i.e., to lower position. Then the static pressure at the boiling point increased, causing saturation temperature, T_{sat}, to increase. Therefore, the geysering period, τ, increased as indicated in equation (2). Figure 14 shows the maximum and minimum temperature distributions during the geysering one cycle. This figure confirmed the above explanations. As the inlet velocity decreases, the saturation temperature at boiling point increases. Whereas, the liquid temperature after refilling stage, T_l, is nearly same in the both case. According to equation (2), the geysering period increases with decreasing the inlet velocity.

Heater length, Riser length In both the experiment and the simulation, as the heater length and/or the riser length increased, the geysering period increased. With increasing heater length and/or riser length, the static pressure at the boiling point increased, because the boiling point moved to upstream region. Therefore the saturation temperature, T_{sat}, increased, resulting in the geysering period, τ, to increase.

Heat flow rate In both the experiment and the simulation, as the heat flow rate increased, the geysering period increased. With increasing the heat flow rate, the boiling point moved to upstream, resulting in T_{sat} to increase. Because the increasing ratio of $(T_{sat} - T_l)$ was larger than that of q''', the geysering period, τ, increased according to equation (2).

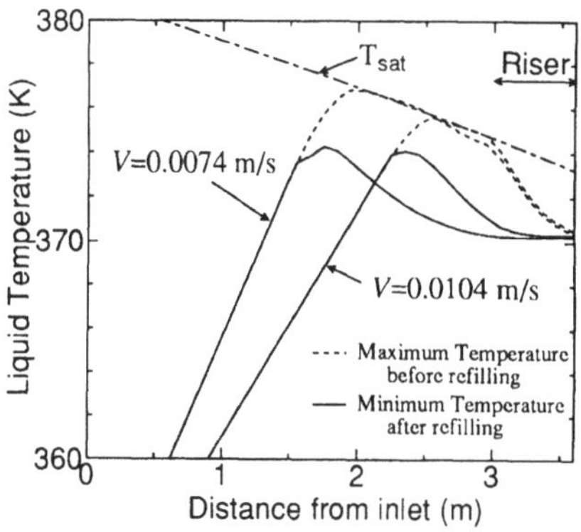

Figure 14. Simulated temperature distribution ($Q = 2$ kW, $\Delta T_{sub/inlet} = 23$ K)

The geysering period could be predicted by equation (2), however, it is difficult to evaluate the T_l In order to simulate the T_l correctly, the correct simulation of the mixing rate in the refilling stage is needed.

CONCLUSION

(a) The geysering was experimentally investigated using large subcooled water in an expansion tank. As the heat flow rate increased, the geysering period decreased. However, as the subcooled temperature in an expansion tank decreased, the geysering period increased.

(b) The Mini-TRAC code could simulate correctly the experimental relations between the geysering period and macro parameters.

(c) By evaluating the saturation temperature at the boiling point, the relations between the geysering period and macro parameters could be explained.

(d) In order to predict the geysering period correctly, the prediction of the boiling point and the mixing ratio at refilling stage is needed.

ACKNOWLEDGMENT

The authors wish to thank Dr. Murao in JAERI for providing MINI-TRAC code for us, and to do Mr.Yamada for supporting the thermocouples.

REFERENCES

Akimoto, H., et al., 1991, "MINI-TRAC code: A driver program for assessment of constitutive equations of two-fluid model", JAERI-M 91-086, (In Japanese)

Aritomi, H., et al., 1991, Transient behavior of natural circulation for boiling two-phase flow; 2nd report, Mechanism of geysering, Proc. 1st JSME/ASME Int. Conf. on Nuclear Eng., **1**, 87

JSME, 1991, "Hand Book of Gas-Liquid Two-Phase Flow Technology", 1st ed., Corona pub. co., 179 (In Japanese)

Liles, D. R., et al., 1986, "TRAC-PF1/MOD1: An advanced best-estimate computer program for pressurized water reactor thermal-hydraulic analysis", NUREG / CR-3858, LA-10157-MS

Morioka, M., et al., 1986, Simulation of the geysering phenomenon in LNG pipes, Adv Cryog Eng **31**, 1111

Nakanishi, S., et al., 1978, Flow Instability in boiling tube; 2nd report, Geysering, Trans. JSME ser. B, **44**-388, 4152 (In Japanese)

NOMENCLATURE

Cp	Specific heat of liquid	(J/kg K)
T_l	Temperature at boiling point after refilling stage	(K)
T_{sat}	Saturation temperature at boiling point	(K)
Q	Heat flow rate	(W)
q	Heat flux	(W/m^2)
q'''	Volumetric heating ratio	(W/m^3)
$\Delta T_{\mathrm{sub}/tank}$	Subcooled temperature in an expansion tank	(K)
$\Delta T_{\mathrm{sub}/inlet}$	Inlet subcooled temperature	(K)
ρ	Density of liquid	(kg/m^3)
τ	Geysering period	(s)

DYNAMICS OF A REDUCED MODEL OF TWO-PHASE FLOW IN A BOILING CHANNEL

C.J. Aldridge and A.C. Fowler

Mathematical Institute
24-29 St Giles'
Oxford OX1 3LB
England
UK

ABSTRACT

A model is presented for the flow of two-phase steam-water flow in a boiling channel. The motivation was to study the dynamics of a boiler which constituted a part of a steam engine. Particular features of the system were that the boiling tube was coiled, and that the externally supplied heat was generated by an exothermic chemical reaction. The reactant supply was controlled so as to maintain an exit temperature of 540°C at an operating pressure of 50—100 bars. The inlet temperature of sub-cooled water was 10°C.

The model presented here considers the flow in the boiler to consist in general of three separate regions: sub-cooled liquid, two-phase steam and water, and superheated steam. These regions are separated by moving boundaries termed the boiling boundary and the superheat boundary. The flow is modelled in each region by one-dimensional, averaged equations of conservation of mass, momentum and enthalpy, together with an equation of state and suitable constitutive relations. In particular, the two-phase region is modelled using 'two-fluid' averaged equations, so that each phase carries its own conservation laws; the enthalpies are not necessarily those of thermodynamic equilibrium, and are related by an averaged jump equation derived from the Stefan condition. Separate constitutive relations, particularly for interfacial friction, must be prescribed for separate regions of bubbly, slug, churn and annular flow.

A much simpler model which retains the accuracy of the original can be derived using the methodology of non-dimensionalisation, scaling and asymptotic analysis. This model facilitates numerical solution, since it also filters out fast space and time scales, which manifest themselves numerically by the requirement of small space and time scales. Here we describe a simple numerical method, and illustrate its use with some typical sample calculations. Some further analysis is possible using the often realistic result that the two phase flow speeds are much larger than the inlet flow speed. In this case the boiling boundary responds quasi-statically to fluctuations in inlet velocity, and the dynamics of the system is then controlled by the delayed feedback between the two.

Instabilities in Multiphase Flows, Edited by G. Gouesbet
and A. Berlemont, Plenum Press, New York, 1993

1. INTRODUCTION

The motivation for this study came from a commercial contract involving design modelling of a steam engine, comprising a pump, boiler, turbine and condenser in series. The specific contractual requirement was for a fast but accurate code for the boiler. Since the exit temperature was controlled to give superheated steam, there was thought to be a significant portion of two-phase flow, much of it in the annular régime. The likelihood of widely differing velocities suggests the use of a two-fluid model, and in the final report to the company (Aldridge et al. 1991), a very general such model, due to Drew and Wood (1985) was proposed. In this paper we summarise this model, and show how it can be usefully approximated, using applied mathematical techniques, to a much simplified form which is capable of rapid and efficient numerical solution.

2. MATHEMATICAL MODEL

We consider a vertical tube $0 < z < l$ of small cross section. Turbulent sub-cooled liquid is admitted at the base $z = 0$ and driven by a pressure gradient towards the outlet at the top. The tube is externally heated at a rate Q (per unit length per unit time), and in general there will be two moving boundaries $z = r(t)$ and $z = s(t)$ which demarcate the regions of sub-cooled flow, two-phase flow and superheated flow. We term $r(t)$ the boiling boundary, and $s(t)$ the superheat boundary.

In what follows, we describe separate models for the three regions. Appropriate equations are averaged cross-sectionally and in time, and the averaging process throws up an enormous number of terms which must be constituted in some way. In this paper, we choose the simplest path, treating the simplest boundary conditions, and ignoring certain terms in Drew and Wood's complete formulation. We must emphasise that in the original report (Aldridge et al. 1991) no such liberty was taken, and we show there that the process described here can be applied to the most general two-fluid model, with exactly the same resultant simplification.

2.1 Sub-cooled flow

Supposing that the cross-sectional area A is constant, suitable equations in $0 < z < r$ are

$$\rho_{lt} + (\rho_l u)_z = 0,$$
$$\rho_l(u_t + uu_z) = -p_z - F_{lw},$$
$$\rho_l(h_{lt} + uh_{lz}) = \frac{Q}{A}, \tag{1}$$

where ρ_l is liquid density, u is liquid velocity, p is pressure and h_l is enthalpy. The term F_{lw} represents the turbulent friction at the wall, and is normally written

$$F_{lw} = \frac{2}{d} f_{lw} \rho_l |u| u, \tag{2}$$

where f_{lw} is the friction factor, and is a weak function of the Reynolds number. We have neglected gravity in the momentum equation. In the original problem, this was because the tube was coiled, so that the average $\bar{g}$ could be neglected. More generally, gravity will be 'small' if $\rho_l gl \ll \Delta p$, where Δp is the pressure drop. Since $\rho_l gl \sim 1$ bar if $l \sim 10$ m, this will be the case where $\Delta p > 10$ bars, a normal operating condition.

2.2 Two-phase flow

In a boiling tube, the two-phase flow will normally evolve through a sequence of different régimes as the void fraction α increases: in turn, we see bubbly flow, slug flow, churn flow and annular flow. It is only in the last of these that the gas phase can move in a relatively unimpeded way, hence the fluid velocities tend to be much larger; equivalently, the region of annular flow is longer. For both these reasons, the annular flow is the primary determinant of two phase flow pressure drop, and hence we concentrate on this. Nevertheless, we emphasise that the difference between the models of the different flow régimes lies in the constitutive terms only. We shall come back to this point eventually.

Suitable two-fluid equations for two-phase flow are

$$(\alpha\rho_g)_t + (\alpha\rho_g v)_z = \Gamma,$$

$$(\beta\rho_l)_t + (\beta\rho_l u)_z = -\Gamma, \tag{3}$$

where α and β are gas and liquid volume fractions, and $\alpha + \beta = 1$; ρ_g is the gas density, v is the gas velocity, and Γ is the rate of change of phase with time due to boiling. The momentum equations are

$$(\beta\rho_l u)_t + (\beta\rho_l u^2)_z = -\beta p_z + F_{lw} + F_{li},$$

$$(\alpha\rho_g v)_t + (\alpha\rho_g v^2)_z = -\alpha p_z + F_{gi}. \tag{4}$$

Here, F_{lw} is the wall stress on the liquid phase, F_{li} is the interfacial stress on the liquid, $F_{gi} = -F_{li}$ being that on the gas. In writing (4), we have already assumed that $F_{gw} = 0$, as for annular flow. We have also assumed that various profile coefficients which arise in averaging the flux terms are equal to one, and that the phasic pressures are equal. This last assumption is thought to be realistic, but it does cause problems, in view of the well-known ill-posedness of the basic equations (3) and (4) when $\rho_g = \rho_l = 1$, $\Gamma = F_{lw} = F_{li} = 0$. This ill-posedness can be resolved in a number of ways, for example by differing phasic pressures, or by a profile coefficient other than one (Fowler and Lisseter 1992), but it will not be of relevance here, since in the approximations we make, the issue of well-posedness is not contentious. There are numerous other terms in (4): Reynolds' stresses, pressure work, momentum phase change, etc., which can be demonstrated to be small.

The two-fluid enthalpy equations are (approximately)

$$\beta\rho_l(h_{lt} + uh_{lz}) = \Gamma(h_l - h_{li}) + E_l + Q/A,$$

$$\alpha\rho_g(h_{gt} + vh_{gz}) = \Gamma(h_{gi} - h_g) + E_g, \tag{5}$$

where h_l and h_g are the liquid and gas enthalpies, E_l and E_g are the interfacial energy transports to the liquid and gas phases, and h_{li} and h_{gi} are the average interfacial enthalpies. Particularly for annular flow, we cannot necessarily assume that h_l and h_g are the equilibrium saturation values, and (5) must be supplemented by an averaged interfacial Stefan condition,

$$E_l + E_g + \Gamma(h_{gi} - h_{li}) = 0, \tag{6}$$

which in effect determines Γ, given the other quantities. In addition, the quantities $\rho_g, \rho_l, F_{lw}, F_{li}, E_l, E_g, h_{li}, h_{gi}$ need to be constituted.

2.3 Superheated flow

The relevant equations are

$$\rho_{gt} + (\rho_g v)_z = 0,$$

$$(\rho_g v)_t + (\rho_g v^2)_z = -p_z - F_{gw},$$

$$\rho_g(h_{gt} + vh_{gz}) = Q/A, \tag{7}$$

and must be supplemented by an equation of state for ρ_g, and an expression for F_{gw}.

2.4 Constitutive relations

We take ρ_l = constant, but define

$$\rho_g = \frac{p}{RT_g} \tag{8}$$

where T_g is the gas temperature. F_{lw} is defined by (2), and equivalently,

$$F_{li} = \frac{2}{d} f_{li}\rho_g \left| v - \chi u \right| (v - \chi u), \tag{9}$$

where a value of $\chi = 2$ was suggested by Wallis (1969). Similarly,

$$F_{gw} = \frac{2}{d} f_{gw}\rho_g \left| v \right| v. \tag{10}$$

The friction factors f_{lw}, f_{li}, f_{gw} depend weakly on the respective Reynolds numbers of relevance: we do not pursue this.

The interfacial heat transports can be defined by

$$E_g = H_{gi}(T_{gi} - T_g)/L_s,$$

$$E_l = H_{li}(T_{li} - T_l)/L_s, \tag{11}$$

where H_{gi} and H_{li} are heat transfer coefficients, T is temperature, and L_s^{-1} is the average surface area per unit volume. Depending on flow régime, it is a function of α, and for annular flow,

$$L_s^{-1} = 4\alpha^{1/2}/d. \tag{12}$$

Lastly, we need to prescribe the enthalpies in terms of the temperatures. We assume that the interfacial temperatures T_{gi}, T_{li} are equal to T_{sat}, the local saturation value. This varies somewhat with pressure, but we ignore this here for simplicity, and thus the interfacial enthalpies are constants,

$$h_{gi} = h_{g0}, \quad h_{li} = h_{l0},$$

$$T_{gi} = T_{sat}, \quad T_{li} = T_{sat}. \tag{13}$$

Then the enthalpies are defined by

$$h_l = h_{l0} + c_{pl}(T_l - T_{sat}),$$

$$h_g = h_{g0} + c_{pg}(T_g - T_{sat}), \tag{14}$$

where c_{pl} and c_{pg} are the specific heats.

2.5 Boundary conditions

The simplest boundary conditions to prescribe are that

$$p = p_{out} \quad \text{at} \quad z = l,$$

$$p = p_{out} + \Delta p, \quad T = T_0 \quad \text{at} \quad z = 0; \tag{15}$$

an alternative to the inlet pressure condition is to prescribe the inlet mass flux per unit area

$$\rho_l u = G_l. \tag{16}$$

We shall mostly dwell on the former.

3. NON-DIMENSIONALISATION AND REDUCTION

We wish to choose scales for the variables so that the dimensionless variables are $O(1)$. Since the velocities can be expected to be large in the two-phase region, we choose to use the two phase equations to determine the appropriate scales. Specifically, we define

$$\beta = \epsilon\beta^*, \quad v = Vv^*, \quad u = Uu^*, \quad p = p_{out} + Pp^*, \quad \Gamma = \bar{\Gamma}\gamma,$$
$$h_g = h_{g0} + c_{pg}\Delta T h_g^*, \quad h_l = h_{l0} + c_{pl}\Delta T h_l^*,$$
$$z = lz^*, \quad t = (l/U)t^*, \quad \rho = \bar{\rho}_g\rho_g^*, \tag{17}$$

where

$$\Delta T = T_{sat} - T_0, \tag{18}$$

and the unknown scales $\epsilon, V, U, P, \bar{\Gamma}, \bar{\rho}_g$ are yet to be chosen. We choose them specifically to balance what we infer (by examining typical orders or magnitude) are the largest terms in the equations. That is, we define

$$\bar{\rho}_g V = l\bar{\Gamma},$$
$$\rho_l\epsilon U = l\bar{\Gamma},$$
$$f_{lw}\rho_l U^2 = f_{li}\bar{\rho}_g V^2,$$
$$P = 2lf_{lw}\rho_l U^2/d,$$
$$\bar{\rho}_g = p_{out}/RT_{sat}, \tag{19}$$

so that the flux terms balance the phase change term in (3), $F_{lw} \sim F_{li}$ in $(4)_1$, $p_z \sim F_{gi}$ in $(4)_2$, and $\rho_g^* = O(1)$ in (8). The final relation depends on the choice of whether pressure drop $(15)_2$ or inlet mass flux (16) is prescribed. We either choose

$$P = \Delta p \tag{20}$$

for prescribed pressure drop, or

$$\rho_l\epsilon U = G_l \tag{21}$$

for prescribed inlet mass flux.

3.1 Two phase region, $r < z < s$

The dimensionless two-phase flow equations are then, from (3),(4),(5) and (6),

$$a_1[(1-\epsilon\beta)\rho_g]_t + [(1-\epsilon\beta)\rho_g v]_z = \gamma,$$
$$\beta_t + (\beta u)_z = -\gamma,$$
$$\epsilon_2[(\beta u)_t + (\beta u^2)_z] = -\epsilon p_z - u^2 + S^2(v - \chi a_1 u)^2,$$
$$\epsilon_3[a_1\{(1-\epsilon\beta)\rho_g v\}_t + \{(1-\epsilon\beta)\rho_g v^2\}_z] = -(1-\epsilon\beta)p_z - S^2(v - \chi a_1 u)^2,$$
$$\gamma = q + a_4(1-\epsilon\beta)^{1/2}h_g + a_5\gamma h_l,$$
$$a_5\epsilon_6\beta[h_{lt} + uh_{lz}] = \epsilon_6 q - [(1-\epsilon\beta)^{1/2} - a_5\epsilon_6\gamma]h_l,$$
$$a_7(1-\epsilon\beta)\rho_g[a_1 h_{gt} + vh_{gz}] = -[(1-\epsilon\beta)^{1/2} + a_7\gamma]h_g, \tag{22}$$

wherein we have dropped the asterisks on the variables, and

$$S^2 = \rho_g, \quad \rho_g = \frac{1 + a_8 p}{1 + a_9 h_g}, \tag{23}$$

where

$$q = Q/\bar{\Gamma} A L, \quad L = h_{g0} - h_{l0} \tag{24}$$

is the latent heat, and the scales are defined by

$$\begin{aligned} \epsilon &= (f_{lw}\bar{\rho}_g / f_{li}\rho_l)^{1/2}, \\ U &= G_l/\rho_l \epsilon, \\ \bar{\Gamma} &= G_l l, \\ V &= G_l/\bar{\rho}_g, \\ P &= 2 l f_{li} G_l^2 / \bar{\rho}_g d. \end{aligned} \tag{25}$$

If G_l is prescribed, these determine the scales; if Δp is prescribed, then $P = \Delta p$, and G_l can be defined by $(25)_5$. The parameters $a_1 - a_9$ are defined by

$$\begin{aligned} a_1 &= U/V, \\ \epsilon_2 &= \epsilon d / 2 f_{lw} l, \\ \epsilon_3 &= d/2 l f_{li}, \\ a_4 &= 4 H_{gi} \Delta T / d L \bar{\Gamma}, \\ a_5 &= c_{pl} \Delta T / L, \\ \epsilon_6 &= \bar{\Gamma} L^2 / H_{li} \Delta T, \\ a_7 &= 4 c_{pg} \Gamma / d H_{gi}, \\ a_8 &= P/p_{out}, \\ a_9 &= \Delta T / T_{sat}. \end{aligned} \tag{26}$$

We use typical numerical estimates of the parameters as follows (these were for the particular application which motivated the study):

$$\begin{gathered} \bar{\rho}_g \sim 30\,\text{kg m}^{-3}, \quad \rho_l \sim 760\,\text{kg m}^{-3}, \quad \dot{m}_0 \sim .2\,\text{kg s}^{-1}, \quad d \sim .014\,\text{m}, \\ A = \pi d^2/4 \sim 1.54 \times 10^{-4}\text{m}^2, \quad G_l = \dot{m}_0/A \sim 1.3 \times 10^3 \text{kg m}^{-2}\text{s}^{-1}, \quad l \sim 20\,\text{m}, \\ c_{pg} \sim 4.6\,\text{kJ kg}^{-1}\text{K}^{-1}, \quad c_{pl} \sim 5.2\,\text{kJ kg}^{-1}\text{K}^{-1}, \quad \Delta T \sim 250\,\text{K}, \\ T_{sat} \sim 270\,\text{K}, \quad L \sim 1571\,\text{kJ kg}^{-1}, \end{gathered} \tag{27}$$

and derived flow characteristics

$$\begin{gathered} f_{lw} \sim .004, \quad f_{li} \sim .02, \\ H_{li} \sim 1.7 \times 10^5 \text{W m}^{-2}\text{K}^{-1}, \\ H_{gi} \sim .7 \times 10^4 \text{W m}^{-2}\text{K}^{-1}; \end{gathered} \tag{28}$$

hence

$$\begin{gathered} \epsilon \sim .09, \quad a_1 \sim .44, \quad \epsilon_2 \sim .009, \quad \epsilon_3 \sim .02, \quad a_4 \sim 6, \\ a_5 \sim .76, \quad \epsilon_6 \sim .01, \quad a_7 \sim .13, \quad a_8 \sim .5 \quad (\text{if}\, p_0 \sim 60\,\text{bars}), \\ a_9 \sim 1. \end{gathered} \tag{29}$$

We see that $\epsilon_2, \epsilon_3, \epsilon_6$ are very small, and ϵ and a_7 are quite small.

3.2 Subcooled region, $z < r(t)$

Here the dimensionless equations are (ρ_l being constant)

$$u_z = 0,$$

$$\epsilon_{10}[u_t + uu_z] = -p_z - u^2,$$

$$h_{lt} + uh_{lz} = \epsilon\gamma/a_5, \tag{30}$$

where

$$\epsilon_{10} = \epsilon_2/\epsilon \sim .1.$$

3.3 Superheated region, $z > s(t)$

Here we have

$$a_1\rho_{gt} + (\rho_g v)_z = 0,$$

$$\epsilon_3[a_1\rho_{gt} + (\rho_g v^2)_z] = -p_z - v^2,$$

$$\rho_g[a_1 h_{gt} + vh_{gz}] = \gamma/a_5. \tag{31}$$

3.4 Boundary conditions

At the ends of the tube, we have

$$p = 0 \text{ on } z = 1,$$

$$h_l = -1,$$

$$p = 1 \text{ or } u = \epsilon \text{ at } z = 0. \tag{32}$$

The boiling boundary is defined by

$$h_l = 0 \text{ on } z = r, \tag{33}$$

and the superheat boundary is defined by

$$\beta = 0 \text{ on } z = s, \tag{34}$$

and we require variables to be continuous at these.

3.5 An asymptotic reduction

3.5.1 Sub-cooled region. If significant evaporation occurs, we can expect $\gamma \sim 1$. Since $\beta = 1/\epsilon$ at $z = r$ and βu is continuous, we have that $u \sim \epsilon$ in $z < r$, thus the solution of (30) is

$$u = \epsilon u_0(t), \quad p = p_0(t), \tag{35}$$

where one or other of u_0 and p_0 can be chosen to be 1, and the enthalpy satisfies

$$h_{l\tau} + u_0 h_{lz} = \gamma/a_5, \quad \tau = \epsilon t, \tag{36}$$

whose solution is

$$z = \int_{\tau-\tau_{sc}(1+h)}^{\tau} u_0(\theta)d\theta, \tag{37}$$

where

$$\tau_{sc} = \frac{a_5}{\gamma} \tag{38}$$

is the sub-cooling delay. In particular,

$$r(t) = \int_{\tau-\tau_{sc}}^{\tau} u_0(\theta)d\theta. \tag{39}$$

Note that the boiling boundary will vary *slowly* with t, being a function of the slow time ϵt.

3.5.2 Two-phase region. If we neglect the small terms in $\epsilon_2, \epsilon_3, \epsilon_6$, ϵ and $a_1 a_7 \sim .06$ in (22), then an approximate model is the following:

$$\begin{aligned}
a_1 \rho_{gt} + (\rho_g v)_z &= \gamma, \\
\beta_t + (\beta u)_z &= -\gamma, \\
u &= \sigma v, \\
p_z &= -u^2, \\
h_l &= 0, \\
a_7 \rho_g v h_{gz} &= -(1 + a_7\gamma) h_g, \\
\gamma &= q + a_4 h_g,
\end{aligned} \tag{40}$$

where

$$\sigma = \frac{S}{1 + \chi a_1 S}. \tag{41}$$

The approximations involved in neglecting ϵ_2 and ϵ_3 are potentially singular, hence we may not be able to satisfy conditions of continuity of all of β, u and v at $z = r$. In this case, we might expect short regions in which the variables adjust from their sub-cooled values to the two-phase values determined by (40). Such boundary regions should not affect the large scale dynamics significantly, however, and their removal enables a direct numerical solution to be significantly easier and more efficient. The neglect of ϵ_6 causes no singular behaviour.

3.5.3 Superheat region. We neglect the term in ϵ_3, thus the reduced model is

$$\begin{aligned}
a_1 \rho_{gt} + (\rho_g v)_z &= 0, \\
p_z &= -v^2, \\
\rho_g [a_1 h_{gt} + v h_{gz}] &= \gamma / a_5.
\end{aligned} \tag{42}$$

3.6 Comments

The parameterisation of the two-phase flow equations implicitly assumes annular flow. In fact, near $z = r$, where $\beta = O(1/\epsilon)$ (this is *scaled*, corresponding to a liquid fraction of $O(1)$), we will have regions of bubbly, slug and churn flow. Since we can expect $u \sim v \sim \epsilon$ in these regions, the gas continuity equation suggests that $z - r \sim \epsilon$, so that the non-annular regions are relatively short. Moreover, the liquid momentum equation suggests that these regions can be adequately modelled by simply changing the definition of S in $(22)_3$. Specifically, drift flux models are consistent with choosing $S \gg 1$, but this does not affect σ significantly. Thus, it is plausible that the two-phase model given here is, in an approximate sense, a valid description for flows other than annular.

3.7 Boundary conditions

We need to specify v and βu on $z = r$. We lose the ability to specify u by losing the inertial terms from the liquid momentum equation, although the condition of equal velocities at $z = r$ ($u = v/a_1$) can be recovered by specifying $\sigma = 1/a_1$ there; we choose

$$v = \epsilon u_0/\sigma, \quad \beta u = u_0, \quad p = p_0,$$

$$h_g = 0 \text{ at } z = r. \tag{43}$$

It follows that $h_g \equiv 0$, $\gamma = q$, and only the first four of (3.24) are relevant.

At $z = s$, we take p, v to be continuous, and $h_g = 0$. Finally, at $z = 1$, $p = 0$.

4. NUMERICAL SOLUTION

The reduced model is of different type to the full model, being partially parabolic, and thus some care needs to be taken with the numerical method. In the sub-cooled region, we solve the enthalpy equation $(30)_3$ with a first order, explicit, upwind method. $r(t)$ is then determined by $h = 0$, interpolating between grid points. The enthalpy equation is solved similarly in $z > s(t)$, when there is a superheat region.

In $z < r$, we obtain u by quadrature; for (3.14) this is trivial, though quadrature is also used if ρ_l is variable. If u is prescribed at $z = 0$, the sub-cooled region is then fully determined.

4.1 Pressure/velocity equations

In $z > r$, ρ_g depends on p (and h_g). Focussing on the p dependence, the gas mass and gas momentum equations have the structure of the set

$$p_t + f_z = 0,$$

$$p_z = -f, \tag{44}$$

where $f = \rho_g v$, and we have omitted irrelevant coefficients. This is a *diffusion* equation for p, and thus we require boundary data at both ends. If u_0 is prescribed, then $p = 0$ at $z = 1$ and v is prescribed at $z = r$; but if p is prescribed at $z = 0$, then we have it at $z = r$ also, and u_0 is then determined from (43) by finding v. And then, in fact, we can integrate u backwards through the sub-cooled region. We use a fully implicit method for (44), which can be got by using forward differences (in z) in one equation, and backward differences in the other.

The remaining equation is that for the liquid volume fraction β. We assume that this equation remains hyperbolic, despite the dependence of u on v and hence p. An explicit method can then be used, and the superheat boundary is determined (by interpolation) when $\beta = 0$.

The method described above has been implemented in FORTRAN and runs at Oxford on a VAX 11/780. It is relatively quick, though there appear to be some regions of numerical instability. This may be due to a change of equation *type* with certain parameters.

Detailed results will be presented elsewhere.

5. ANALYSIS

Suppose, for simplicity, that there is no superheat region. Evidently, r varies on the slow time scale $\tau = \epsilon t$, while disturbances propagate through the two-phase region on a time of $O(1)$. It follows that, barring instability in the two phase region, the equations in $z > r$ can be solved quasi-statically. Specifically

$$\rho_g v \approx \gamma(z-r),$$

$$\beta u = u_0 - \gamma(z-r),$$

$$u = \sigma v,$$

$$p_z = -\sigma^2 v^2. \tag{45}$$

Now $\sigma = S/(1+\chi a_1 S)$, and Aldridge et al. (1991) show that Reynolds number dependence leads to

$$S \approx \left[\frac{\rho_g(\beta+b)(\beta u)^{1/4}}{(\rho_g v)^{1/4}}\right]^{1/2}, \quad b \approx 0.14; \tag{46}$$

in particular σ depends on β, $\partial\sigma/\partial\beta > 0$. Evidently, p can be determined from quadrature, and since $-\int_r^1 p_z dz = p_0$ ($= 1$ if pressure is prescribed at the inlet) and σ increases with u_0, it follows that this quadrature determines r as an increasing function of u_0, for example as shown in Fig. 1.

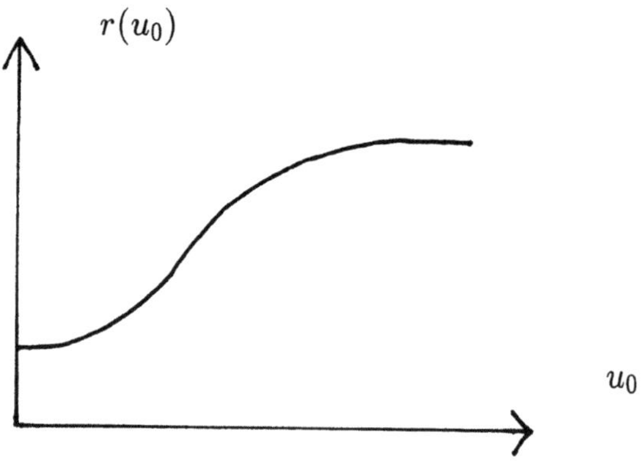

Figure 1

The whole dynamics of the system is then described by the equation

$$r(u_0) = \int_{T-\tau_{sc}}^{T} u_0(\theta)d\theta : \tag{47}$$

this represents a decided simplification of the original model! And yet, no serious quantitative errors have been made.

Steady state

These are simply solutions of

$$r(u_0) = \tau_{sc} u_0, \tag{48}$$

and hysteresis is possible for the sigmoidal curve illustrated in Fig. 1. This is associated with the *Ledinegg instability* (Ledinegg 1938) which occurs on regions of the Δp versus u_0 curve which have negative slope. The transition to instability occurs at the values of u_0 where $r' = \tau_{sc}$.

Stability

Perturbations to a steady solution exist proportional to $\exp(\lambda\tau)$, provided

$$r'(u)\lambda = 1 - e^{-\lambda\tau_{sc}}. \tag{49}$$

For large τ_{sc}, $\lambda \approx 1/r'$ or Re $\lambda < 0$, and Re $\lambda = 0$ only if $r' = \tau_{sc}$. In particular, this reduction does not admit oscillatory solutions. It follows that oscillatory instability must be associated with low values of τ_{sc}. Putting

$$\tau_{sc} = \epsilon\tau_0, \tag{50}$$

then

$$r(t) = \epsilon \int_{t-\tau_0}^{t} u_0(s)ds, \tag{51}$$

and we essentially ignore the sub-cooled region. Instability is then determined by the solution of the two-phase equations, that is

$$\begin{aligned}
\alpha_1\rho_t + (\rho v)_z &= \gamma, \\
\beta_t + (\beta u)_z &= -\gamma, \\
u &= \sigma v, \\
p_z &= -u^2 \\
\rho &= 1 + a_8 p,
\end{aligned} \tag{52}$$

together with

$$\begin{aligned}
p = 1, \quad \beta u = u_0, \quad \rho v \approx 0 \quad \text{at} \quad z \approx 0, \\
p = 0 \quad \text{at} \quad z = 1.
\end{aligned} \tag{53}$$

We must choose u_0, in order that the third condition holds. If σ is independent of β, the equation for β uncouples, and that for ρ and v can be reduced to (with appropriate rescaling)

$$\rho_t = 1 - \frac{\partial}{\partial z}\left[\rho\left\{-\frac{\partial\rho}{\partial z}\right\}^{1/2}\right], \tag{54}$$

with ρ being prescribed at $z = 0$ and $z = 1$ ($\rho_0 > \rho_1$). We have not examined the stability of this equation, but as it is of nonlinear diffusion type, we might expect it to be stable.

If σ depends on β (as in fact it does), then ρ, β, v are all coupled: in such circumstances, instability may be more likely.

A final possibility occurs when heating is low, and the annular flow régime is not reached. We rescale the variables as

$$t \sim 1/\epsilon, \quad u \sim \epsilon, \quad v \sim \epsilon a_1, \quad \gamma \sim \epsilon a_1, \quad \beta \sim 1/\epsilon, \quad p \sim \epsilon^2, \tag{55}$$

so that (in rescaled terms)

$$\begin{aligned}
r &= \int_{t-\tau_{sc}}^{t} u_0(\theta)d\theta, \\
[\rho(1-\beta)]_t + [\rho(1-\beta)v]_z &= \gamma, \\
\beta_t + (\beta u)_z &= 0, \\
p_z &= -u^2,
\end{aligned}$$

$$u = \sigma a_1 v,$$

$$\rho = 1 + a_8 p, \tag{56}$$

in the two-phase region, and the pressure drop condition is

$$1 = u_0^2 r + \int_r^1 u^2 dz. \tag{57}$$

This is almost the homogeneous model (and is if $\sigma a_1 = 1$), which is known to have oscillatory solutions (Davies and Potter 1967, Fowler 1978).

6. CONCLUSIONS

Through the use of non-dimensionalisation and scaling procedures, complicated two-phase flow problems can be reduced to much simpler, yet no less accurate, forms, which greatly facilitate both numerical computations and the ability to carry out analysis.

REFERENCES

Aldridge, C.J., A.C. Fowler and R.L. Molla 1991 Real time modelling of transient steam-water flows in boilers. OCIAM Technical Report No. 1, Mathematical Institute, Oxford.

Davies, A.L. and R. Potter 1967 Hydraulic stability: an analysis of the causes of unstable flow in parallel channels. Paper presented at the symposium on Two-phase Flow Dynamics, Eindhoven EUR 4288e, 1225-1266.

Drew, D.A. and R.T. Wood 1985 Overview and taxonomy of models and methods for workshop on two-phase flow fundamentals. Nat. Bureau of Standards, Gaithersburg, MD.

Fowler, A.C. 1978 Linear and non-linear stability of heat exchangers. J. Inst. Maths. Applics. **22**, 361-382.

Fowler, A.C. and P.E. Lisseter 1992 Flooding and flow reversal in annular two- phase flows. SIAM J. Appl. Math., in press.

Wallis, G.B. 1969 One-dimensional two-phase flow. McGraw-Hill, New York.

BIFURCATIONS IN TWO-PHASE FLAMES

N.-H. Chen and B. Rogg

University of Cambridge, Department of Engineering
Trumpington Street, Cambridge CB2 1PZ, Great Britain

INTRODUCTION

Ordinary droplet and spray combustion arise, for instance, by injection of a liquid fuel into a hot environment. After injection the liquid fuel eventually breaks up into small droplets that evaporate, ignite and burn. The Diesel engine provides a practical example of such a process. The interaction of droplets and sprays with a gaseous phase are particularly complex, if the gaseous phase is a flame that is self-sustained by gaseous fuel and oxidizer. Such flames are termed "two-phase flames", and they are the objective of the present study. Suppression of explosion or extinction of fire by water sprays provides practical examples of the occurrence of two-phase flames. The two-phase flames modelled and simulated in the present work are counterflow flames, i.e., flames in a flow field generated by directing two gaseous streams towards each other. One of the gaseous streams carries a combustible fuel-air mixture and the water droplets, the other stream carries a flow of hot combustion products.

The purpose of the present paper is to go beyond previous numerical studies such as the one by Continillo and Sirignano (1990) and Chen et al (1992). Specifically, in the present paper laminar, premixed, strained methane-air flames with water droplets added to the fresh, unburnt mixture are considered. Attention is focussed on the determination of bifurcation points, i.e., points of two-phase flame ignition and extinction, in a suitably defined heat release - number density plane. In the numerical simulations detailed models for thermodynamics, molecular transport and chemistry are taken into account.

GOVERNING EQUATIONS

Gas-Phase Governing Equations

Low Mach-number flow of an ideal-gas mixture is considered. Therefore, the pressure is taken as thermochemically constant and effects of viscous dissipation are neglected. Furthermore, effects of radiative heat transfer, thermal diffusion, diffusion caused by pressure gradients, and external forces are neglected. By virtue of an accumulative convective transport operator L,

Instabilities in Multiphase Flows, Edited by G. Gouesbet
and A. Berlemont, Plenum Press, New York, 1993

$$L(\phi) \equiv \frac{\partial(\rho\phi)}{\partial t} + \frac{1}{x^j}\frac{\partial(x^j \rho u \phi)}{\partial x} + \frac{\partial(\rho v \phi)}{\partial y}$$

for any dependent variable ϕ, where $j = 0$ for planar and $j = 1$ for axisymmetric flow, the gas-phase boundary-layer equations describing the conservation of overall mass, momentum, energy and species mass in a laminar, chemically reacting, stagnation point flow can be written as

$$L(1) = S_v \,, \tag{1}$$

$$L(u) = -\frac{\partial p}{\partial x} + \frac{\partial}{\partial y}(\mu \frac{\partial u}{\partial y}) + S_m \,, \tag{2}$$

$$L(T) = \frac{1}{c_p}\frac{\partial}{\partial y}(\lambda \frac{\partial T}{\partial y}) - \frac{\partial T}{\partial y}\sum_{i=1}^{N} \frac{c_{pi}}{c_p}\rho Y_i V_{iy} - \frac{1}{c_p}\sum_{i=1}^{N} h_i w_i + \frac{S_e}{c_p} \,, \tag{3}$$

$$L(Y_i) = -\frac{\partial(\rho Y_i V_{iy})}{\partial y} + w_i + (\delta_{id} - Y_i) S_v \,, \tag{4}$$

$i = 1, ..., N$. In Eqs. (1) – (4), t is the time, x and y are the spatial coordinates, and u and v the respective velocity components; ρ is the density, p the pressure, T the temperature; Y_i is the mass fraction of species i, and V_i its diffusion velocity; c_{pi} and h_i denote the specific heat capacity and specific enthalpy, respectively, of species i, and w_i is its mass rate of production; c_p, μ and λ are the mixture's specific heat capacity, dynamic viscosity and thermal conductivity, respectively. The δ_{id} appearing in Eq. (4) is the Kronecker delta, where here and below the subscript d is used to identify quantities associated with the one-component liquid-fuel droplets. The terms S_v, S_m and S_e appearing in Eqs. (1) – (4) represent sources of mass, momentum and energy; they are due to the presence of the liquid phase. The source terms can be written as

$$S_v \equiv \lim_{K\to\infty} \sum_{k=1}^{2K} S_{v,k} \,, \qquad S_{v,k} \equiv (n\dot{m})_k \,, \tag{5}$$

$$S_m \equiv \lim_{K\to\infty} \sum_{k=1}^{2K} S_{m,k} \,, \qquad S_{m,k} \equiv \left(S_v u_d - n m_d \frac{du_d}{dt} \right)_k \,, \tag{6}$$

$$S_e \equiv \lim_{K\to\infty} \sum_{k=1}^{2K} S_{e,k} \,, \qquad S_{e,k} \equiv \left[-n\left(\dot{q} + \dot{m}(c_{pF}(T - T_s) + L) \right) \right]_k \,. \tag{7}$$

Prior to defining the various terms appearing in Eqs. (5) – (7) some comments are in order. The sum on the r.h.s. of the first equation in (5) reflects the phenomenon of multiple crossings of droplets through the flame. This phenomenon has been termed "push-back effect" and is described in detail in reference (Chen et al., 1992). Furthermore, in Eqs. (5) – (7), n denotes the droplet number density; m_d and $\dot{m}$ are the mass and mass vaporization rate, respectively, of a single droplet; u_d is the x component of the droplet velocity; $\dot{q}$ is the total heat transferred to the droplet interior; c_{pF} is the specific heat capacity of of the liquid-fuel vapour, T_s the droplet-surface temperature, and L the latent heat. Subsequently, variables and properties of the fuel vapour will be identified by the subscript F.

Provided the quantities appearing on the r.h.s.'s of Eqs. (5) – (7) are known, the system of Eqs. (1) – (4) is closed by the ideal-gas equation of state

$$p/\rho = R^0 T \sum_{i=1}^{N} Y_i/W_i \,. \tag{8}$$

Gas-Phase Models for Molecular Transport

Recent work (Puri and Libby; 1991) has shown that a detailed description of molecular transport in the gas phase surrounding a droplet is a prerequisite for confident numerical predictions of the combustion of droplets and hence sprays. In the present work detailed models are employed for molecular transport chemistry. Specifically, the specific heat capacities c_{pi} and enthalpies h_i are taken from the JANAF tables. The transport coefficients μ_i and λ_i of the pure species and the binary diffusion coefficients D_{ij} are calculated from molecular data. Using the Hirschfelder-Curtiss approximation (Hirschfelder and Curtiss; 1954), the diffusion velocities V_i are expressed as $V_i = -(D_i/X_i)(\partial X_i/\partial y)$, where the average diffusion coefficient D_i is given in terms of the D_{ij}, viz., $D_i = (1 - Y_i)/(\sum_{j=1}^{N} X_j/D_{ij})$. Soret effects are included but Dufour effects neglected, since it has been shown (García–Ybarra et al.; 1984) that, the latter is generally unimportant, the former can influence flame behaviour through its influence on the temperature at the reaction zone. For the dynamic viscosity and the thermal conductivity of the gas mixture the widely accepted empirical formulas

$$\mu = \frac{1}{2}\left(\sum_{i=1}^{N} X_i\mu_i + \frac{1}{\sum_{i=1}^{N} X_i/\mu_i}\right), \tag{9}$$

$$\lambda = \frac{1}{2}\left(\sum_{i=1}^{N} X_i\lambda_i + \frac{1}{\sum_{i=1}^{N} X_i/\lambda_i}\right) \tag{10}$$

are adopted.

Liquid-Phase Governing Equations

The model for the liquid phase is essentially that of Continillo and Sirignano (1990), except that in the formulation lift forces were taken into account which according to estimates (Puri and Libby; 1989) may be important. Thus, this section is kept rather short; for details the previously published work (Continillo and Sirignano, 1990; Puri and Libby, 1989 Abramzon and Sirignano, 1987) should be consulted.

A one-component, spherically symmetric droplet, surrounded by a quasi-steady spherically symmetric film is considered. The local free-stream conditions are those of the gas phase surrounding the droplet. The thermophysical properties are evaluated at average temperature and concentrations in the film except for the density appearing in the Reynolds number, which is the local gas phase density (Abramzon and Sirignano, 1987). Droplet mass conservation is

$$\frac{dm_d}{dt} = -\dot{m} \,, \tag{11.a}$$

where the droplet mass and initial mass are given by

$$m_d = \rho_d \frac{4}{3}\pi R^3 \,, \qquad m_{d0} = \rho_d \frac{4}{3}\pi R_0^3 \,. \tag{11.b}$$

The mass evaporation rate $\dot{m}$ appearing on the r.h.s. of Eq. (11.a) can be written as

$$\dot{m} = 2\pi\rho_f D_f R \cdot \widetilde{Sh} \cdot \ln(1 + B_M), \tag{12.a}$$

where the modified Sherwood number is given by

$$\widetilde{Sh} = 2 + [(1 + Re \cdot Sc)^{\frac{1}{3}} \max(1, Re)^{0.077} - 1] \cdot \frac{B_M \cdot (1 + B_M)^{-0.7}}{\ln(1 + B_M)}, \tag{12.b}$$

$$B_M = \frac{Y_{Fs} - Y_{F\infty}}{1 - Y_{Fs}}, \quad Re = \frac{\rho_\infty v_r (2R)}{\mu_f}, \quad Sc = \frac{\nu_f}{D_f}, \tag{12.c}$$

$$Y_{Fs} = \frac{W_F X_{Fs}}{W_F X_{Fs} + (1 - X_{Fs})\overline{W}}, \quad X_{Fs} = \frac{1}{p} exp(-\frac{C_{1F}}{T_s} + C_{2F}), , \tag{12.d}$$

$$C_{1F} = \frac{L_{boil}}{R_F^0}, \quad C_{2F} = ln(p_{F0}) + \frac{C_{1F}}{T_{F0}}, \tag{12.e}$$

and where the other symbols have their usual meaning. The droplet equations of motion are

$$m_d \frac{d^2 x_d}{dt^2} = \frac{1}{2}\rho_\infty v_r \pi R^2 [-C_L(v - v_d) + C_D(u - u_d)] + m_d g_x, \tag{13}$$

$$m_d \frac{d^2 y_d}{dt^2} = \frac{1}{2}\rho_\infty v_r \pi R^2 [C_L(u - u_d) + C_D(v - v_d)] + m_d g_y, \tag{14}$$

where C_L and C_D denote the lift and drag coefficient, respectively. The droplet energy equation is

$$\frac{\partial T_d}{\partial t} = \alpha_d \frac{1}{r^2} \frac{\partial}{\partial r}(r^2 \frac{\partial T_d}{\partial r}), \tag{15}$$

which must be solved subject to the initial and boundary conditions

$$\begin{aligned} T_d(r, 0) = T_{do}, \quad & \frac{\partial T_d}{\partial r}|_{r=0} = 0, \\ \frac{\partial T_d}{\partial r}|_{r=R(t)} = & \frac{\dot{q}}{4\pi R^2 \alpha_d \rho_d c_{pd}}. \end{aligned} \tag{16}$$

The heat $\dot{q}$ transferred into the droplet, which appears on the r.h.s. of the third of equations (16), can be written as

$$\dot{q} = \dot{m}[\frac{c_{pF}(T_\infty - T_s)}{B_T} - L], \tag{17.a}$$

where

$$L = L_{boil}(\frac{T_{crit} - T_s}{T_{crit} - T_{boil}})^{0.38}, \quad B_T = (1 + B_M)^\phi - 1, \quad \phi = \frac{c_{pF}}{c_{pf}} \cdot \frac{\widetilde{Sh}}{\widetilde{Nu}} \cdot \frac{1}{Le} \tag{17.b}$$

The modified Nusselt number appearing in the third of Eqs. (17.b) is defined as

$$\widetilde{Nu} = 2 + [(1 + Re \cdot Pr)^{\frac{1}{3}} \max(1, Re)^{0.077} - 1] \cdot \frac{B_T \cdot (1 + B_T)^{-0.7}}{\ln(1 + B_T)}. \tag{18}$$

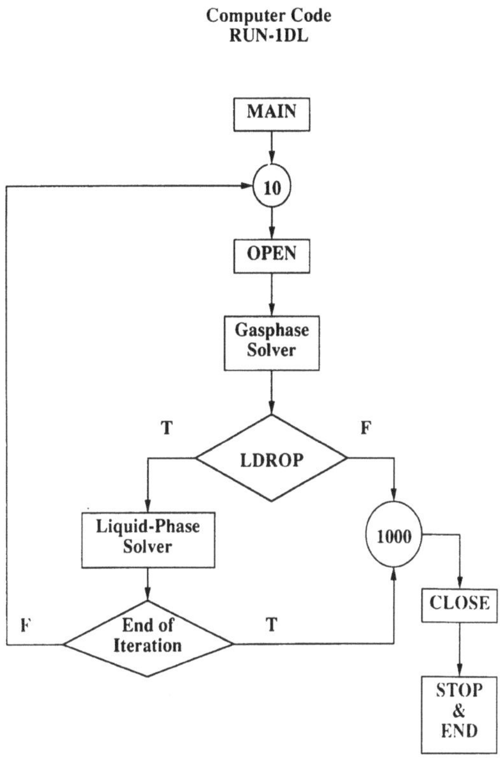

Figure 1. Simplified flow chart of the Cambridge laminar-flame code RUN-1DL.

The Lewis and Prandtl number appearing in Eqs. (17) and (18) are defined as usual; they are to be evaluated at film conditions. Under steady-state conditions the droplet velocity field $\vec{V_d}$ is given by $\nabla \cdot (n\vec{V_d}) = 0$. This equation can be integrated once to give, after rearrangement, the number density n, viz.,

$$n = \frac{n_0 x_{d0} v_{d0}}{x_d v_d} . \tag{19}$$

Models for Mass, Heat and Momentum Transfer Between Liquid Phase and Gas Phase

The models for mass, heat and momentum transfer between the liquid and the gaseous phase, some of which are already incorporated in Eqs. (12) and (17), are too comprehensive to be discussed here. For details as well as for a general overview, we refer to the detailed work by Sirignano and coauthors (Abramzon and Sirignano, 1987; Continillo and Sirignano, 1990).

NUMERICAL-SOLUTION PROCEDURE

The hybrid Eulerian-Lagrangian system of governing equations is solved with the Cambridge laminar-flame code RUN-1DL (Rogg, 1991). Shown in Fig. 1 is a flow chart of the code. Since the gas-phase and liquid-phase equations, Eqs. (1), (2), (3), (4), (11), (13), (14) and (15), are strongly coupled, RUN-1DL employs an iterative approach to determine the solution to the full problem. Initially, assumed profiles are assigned to the droplet source terms S_v, S_m and S_e that appear in the gas-phase conservation equations. The latter equations are then solved by a gas-phase driver subroutine, with the profiles of S_v, S_m and S_e held fixed. Then, with the gas-phase solution just obtained kept fixed, the liquid-phase governing equations are solved by a liquid-phase driver subroutine. The source terms S_v, S_m and S_e are then calculated and fed back into the gas-phase equations, which are solved again, and so forth. In this way, the two sets of equations are integrated in sequence until the "steady-state" solution to the overall problem is obtained. Since the droplet contributions are to be evaluated with the updated gas-phase variables, there is no point in calculating at each cycle of the outer, main iteration loop an accurate gas-phase solution with still inaccurate droplet source terms. Therefore, the convergence criterion for the iteration loop inside the gas-phase solution procedure is chosen not so severe as the overall criterion. Only as the overall solution is iteratively approached, a more severe convergence criterion for the gas-phase solution is employed.

The gas-phase governing equations are discretized on a non-uniform mesh using finite differences. The discretized equations are solved with a damped, modified Newton method (Deuflhard, 1974; Smooke 1983). To solve the gas-phase governing equations, procedures and criteria for the adaptive selection of grid points are of critical importance to the efficiency of the overall solution algorithm. In particular, a strategy is required that places the grid points where they are needed in order to bound the local space-discretization error. RUN-1DL employs a static self-adaptive gridding procedure (Giovangigli and Smooke, 1987) and is described in detail by Rogg (1990).

After discretization of the droplet energy equation (15) with respect to the droplet spatial coordinate r, the liquid-phase governing equations result in a system of differential algebraic systems. The latter is solved by RUN-1DL with an extrapolation method developed for the numerical solution of stiff systems (Deuflhard et al., 1985; Deuflhard and Nowak, 1985).

RESULTS AND DISCUSSION

Results are presented and discussed for laminar, strained, premixed fresh-counter-to-burnt two-phase flames with droplets added to the fresh, unburnt mixture. Such flames can be realized, for instance, by directing a stream of fresh, gaseous reactants towards a stream of hot combustion products whose thermochemical state corresponds to the fresh reactant's chemical-equilibrium composition at the adiabatic flame temperature. For the present computations a fresh, stoichiometric mixture of methane and air at a temperature of 300 K and a pressure of 1 bar has been selected. The liquid added to the fresh mixture has been taken as water at 300 K and 1 bar. The

Table 1. Short mechanism of elementary reactions for premixed CH_4-air flames. Units are cm, mole, s, kJ and K; third-body efficiencies were taken from Warnatz (1984).

No.	‡			Reaction			A	α	E
1	1	O_2	+ H	→ OH	+ O		2.2000×10^{14}	0.00	70.30
2b	2	OH	+ O	→ O_2	+ H		1.7244×10^{13}	0.00	3.52
3	3	H_2	+ O	→ OH	+ H		5.0600×10^{4}	2.67	26.30
4b	4	OH	+ H	→ H_2	+ O		2.2199×10^{4}	2.67	18.28
5	5	H_2	+ OH	→ H_2O	+ H		1.0000×10^{8}	1.60	13.80
6b	6	H_2O	+ H	→ H_2	+ OH		4.3137×10^{8}	1.60	76.47
7	7	OH	+ OH	→ H_2O	+ O		1.5000×10^{9}	1.14	0.42
8b	8	H_2O	+ O	→ OH	+ OH		1.4749×10^{10}	1.14	71.11
9	11	H	+ OH	+ M'→ H_2O	+ M'		2.2000×10^{22}	-2.00	0.00
10	15	H	+ O_2	+ M'→ HO_2	+ M'		2.3000×10^{18}	-0.80	0.00
11	19	HO_2	+ H	→ H_2	+ O_2		2.5000×10^{13}	0.00	2.90
12	25	HO_2	+ OH	→ H_2O	+ O_2		6.0000×10^{13}	0.00	0.00
13	38	CO	+ OH	→ CO_2	+ H		4.4000×10^{6}	1.50	-3.10
14b	39	CO_2	+ H	→ CO	+ OH		4.9735×10^{8}	1.50	89.75
15	59	CHO	+ O_2	→ CO	+ HO_2		3.0000×10^{12}	0.00	0.00
16	61	CHO	+ M'	→ CO	+ H	+M'	7.1000×10^{14}	0.00	70.30
17	68	CH_2O	+ H	→ CHO	+ H_2		2.5000×10^{13}	0.00	16.70
18	72	CH_2O	+ OH	→ CHO	+ H_2O		3.0000×10^{13}	0.00	5.00
19	82	CH_3	+ H	→ CH_4			1.9000×10^{36}	-7.00	38.00
20	84	CH_3	+ O	→ CH_2O	+ H		7.0000×10^{13}	0.00	0.00
21	86	CH_3	+ CH_3	→ C_2H_6			1.7000×10^{53}	-12.00	81.20
22	90	CH_3	+ CH_3	→ C_2H_4	+ H_2		1.0000×10^{16}	0.00	134.00
23	92	CH_4	+ H	→ H_2	+ CH_3		2.2000×10^{4}	3.00	36.60
24	96	CH_4	+ OH	→ H_2O	+ CH_3		1.6000×10^{6}	2.10	10.30
25	112	C_2HO	+ O	→ CO	+ CO	+H	1.0000×10^{14}	0.00	0.00
26	115	C_2H_2	+ O	→ C_2HO	+ H		4.3000×10^{14}	0.00	50.70
27	133	C_2H_3	+ O_2	→ CH_2O	+ CHO		1.5000×10^{12}	0.00	0.00
28	134	C_2H_3		→ C_2H_2	+ H		1.6000×10^{32}	-5.50	193.50
29b	135	C_2H_2	+ H	→ C_2H_3			8.3508×10^{31}	-5.50	4.29
30	147	C_2H_4	+ OH	→ C_2H_3	+ H_2O		3.0000×10^{13}	0.00	12.60
31	167	C_2H_5	+ H	→ CH_3	+ CH_3		3.0000×10^{13}	0.00	0.00
32b	168	CH_3	+ CH_3	→ C_2H_5	+ H		3.6583×10^{12}	0.00	37.36
33	171	C_2H_5	+ O_2	→ HO_2	+ C_2H_4		2.0000×10^{12}	0.00	20.90
34	175	C_2H_5		→ C_2H_4	+ H		1.0000×10^{43}	-9.10	224.10
35	177	C_2H_6	+ H	→ H_2	+ C_2H_5		5.4000×10^{2}	3.50	21.80
36	179	C_2H_6	+ O	→ OH	+ C_2H_5		3.0000×10^{7}	2.00	21.40
37	181	C_2H_6	+ OH	→ H_2O	+ C_2H_5		6.3000×10^{6}	2.00	2.70

variable parameters in the calculations are the initial number density n_0 and, alternatively, the initial water volume fraction χ_0, $\chi_0 = \pi D_0^3 n_0/6$, where D_0 denotes the initial droplet diameter which has been kept at 20μm in all calculations.

The calculations were performed with detailed gas-phase chemistry. The mechanism of elementary reactions is shown in Table 1. This mechanism was derived by Wang (Wang and Rogg, 1991) from a comprehensive mechanism recommended by Warnatz (1984) such that at atmospheric pressure both structures and burning velocities calculated with this mechanism are in excellent agreement with those calculated

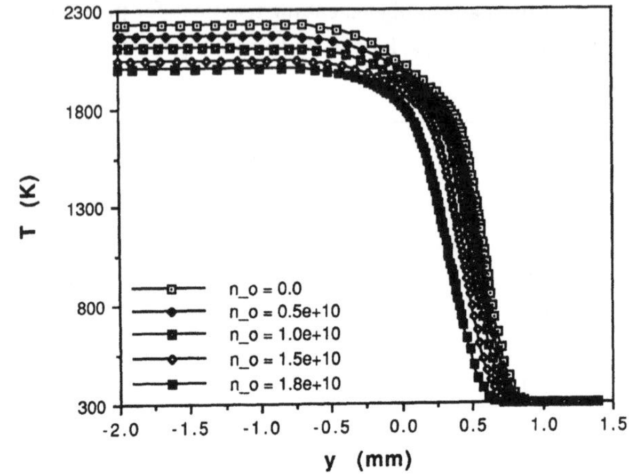

Figure 2a.

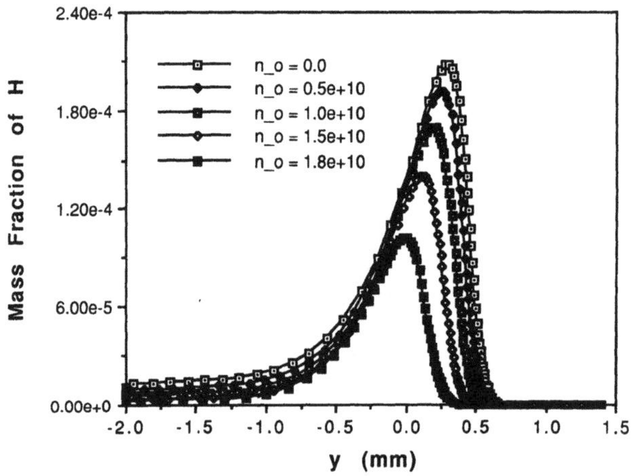

Figure 2b

Figure 2. Profiles of temperature (2a) and H-atom mass fraction (2b) through a two-phase counterflow flame for various values of the initial number density n_0.

from the comprehensive mechanism over the entire range of stoichiometries. Shown in Fig. 2 are the profiles of temperature (2a) and H-atom mass fraction (2b) for a strain rate of $a = 1500\text{s}^{-1}$ (Note that the case $n_0 = 0$ corresponds to a single water droplet which does not affect the structure of the gaseous flame.) As it is to be

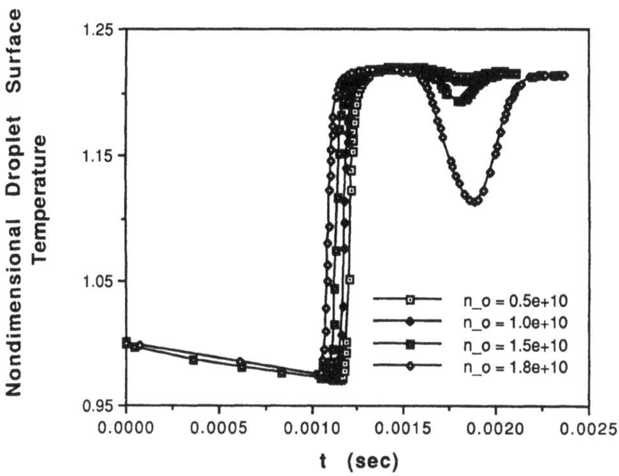

Figure 3. History of nondimensional surface temperature of water droplets in the flames of Fig. 2

expected on physical grounds, is seen that an increasing liquid-water content in the fresh mixture leads to a decrease in the final gas temperature and in the peak values of the H-atom concentration indicating the onset of flame extinction for sufficiently large values of n_0. Shown in Fig. 3 is the time history of the nondimensional droplet-surface temperature for the same non-zero initial number densities as in Figs. 2a and 2b. It is seen that initially evaporation leads to cooling of the droplet surface which, once the droplets enter the flame zone, is followed by a quick rise in the droplet-surface temperature. The oscillations in surface temperature observed for some of the values of n_0 are the consequence of the push-back effect which is discussed in detail in by Chen et al. (1992).

It is seen from Fig. 4 that for suitable values of n_0 and hence χ_0, flame extinction and ignition indeed occurs. Plotted in Fig. 4a is the nondimensional heat-release rate integrated through the flame Q,

$$Q \equiv - \int_{-\infty}^{+\infty} \frac{\sum_{i=1}^{N} h_i w_i}{c_p(T_u - T_{ad})} dy \,, \tag{20}$$

for three different values of the strain rate. The solid lines represent the computational results, the dashed lines represent interpolation. The results clearly exhibit the "classical" reflected image of an S-shaped curve indicating that in a certain range of χ_0 there exist up to three solutions for a single value of χ_0. In the present context, solutions on the upper branch of the reflected image of the S-shaped curve correspond to negligible heat release, i.e., to extinguished flames, whilst the solutions on the lower branch correspond to regularly, steadily burning counterflow two-phase flames. Solutions on the middle branch, which herein were not obtained numerically,

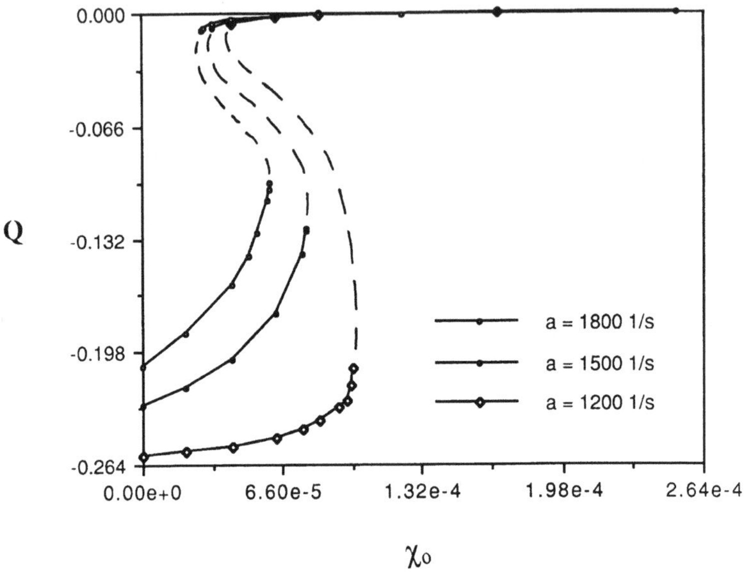

Figure 4a. Integrated heat-release rate Q versus initial water volume fraction for different values of the strain rate. Solid lines and symbols: computed; dashed lines: interpolated.

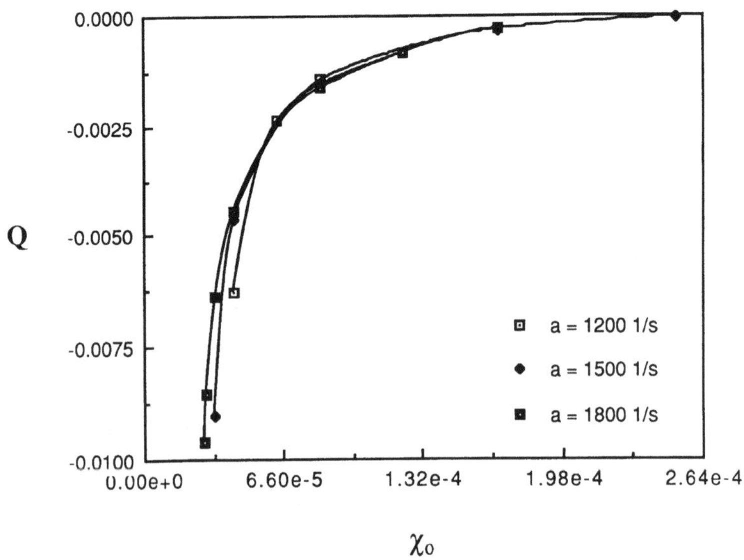

Figure 4b. Magnified view of the ignition area in the Q-χ_0 plane of Fig. 4a.

are physically meaningless. Bifurcation of the solution occurs at the two points in the Q-χ_0 plane where the tangent to the curve is vertical, i.e., at an ignition-point at the transition from the upper to the middle branch and at an extinction point at the transition from the lower to the middle branch. Figure 4b shows an enlarged view of the ignition region in the Q-χ_0 plane of Fig. 4a.

CONCLUSIONS

In the present paper we have modelled laminar two-phase counterflow flames employing detailed models for evaporation and for heat, mass and momentum transfer between the gaseous and the liquid phase. Also detailed models for chemistry and molecular transport have been employed. Specifically, strained methane-air counterflow premixed flames with water droplets added to the fresh, unburnt mixture were studied. Depending on the initial water-droplet volume fraction multiple solutions have been found. For various rates of strain, ignition points, at which the frozen solution bifurcates into the solution of a steadily burning or, alternatively, physically meaningless flame, have been identified as well as extinction points, where a bifurcating transition takes place from steady burning to inert or, alternatively, physically meaningless flow.

ACKNOWLEDGEMENTS

The authors are grateful to Professor K.N.C. Bray for stimulation and helpful comments. This study was supported in part by British Gas, UK.

REFERENCES

Abramzon, B. and Sirignano, W.A., 1987. Approximate Theory of a Single Droplet Vaporization in a Convective Field: Effects of Variable Properties, Stefan Flow and Transient Liquid Heating. In *Proceedings of the 2nd ASME-JSME Thermal Engineering Joint Conference*, volume 1, pages 11–18, Honolulu, Hawaii, March 1987.

Chen W.A., Rogg, B., and Bray, K.N.C., 1992. Modelling Laminar Two-Phase Counterflow Flames with Detailed Chemistry and Transport. In *Twenty-Fourth Symposium (International) on Combustion*, Sydney (Australia), submitted, 1992. The Combustion Institute.

Continillo, G. and Sirignano, W.A., 1990. Counterflow Spray Combustion Modeling. *Combust. Flame*, 81:325–340.

Deuflhard, P., 1974. A modified newton method for the solution of ill-conditioned systems of nonlinear equations with application to multiple shooting. *Numer. Math.*, 22:289.

Deuflhard, P., Hairer, E. and Zugck, J., 1985. One-step and extrapolation methods for differential-algebraic systems. Technical Report Preprint No. 318, Universität Heidelberg, Sonderforschungsberich 123.

Deuflhard, P. and Nowak, U., 1985. Extrapolation integrators for quasilinaer implicit odes. Technical Report Preprint No. 332, Universität Heidelberg, Sonderforschungsberich 123.

García–Ybarra, P. Nicoli, C. and Clavin, P., 1984. Soret and Dilution Effects on Premixed Flames. *Combust. Sci. and Tech.*, 42:87–109.

Giovangigli, V. and Smooke, M.D., 1987. Adaptive continuation algorithms with application to combustion problems. Technical Report Report ME-102-87, Yale University.

Hirschfelder, J.O. and Curtiss, C.F., 1949. Theory of Propagation of Flames. Part I: General Equations. In *Third Symposium on Combustion and Flame and Explosion Phenomena*, pages 121–127, Baltimore, 1949. Williams and Wilkins Co.

Puri, I.K. and Libby, P.A., 1989. Droplet Behavior in Counterflowing Streams. *Combust. Sci. and Tech.*, 66:267–292.

Puri, I.K. and Libby, P.A., 1991. The Influence of Transport Properties on Droplet Burning. *Combust. Sci. and Tech.*, 76:67–80.

Rogg, B., 1990. Adaptive computational methods for time-dependent problems in combustion engineering. In Pande G.N. and Middleton J., editors, *NUMETA 90, Numerical Methods in Engineering, Vol.2*, pages 1129–1140, London-New York, 1990. Elsevier.

Rogg, B., 1991. RUN-1DL: A Computer Program for the Simulation of One-Dimensional Chemically Reacting Flows. Technical Report CUED/A-THERMO/TR39, University of Cambridge, Department of Engineering, April 1991.

Smooke. M.D., 1983. An Error Estimate for the Modified Newton Method with Application to the Solution of Nonlinear Two-Point Boundary Value Problems. *J. Opt. Theory and Appl.*, 39:pp. 489, 1983.

Wang, W. and Rogg, B., 1991. On the Systematical Reduction of Detailed Kinetic Mechanisms. Technical Report CUED/A-THERMO/TR42, University of Cambridge, Department of Engineering, June 1991.

Warnatz, J., 1984. Rate Coefficients in the C/H/O/System. In W.C. Gardiner, Jr., editor, *Combustion Chemistry*, pages 197–360. Springer, New York.

INFLUENCE OF THE SURFACE DEFORMABILITY ON CONCENTRATION-DEPENDENT, OSCILLATORY AND STATIONARY CONVECTION IN A LIQUID LAYER

Zhasmina D. Kozhoukharova

Institute of Mechanics and Biomechanics
Acad. G. Bonchev St. Bl. 4, 1113 Sofia, Bulgaria

INTRODUCTION

The main goal of the present paper is to lay stress on the influence of the free surface deformability on the Marangoni instability in a horizontal liquid layer. Here, by means of the small disturbance analysis, the conditions for presence of stationary and oscillatory convection due to physical absorption are discussed.

When a gaseous surface-active solute absorbs or desorbs on a free liquid surface convective motions can take place. The fluid instability is caused by concentration gradients along a free surface due to dependence of the surface tension on surfactant concentration.

Cell patterns due to the mass transfer trough the interface are observed by Orel and Westwater (1961) and Linde and Schwarz (1963). First, Sternling and Scriven (1959) explain the concentration-capillary mechanism of instability (named also Marangoni instability). Their theoretical work is generalized by many authors. For instance, Brian (1971), Brian and Smith (1972), Palmer and Berg (1972), Chu and Chen (1986), Elenin et al. (1987), Chu and Velarde (1989), Kozhoukharova (1991), and Kozhoukharova and Slavchev (1991, 1992) investigate the Marangoni instability in a horizontal liquid layer with a free surface.

The surface deformability influences the system instability in two different ways. While the gravity waves have a stabilizing effect, the capillary waves destabilize the gas-liquid system.

Instabilities in Multiphase Flows, Edited by G. Gouesbet
and A. Berlemont, Plenum Press, New York, 1993

For Pearson's problem of thermocapillary convection (Pearson, 1958) Vidal and Acrivos (1966) verify that the instability sets in only as a stationary convection under condition of the non-deformable free liquid surface. In the case of a heated thin liquid layer the effect of the surface deflection on the existence of oscillatory modes is studied by Takashima (1981b), Benguria and Depassier (1989), Gouesbet et al. (1990), and Perez-Garcia and Carneiro (1991).

The influence of Gibbs adsorption on stationary and oscillatory Marangoni instability in a thin liquid layer with a flat free surface is analyzed by Brian (1971), and Brian and Smith (1972). These authors conclude that the only oscillatory modes found correspond to no physical meaning situations. This motivates further extension of the linear analysis in respect to influence of surface deformability on the system instability and to searching some oscillatory modes.

FORMULATION OF THE PROBLEM

Consider a physical absorption for an infinite liquid layer placed at a horizontal rigid wall and opened to the ambient gas of a component A. This component has a surface-active property. Both kinds of surface-active substances lowering or raising the surface tension will be considered. If the bulk surfactant concentration c_i at the surface is larger than the concentration c_g in the ambient gas there is a desorption process. Otherwise, the surface-active agent absorbs on the surface dissolving in the liquid simultaneously.

Restricting our analysis to dilute solutions, the fluid viscosity μ and density ρ are assumed unchangeable. The physical quantities that vary are the bulk concentration c, the surface excess concentration Γ, and the surface tension σ. In the equilibrium state σ depends linearly on Γ. Considering very thin layers the buoyancy is neglected.

In the equilibrium state liquid is at rest and the concentration gradient ($-\beta$) across the layer is constant being positive at absorption (e.g. $\beta < 0$) and negative at desorption ($\beta > 0$)

$$v_o = 0, \quad p_o = p_g + \rho g (d - z), \quad h = d, \tag{1}$$

$$c_o = c_w - \beta z. \tag{2}$$

Here $v_o = (u_o, v_o, w_o)$ is the fluid velocity, p_o the hydrostatic pressure, p_g the gas pressure above the layer, d the liquid depth, c_w solute concentration at the bottom of the layer, h=h(x,y,t) describes the free liquid surface. Cartesian coordinates (x,y,z) are employed where the vertical coordinate z is measured upward from the bottom of the liquid layer.

Let suppose the equilibrium state to be perturbed by disturbances denoted by primes, i.e.

$$v = v' \quad , p = p_o + p' \quad , c = c_o + c'. \tag{3}$$

Assuming the bulk diffusion controlled transfer in the liquid phase an instantaneous equilibrium between the perturbed adsorption and concentration at the sublayer is considered, e.g. the surface excess concentration is expressed as the sum

$$\Gamma = \Gamma_o + \Gamma', \quad \Gamma = \delta \, . \, c_i \, , \tag{4}$$

where δ is the "Gibbs depth". According to the linear stability theory the vertical component of the velocity w' and the perturbed concentration c' satisfy the following linear equations (prime are omitted)

$$\left(\frac{\partial}{\partial t} - \upsilon \nabla^2 \right) \nabla^2 w = 0 \quad , \tag{5}$$

$$\left(\frac{\partial}{\partial t} - D_L \nabla^2 \right) c = \beta \, w, \tag{6}$$

where t is the time, D_L the diffusion coefficient for the solute in the liquid, ∇ the gradient operator, ∇^2 the Laplacian expressed in the form

$$\nabla^2 = \frac{\partial^2}{\partial z^2} + L^2; \quad L^2 = \frac{\partial^2}{\partial x^2} + \frac{\partial^2}{\partial y^2}. \tag{7}$$

The perturbed free surface is presented by the equation

$$h = d + \zeta \, (x,y,t) \quad . \tag{8}$$

Since ζ is small compared to d, the boundary conditions at z=h are linearized and written at z = d by Taylor expansions. The kinematic condition is reduced to the equation

$$w = \frac{\partial \zeta}{\partial t} \, . \tag{9}$$

Neglecting the gas motion above the layer and excluding the hydrodynamic pressure the normal and tangential force balances take the forms

$$\Gamma_o \frac{\partial}{\partial t} (L^2 \, w) = \rho \left[\frac{\partial}{\partial t} - \upsilon \left(\frac{\partial^2}{\partial z^2} + 3L^2 \right) \right] \frac{\partial w}{\partial z} - (\rho \, g - \sigma \, L^2) L^2 \zeta, \tag{10}$$

$$- \Gamma_o \frac{\partial}{\partial t} \left(\frac{\partial w}{\partial z} \right) = \rho \upsilon \left(\frac{\partial^2}{\partial z^2} - L^2 \right) w - \frac{\alpha}{\delta} \, L^2 \, \Gamma', \tag{11}$$

where $\alpha = - d\sigma/dc$. The solute material balance is expressed as follows

$$- D_L \frac{\partial c}{\partial z} - Hk_G(c - \beta z) = \frac{\partial \Gamma}{\partial t} - \Gamma_O \frac{\partial w}{\partial z} - D_S L^2 \Gamma \quad , \tag{12}$$

In the above presented formulae H is the Henry's law constant, k_G is the gas-phase mass transfer coefficient, D_S the diffusivity of the solute in the interface.

The boundary conditions at the solid wall (z=0) are given for the velocity by

$$w = \frac{\partial w}{\partial z} = 0 \tag{13}$$

and for the concentration by

$$D_L \frac{\partial c}{\partial z} = \lambda_S c \ , \tag{14}$$

where λ_S is the solid-liquid mass transfer coefficient.

A solution of equations (5), (6) with boundary conditions (9)-(14) is sought by the usual way of decomposing it in normal modes (see the book of Chandrasekhar (1961)), i.e.

$$\left[w \ , \ c, \ \zeta\right] = \left[(D_L/d)W(z), \ (\beta d)C(z), \ dZ\ \right] \Phi(x,y).\exp \omega t \tag{15}$$

Here W and C are the non-dimensional velocity and concentration respectively, ω the time constant which is in generally a complex quantity ($\omega=\omega_r+i\omega_i$), function $\Phi(x,y)$, describing a periodic variation in the x and y direction, satisfies the Helmholtz equation

$$(L^2 + l^2)\ \Phi = 0, \quad l^2 = l_X^2 + l_Y^2 \ . \tag{16}$$

The governing equations and boundary conditions are rewritten in a dimensionless form using d^2/D_L as scaling quantity for time (D=d/dz)

$$\left[\ Sc^{-1}\omega - (D^2 - l^2\ )\right]\ (D^2 - l^2\ )W = 0 \quad , \tag{17}$$

$$\left[\ \omega - (D^2 - l^2\ )\right]\ C = W \tag{18}$$

with the boundary conditions at z=1

$$W - \omega\ Z = 0 \quad , \tag{19}$$

$$- \omega\gamma CrSc^{-1}l^2W = Cr\left[\ Sc^{-1}\omega - (D^2 - 3l^2)\right]DW + (Bo + l^2)l^2Z, \tag{20}$$

$$(D^2+l^2)W + Mal^2 (C - Z) + \omega\gamma\, Sc^{-1} DW = 0, \tag{21}$$

$$DC - NaDW + (L+l^2 S + \eth\omega)(C - Z) = 0, \tag{22}$$

and at z=0

$$W = DW = 0, \tag{23}$$

$$DC = Bi\ C\ . \tag{24}$$

In the equations and the boundary conditions the following dimensionless parameters appear: $Ma=\alpha\beta d^2/\rho\upsilon D_L$ the Marangoni number, $Na=\Gamma_0/\beta d^2$ the adsorption number, $Cr=\rho\upsilon D_L/\sigma_0 d$ the Crispation number, $Sc=\upsilon/D_L$ the Schmidt number, $Bo=\rho g d^2/\sigma_0 d$ the Bond number, $Bi=\lambda_S d/D_L$ the diffusion solid-liquid Biot number (briefly, Biot number), $L=Hk_G d/D_L$ the diffusion gas-liquid Biot number (briefly, modified Biot number), $S=D_S\delta/D_L d$, $\eth=\delta/d$, $\gamma=\Gamma_0\rho/d$.

The Marangoni number represents the ratio between surface and viscous forces. It enters the problem through the tangential stress boundary conditions at the surface and this dimensionless group is the most important one in the problem of capillary convection. The deformability of the fluid surface is characterized by the Crispation number as well as by the Bond number. The Crispation number is negligibly small only for thick layers or/and interface with a high surface tension. The adsorption number and the parameter S account for the surface convection and diffusion, respectively. The modified Biot number stands for the ratio between the liquid phase mass transfer resistance and to the gas phase resistance. The value Bi=0 corresponds to the so called "insulating" case and $Bi=\alpha$ to the "conducting" one. Following Brian (1971), one could imagine that the first case is physically realized if the bottom is treated as a semipermeable membrane through which the solute diffuses from the bulk beneath the membrane. The second limiting case takes place when the resistance of the membrane is quite high. The parameter γ is the ratio of the solute mass of the surface unit element to the liquid mass of the column of the layer beneath this element.

SOLUTION AND RESULTS

The eigenvalue problem (17-24) has been solved analytically for both cases of $\omega=0$ and $\omega\neq0$. In general case the Marangoni number is presented in form:

$$Ma = F(l,\omega,Na,Cr,Bo,Bi,L,Sc,S,\gamma,\eth), \tag{25}$$

where F is a real-valued function of the parameters in parentheses.

In the case of the oscillatory convection Ma is, in general, a complex function for arbitrary values of the enumerated parameters, e.g. $Ma=Ma_r+iMa_i$. But, the solution of the problem is only reasonable for real values of Ma. A numerical search has been conducted to find the values of ω_i for which the imaginary part of Ma vanish.

Note that in the case of stationary convection ($\omega=0$) the Schmidt number, $\mathfrak{F}$ and γ do not influence the neutral stability curves and consequently, the values of the critical Marangoni number.

The determined formulae for a stationary case are already presented in (Kozhoukharova and Slavchev, 1991) and for an oscillatory one in (Kozhoukharova, 1991, and Kozhoukharova and Slavchev, 1992). For the sake of brevity only the formula expressed the stationary Marangoni number noted by Ma_s in the conducting case is given here

$$Ma_s=\frac{8l(\cosh l.\sinh l-1)[l.\cosh l+(L+l^2 S)\sinh l]}{\dfrac{8l^5 Cr}{Bo+l^2}.\cosh l-4.Nal^2\sinh l(\sinh^2 l-l^2)+(\sinh^3 l-l^3\cosh l)} \tag{26}$$

The behaviour of the stationary neutral curves for small wave numbers depends considerably on the Crispation number. In the vicinity of $l=0$ the Marangoni number for the conducting case is presented by Taylor series

$$Ma_s=\frac{2Bo}{3Cr}(1+L)+\frac{2}{3Cr}\left\{\frac{Bo^2}{6Cr}(1+L)(Na-\frac{1}{20})+(1+L)(1-\frac{2}{15}Bo)\right.$$
$$\left.(S+\frac{1}{3})Bo\right\}l^2+O(l^4)\ . \tag{27}$$

In particular, in the case of non-deformable free boundary the Crispation number disappears as well as the Bond number. The determined formulae are reduced to Brian's (1971) and Brian & Smith's ones (1972) in an explicit form. In the conducting case when γ, $\mathfrak{F}$, S and Na are equal to zero the obtained results formally coincide with those for Pearson's problem of thermocapillary convection presented by Takashima (1981a, 1981b), Gouesbet et al. (1990) and Perez-Garcia and Carneiro (1991) provided the buoyancy force are excluded. A perfect agreement is established by comparing our results with those in previous works.

When Gibbs adsorption is assumed, Ma and Na always have the same sign coinciding with the sign of the product $\alpha\beta$ (Brian, 1971, Kozhoukharova and Slavchev, 1991). It means that any solution of the problem obtained for Ma and Na with different signs has not a physical sense. So, the layer is stable in

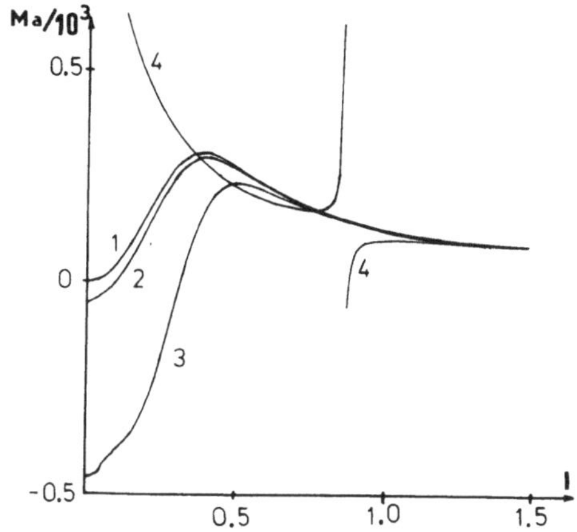

Figure 1. Neutral stability curves for the conducting case, Na=-0.0004, Cr=0.00016, L=0, and various values of the Bond number: Bo=0 (label 1), Bo=-0.01 (label 2), Bo=-0.1 (label 3), Bo=-1. (label 4).

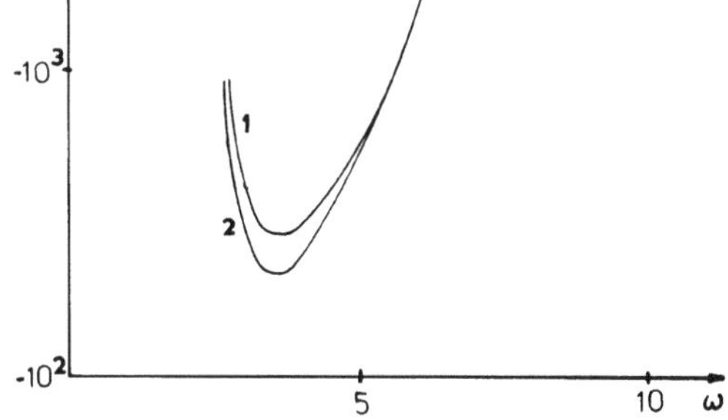

Figure 2. Curve of overstability as a function of frequency, conducting case, Bo=-0.1, Cr=0.1, Sc=0.5, L=0, and various values of Na: Na=0.0004 (label 1), Na=0.004 (label 2).

that case. Both numbers α and β are positive in the case of desorbing a surface-tension-lowering solute (when $\alpha > 0$, $\beta > 0$), or absorbing a solute which increases the surface tension ($\alpha < 0$, $\beta < 0$). They are negative when a surface-tension absorbs ($\alpha > 0$, $\beta < 0$), or surface-tension-increasing solute desorbs ($\alpha < 0$, $\beta > 0$).

In this paper our attention is focused on the case when the liquid layer is suspended from a rigid ceiling, e.g. the Bond number is negative and only the conducting case is considered.

It is easy to show when l goes to infinity the stationary Marangoni number tends to the value -2/Na. So, there is a wave number l_F above which Ma_S and the adsorption number have different signs. Hence, the liquid system is always stable with subject to very small-length wave disturbances.

For non-negative values of Cr, Bi and Bo it is shown (Kozhoukharova and Slavchev, 1991) that the stationary Marangoni number is non-negative for $Na \leq 0$. In the case of the suspended liquid layer from a ceiling Takashima (1981a) shows that of all neutral stability curves cut across the l-axis at $l_I = (-Bo)^{1/2}$. In our investigation because Ma_S is strongly influenced by the surface convection the wave number l_S different from l_I exists and $Ma_S(l_S)$ is also equal to zero for the large enough Bond number(see Fig. 1). When $l > (-Bo)^{-1/2}$ all curves lie in the positive region Ma. The region below each curve represents stable state.

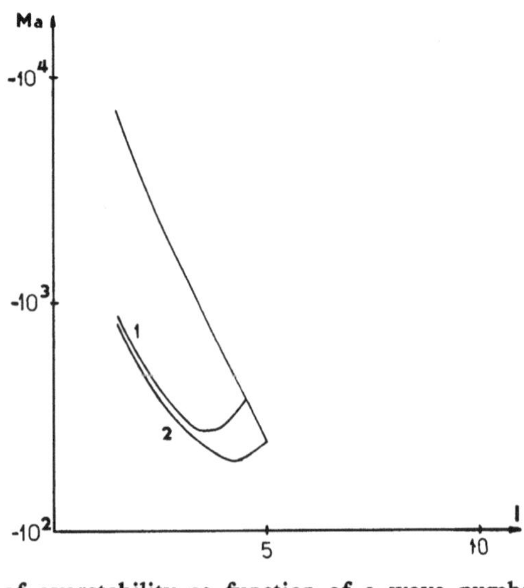

Figure 3. Curve of overstability as function of a wave number conducting case, Bo=-0.1, Cr=0.1, Sc=0.5, L=0, and various values of Na: Na=-0.004 (label 1), Na=-0.0004 (label 2).

From provided computations it is established in the conducting case if both the adsorption and the Bond number are negative there exist some conditions under which the oscillatory Marangoni number (Ma_o) also is negative. So, in the case under consideration the liquid system may lose its stability with respect to an oscillatory mode. The dependence of the Marangoni number on the frequence for various values of the adsorption number is presented in Fig. 2 and on the wave number on Fig. 3. It is seen from Fig. 2 and Fig. 3 that the every neutral overstability curve has two different frequency and wave number branches, respectively. The one of the weakly depends on the adsorption number.

When the conducting case with the negative adsorption number is under consideration the liquid system stability strongly depend of type of the liquid layer. In the first case when it is placed at the rigid wall ($Bo > 0$) it is always stable with respect to stationary modes (see Kozhoukharova and Slavchev, 1991) but it may lose its stability with respect to an oscillatory mode (see Kozhoukharova and Slavchev, 1992). In the second case, when a liquid layer is suspended from a rigid ceiling two type of stability modes may exist simultaneously (see Fig. 1 and Fig. 2).

REFERENCES

Benguria, R.D., and Depassier, M.C., 1989, On the linear stability theory of Benard-Marangoni convection, *Phys. Fluids A.* 1:1127.

Brian, P.L.T., 1971, Effect of Gibbs adsorption on Marangoni instability, *A.I.Ch.E. J.* 17:765.

Brian, P.L.T., and Smith, K.A., 1972, Influence of Gibbs adsorption on oscillatory Marangoni instability, *A.I.Ch.E. J.* 18:231.

Chandrasekhar, S., 1961, "Hydrodynamic and Hydromagnetic Stability, Oxford Univ. Press, Oxford.

Chu, X.-L., and Chen, L.-Y., 1986, Stability of two-component Benard fluid with surface adsorption, *Commun. Theor. Phys. (Beijing)* 6:237.

Chu, X.-L., and Velarde, M.G., 1989, Benard-Marangoni convection in liquid layers with a deformable open surface: note on the role of solute accumulation at the air-liquid interface, *J.Coll. Int. Sci.* 127:205.

Elenin, G.G., Kalachinskaya, I.S., and Rabinovich, L.M., 1987, Numerical modelling of interface instability in a two-dimensional layer of reacting liquid, *Fluid Dynamics, Consults Bureau.* 22:927 (translated from Russian).

Gouesbet. G., Maquet, J., Roze, C., and Darrigo R., 1990, Surface tension- and coupled buoyancy-driven instability in a horizontal liquid layer. Overstability and exchange of stability, *Phys. Fluids A.* 2:903.

Kozhoukharova, Zh., 1991, Influence of the surface deformability on the instability in an insulating layer, *J.Teor. Apll. Mech.* 22, No. 4:50.

Kozhoukharova, Zh., and Slavchev, S., 1991, Influence of the surface deformability in a liquid layer with surface chemical reaction. I. Stationary convection, *J. Coll. Int. Sci.* 148:42.

Kozhoukharova, Zh., and Slavchev, S., 1992, Influence of the surface deformability in a liquid layer with surface chemical reaction. II. Overstability, *J. Coll. Int. Sci.* 152:473.

Linde, H., and Schwarz, E., 1963, Unterschungen zur Charakteristik der freien Grenzflachankonvektion beim Stoffubergang an fluiden Grenzen, *Z. Phys. Chem.* 224:331.

Orell, A., and Westwater, J.W., 1961, Natural convection cells accompanying liquid-liquid extraction, *Chem. Eng Sci.* 16:127.

Palmer, H.J., and Berg, J.C., 1972, Hydrodynamic stability of surfactant solution heated from below, *J. Fluid Mech.* 51:385.

Pearson, J.R.A., 1958, On convection cells induced by surface tension, *J.Fluid Mech.* 4:489.

Perez-Garcia, C., and Carneiro, G., 1991, Linear stability analysis of Benard-Marangoni convection in fluids with a deformable free surface, *Phys. Fluids A*, 3:292.

Smith, K.A., 1966, On convective instability induced by surface-tension gradients, *J. Fluid. Mech.* 24:401.

Sterling, C.V., and Scriven, L.E., 1959, Interfacial turbulence: hydrodynamic instability and the Marangoni effect, A.I.Ch.E. J. 5:514.

Takashima, M., 1981a, Surface tension driven instability in a horizontal liquid layer with a deformable free surface. I. Stationary convection, *J. Phys. Soc. Japan* 50:2745.

Takashima, M., 1981b, Surface tension driven instability in a horizontal liquid layer with a deformable free surface. II. Overstability, *J. Phys. Soc. Japan* 50:2751.

Vidal, A., and Acrivos, A., 1966, Nature of the neutral state in surface-tension driven convection, *Phys. Fluids* 9:615.

INSTABILITIES, MASS-TRANSFER AND SELF-ORGANIZATION OF HEAT CYCLES BY INTERACTION OF LASER RADIATION WITH LIQUIDS

A.T.Sukhodolsky

Institute of General Physics
Russian Academy of Sciences
117942, Moscow, Russia

INTRODUCTION

Interest in research on processes of self-organization of open systems increased in recent years [1]. The main problem in this sphere of modern physics knowledge is an entropy problem [2]. This paper gives a review of some phenomena of laser radiation interaction with absorbing liquids. These phenomena are united on the basis of an original approach to the problem of self-organization introduced in the context of classical thermodynamics of heat cycles in heat engines. This approach could be justified by an attempt to merge new experimental data to the sphere of instabilities in multiphase flows caused by laser radiation interaction with liquids and solutions.

It is common knowledge that any amount of work can be transformed into heat. On the contrary, a reverse process of transformation of heat into work can not arise "by itself". At the same time a process like this one can take place in special artificial mechanisms (engines) made by a human being. So if a motor already exists then the problem of negative change of entropy inside this system will be simply solved: part of the heat coming from the source is transformed into work in a cycle. During a part of this process some part of the energy of expanding gas is stored in the engine, for instance in the kinetic energy of a flywheel. Then during the other part of the cycle this stored energy covers movement in the system with negative change of entropy. Something like this will be showed below for simplest nonequilibrium fluid systems.

The main idea would be kept in mind by considering below: are there in nature any nonartificial systems transforming heat into work "by itself" ? It should be noted that we take into consideration a modest goal: " How can some natural systems function by themselves transforming heat into work ?" instead of much more complicated: " How does a self-organised system arise by itself ?". In all the processes which will be described below a laser beam is in service as a source of energy.

Instabilities in Multiphase Flows, Edited by G. Gouesbet
and A. Berlemont, Plenum Press, New York, 1993

After absorption, the energy of radiation transforms into thermal energy. It should be noted that radiation heating differs from heating by thermoconductivity because of its volumetrical character . Namely laser radiation being absorbed in fluid matter creates condition for observation of a lot of phenomena which will be united by the idea of heat cycle self-organization.

LIGHT-PHORESIS(Circulator 1)

Recently [3], the phenomenon of so-called laser-channeled lightphoresis has been revealed, when an Ar-laser beam has been used for illumination of microemulsion. The phenomenon was found to be observed as capturing a particle which crossed over the laser beam and its following pulling along the beam in the direction opposite to the flow of light. The velocity of the particles was measured to go as high as $10cm/s$. In this chapter photophoresis is considered on the basis of a self-organized heat cycle in which the interface separating two phases has been brought into use as a working fluid. To obtain an nonequilibrium analogy of the classical $p-v$-diagram the conception of both the interface source and the interface sink as well as surface flow was introduced [4]. The idea of self-organization has been developed for explanation of exceeding lightphoresis velocity compared to characteristic velocity of particles in usual thermodiffusion.

Hydrodynamic problem

Let us consider an opaque spherical liquid drop of radius R embedded into a transparent nonmixed fluid. The particle was illuminated with homogeneous radiation along z axes. The liquid drop has an absorption coefficient $\alpha \gg 1/R$. When the particle gets into the beam, the light excites thermocapillary convection inside and outside of the drop due to the gradient of the interface tension between the illuminated and the dark side of the drop. The particle as a whole moves thanks to convection.

To obtain the value of the velocity of lightphoresis $\vec{u}$ we are going to consider a drop of spherical form with the linear dependence of surface tension $\sigma = \sigma_0 + \Delta \cos \Theta$. The angle Θ is assigned in the spherical system of coordinates associated with the center of the drop. We have to note that the value Δ depends generally on the velocity, therefore, the following solution of the hydrodynamics problem can be considered as a rough estimation to develop a thermodynamical approach. The initial hydrodynamics equations consist of linear Navies-Stokes and continuity equations [5]

$$\eta_1 \nabla^2 \vec{v}^{(2)} = \nabla p^{(1)} \qquad \eta_2 \nabla^2 \vec{v}^{(2)} = \nabla p^{(2)} \tag{1}$$

$$\mathrm{div}\vec{v}^{(1)} = 0 \qquad \mathrm{div}\vec{v}^{(2)} = 0 \tag{2}$$

Where η_1 and η_2 are dynamic viscosity inside and outside the drop. For nonmixed fluids the boundary conditions are held

$$\vec{v}^{(1)} \mid_{r=R} = 0 \qquad \vec{v}^{(2)} \mid_{r=R} = 0 \tag{3}$$

$$\vec{v}^{(1)}_{\Theta} \mid_{r=R} = \vec{v}^{(2)}_{\Theta} \mid_{r=R} \tag{4}$$

$$\vec{p}^{(1)}\,|_{r=R} - \frac{\partial\sigma}{R\partial\Theta} = \vec{p}^{(2)}_{\Theta r}\,|_{r=R} \tag{5}$$

Solution of equations (1) and (2) is found in the form of

$$\vec{v}^{(1,2)} = \text{rot rot} f^{(1,2)}\vec{u} + \vec{u} \tag{6}$$

where

$$f^{(1)} = a \cdot r + \frac{b}{r}, f^{(2)} = \frac{A}{4}r^2 + \frac{B}{8}r^4$$

Substituting this into (1) (6) gives the system of equations for finding a, b, A, B [5]. In the moving coordinate system fixed in the center of the drop the whole force has to be equal to zero. Using (1) (2) together with (4)-(5) the force can be obtained by $F = 8\pi\eta_1 a$. Then, the velocity of the particle is calculated from equation

$$a = \frac{9R\eta_1 u + 6R\eta_2 u + 2\Delta R}{12u(\eta_1 + \eta_2)}$$

At last this gives expression for the velocity of propulsion

$$u = \frac{2\Delta}{3(3\eta_1 + 2\eta_2)} \tag{7}$$

Thermodynamic approach

Let us use a thermodynamic principle to describe the moving process. First of all we consider an element of boundary region separating the phase i and e with area ΔS. This domain of thickness h is the confined region of existence of the third intermediate phase s with properties differing from both phase i and e. According to Gibbs[6] in region s can be set a separating plane with infinitly small thickness. This Gibbs'es plane (GP) can serve as a pure boundary of bubble in conditions (3)-(4). In this case the gradient of the interface tension properties of region s can not be described by the equilibrium thermodynamics at all.

In fact, it is necessary to bring into use attributes of nonequilibrium thermodynamics. First, we introduce a principle of local equilibrium which in this case can be formulated as the principle of local equilibrium (PLE) of GP. Its essence consists in local dynamic equilibrium of any piece of boundary s with volumes of both phases i and e. This supposition can be quite true because the characteristic speed of interchange of molecules between the boundary s and both phases i and e is considerably faster than the hydrodynamics velocity along GP.

On the basis of PLE, we can introduce the GP flux. Let us consider an element of interface with the area ΔS. Suppose that hydrodynamic velocity v of a certain boundary point is the velocity of the corresponding point of GP. Then we can introduce GP flux as flux of GP through an element of segment $\vec{dl}$, restricting ΔS. The GP flux is equal to $\rho_s \vec{v}_s \vec{dl}$, where $\vec{dl}$ is directed along the external normal, ρ_s is the density of GP which can be simply presented as an average number of mutually adsorbed molecules per unit area of the interface. The total amount of GP which flows through the boundary ΔS is equal to $\oint \rho_s \vec{v}_s \vec{dl}$. On the other hand the GP flux through the boundary ΔS can be determined by the sources and sinks of GP due to departure of molecules from GP into volumes i and e and their appearance on GP. Full flux is equal to $\int_s \xi ds$, where ξ is the source or sink density. Besides, the change of the whole GP inside of ΔS is equal to $-\frac{\partial}{\partial t}\int\limits_s \rho_s ds$. The continuity equation for two-dimensional GP flux is derived from the mass balance on ΔS as follows,

$$(input - output)/dt = accumulation/dt,$$

$$\int_s \xi ds - \frac{\partial}{\partial t_s} = \oint \rho_s \vec{v}_s d\vec{l}$$

Using conversion of integral $l \rightarrow s$ the last expression yields

$$\rho_s \text{div}_s \vec{v}_s - \vec{v}_s \text{grad} \rho_s = \xi - \frac{\partial \rho_s}{\partial t}$$

The important question arises about the terms $\text{grad}\rho_s$ and $\frac{\partial \rho_s}{\partial t}$ in this equation. The role of these terms is associated with the problem of compressibility of GP. By comparing the preceding two expressions we obtain the continuity equation for GP

$$\rho_s \text{div}_s \vec{v}_s + \vec{v}_s \text{grad} \rho_s - \xi = 0$$

which for the case of incompressible surface $\rho = \text{const}$ shows as

$$\xi = \rho_s \text{div}_s \vec{v}_s \tag{8}$$

Now we can introduce the nonequilibrium powerful analogy of classical $p - v$ diagram[7]. First, we can notice that surface tension is the reversible work of isothermic formation of an unit of GP. Positive work $(\sigma\xi/\rho_s)$ associated with GP is done in the region of positive ξ. Oppositely, negative one is done in domain of GP where ξ is negative. The difference between them is exploited for the useful work of drop's movement. Hence in accordance with classical approach we see here all the symptoms of a reversible thermodynamic cycle. But in our case useful power can be found by integration over all sources and sinks

$$N = \int \frac{\sigma\xi}{\rho} d\xi \tag{9}$$

For the lightphoresis of the liquid drop from hydrodynamic solution follows

$$\xi(\Theta) = 3(\frac{u}{R}) \cos\Theta \tag{10}$$

Discussion

The dependence of surface tension $\sigma(\Theta)$ from so-called "integer" surface $S(\Theta)$ whose differential is equal to $(\xi(\Theta)/\rho_s)ds$ can serve as nonequilibrium analogy of $p - v$-diagram in case of the lightphoresis. As it was illustrated in [4] this relationship was, obtained from eq.(10) , taking into account eqs.(10) and (9). We should note that the curve motion corresponds to a spatial distribution of the surface region, making positive and negative work, rather than temporal sequence of thermodynamic cycle performed by the working fluid. A total "presented" area at the turning of angle Θ from 0 to π is equal to zero. This is satisfied if the area of the drop surface is constant. An analysis conducted makes it possible to conclude that the useful power of the $E \rightarrow N$ conversion in the case of lightphoresis is determined mainly by a difference between maximum and minimum values of stationary Δ.

THERMOCAPILLARY LENS OSCILLATION(circulator 2)

An interface between two phases will be considered as working fluid in another thermocapillary phenomenon so-called thermocapillary lens oscillation or heartbeats (HB)[8]. This phenomenon is produced when a CW laser beam travels in an absorbing liquid near the free interface. HB-effect was observed with several experimental configuration [9, 10, 11, 12, 13]. Seemingly, the first observation of thermal lens oscillations has been reported in a short letter by Jakeman et al but without any further investigation [14]. The phenomenon has been systematically investigated by G.Gouesbet[8, 15, 16]. The reader may consult a review paper in which many references up to 1990 are quoted [17]. Basically, thermal lens oscillations may occur when a laser beam travels horizontally below the free surface of an absorbing liquid (so-called optical heart-beat 1, HB1) or vertically upwards (HB2), leading to oscillations of the beam going out of the liquid cell, associated with oscillatory convection in the bulk and propagating waves at the free surface. Also, a new HB3 has been designed which, to some extent, combines HB1 and HB2 features[18].

Experiment

Multimode radiation from an argon laser with a radiated power up to 1 W was used for the experiment[13]. The laser beam was directed downward through a lens with a focal length of ~15 cm onto a Petri dish containing an absorbing fluid. The absorbing substances were different oily fluids: hypoid oil and transformer oil with absorption coefficients of $4-7cm^{-1}$ at the wavelength of the argon laser. Volume and surface capillary thermal lenses were formed under laser heating, causing appreciable defocusing of the laser beam. The motion of the fluid was observed in a microscope for the purpose of which the particles of aluminum powder were introduced into the oil. A photodiode with an iris placed in the cone of laser light transmitted through the fluid layer was used to analyze the temporal dynamics of the transmitted beam; the output signal of the photodiode was sent to a graphic recording instrument and a frequency meter.

The onset of instability of the fluid motion manifested itself in the transition from steady-state motion to an ordered periodic motion at a certain threshold thickness of the layer (and/or a threshold pump power). The measured frequency stability under fixed experimental conditions was better than 5%, and the vibrational process exhibited the characteristic behavior of nonlinear oscillation.

Simple model

We interpret the experiment on the basis of the dissipative structure or "circulator" model; this model describes the main components of heat and mass transfer under the action of gravity and thermocapillary mechanisms without completely solving the problem, which entails considerable mathematical difficulties.

The analysis is based on the model of a continuous stream tube of finite length and radius r with a constant velocity v over the entire length at a given time. The vertical section of the tube of length L_1 is subjected to a buoyant force proportional to the integral of the temperature over the vertical distance:

$$F_b = C_4 \int_0^{L_1} T dz$$

The horizontal part of the tube, which corresponds to its motion over the surface, is acted upon by the Marangoni force, which is proportional to the temperature difference at the point of impingence of the jet on the surface and on the edge:

$$F_N = C_5(T_2 - T_3)$$

The tube is also acted upon by the Newtonian friction force

$$F_f = -C_6 v$$

If the "mechanical" part of the problem is augmented with the heat-balance equation, which takes into account continuous heating from below, convective heat transfer and heat transfer into the surrounding medium, we obtain the basis system of equations

$$\frac{\partial T}{\partial t} = C_1 e^{-C_2 z} - v\frac{\partial T}{\partial z} - C_3 T \tag{11}$$

$$\frac{\partial v}{\partial t} = C_4 \int_0^{L_1} T dz + C_5(T_2 - T_3) - C_6 v \tag{12}$$

in which

$$C_1 e^{C_2 z} C_p \rho \pi r^2$$

is the power of the heat sources per unit length of the tube, and $-C_3 T$ is the heat loss into the surrounding volume. For the coefficients in Eqs.(11) and (12) we have the estimates (in the SI system units).

$$C_1 \sim \frac{I_o}{aC_p\rho} \sim 10^2; \qquad C_2 \sim \frac{1}{a} \sim 10^3; \qquad C_3 \sim \frac{\chi}{R^3} \sim 1 - 10$$

$$C_4 \sim \frac{g\beta}{R^3} \sim 1 - 10; C_5 \sim \frac{\alpha\sigma R}{L\rho r^2} \sim 10^{-3}; \qquad C_6 \sim \frac{\nu}{r^2} \sim 1 - 10$$

where

$$\beta = -\frac{\partial \rho}{\rho \partial T}; \qquad \alpha = -\frac{\partial \sigma}{\sigma \partial T};$$

I_o is the laser power density, $1/a$ is the absorption coefficient, C_p is the heat capacity, ρ is the density, χ is the thermal diffusivity, σ is the surface tension coefficient, L_1 is the characteristic length of the stream tube, r is its transverse radius, R is a characteristic radial length of heat transfer away from the stream tube and ν is the dynamic viscosity.

Equations (11) and (12) were computed by a fourth-order Runge-Kutta procedure, and the entire active section of the tube from T_1 to T_3 was partitioned into 19 zones, the temperatures T_1, T_2 and T_3, corresponding to the temperatures of zones 1,

10 and 19. In [13] was shown the time dependence of the fluid velocity along with a phase diagram v as a function of T_2. So it appears that the solution of Eqs.(11) and (12) enters a stable cycle, as is corroborated by the choice of more or less arbitrary initial conditions.

Discussion

If we compare the experimental and calculated results, we see that our model, which reflects the main mechanisms responsible for fluid motion provides a good description of the experiment. The essence of the ordering of the fluid motion and self-organization of heat cycle can be traced qualitatively in the phase diagram [13]. The initial phase of the motion is associated with an increase of the velocity and temperature on the surface under the combined action of gravity and capillary forces. An abrupt drop of the surface temperature at the center takes place along path of diagram as a result of the emergence at the surface of the fluid from the depths which has not had time to heat up. Latest path of heat cycle is associated with a change of the Marangoni force which changes the direction of motion.

SELF-ORGANIZATION HEAT CYCLE BY THERMOCAVITATION (Cavitator)

In the present chapter we discuss the observation of the effect of self-organization of transformation $H \rightarrow A$ in process so-called thermocavitation of a single gas bubble. The process of heat transformation into work arising in the continuous laser heating of a liquid proceeds without the use of any devices - "by itself". An interpretation of the effect is given on the basis of a model of the dissipative structure, called a "cavitator", through a solution of the system of nonlinear differential equations describing the dynamics of the supply and dissipation of heat and the hydrodynamics of the cavitating single bubble.

Simple experimental setup

The experiment was conducted using an argon laser radiation, which was focused on the flat window of a cell containing an absorbing liquid[19]. Part of the radiation reflected from the interface between the window and the liquid was recorded with a photodiode. The acoustic signal was recorded simultaneously with a hydrophone placed in the liquid. Solutions of various organic dyes in water and ethanol with an absorption coefficient of the order of $100cm^{-1}$ were used as absorbing liquids.

At the pumping radiation densities higher than $10^6 W/cm^2$ the liquid boiled. The boiling was accompanied by the formation of a set of small bubbles and a characteristic random crackling. With softer focusing, however, at pumping power densities of the order of $(0.4-1)10^4 W/cm^2$ we observed self-organization of heat transformation into mechanical energy. A single bubble, whose maximum size reached 1 mm, formed and collapsed in the liquid periodically with a repetition frequency of 20-1000 Hz. The lifetime of the bubble was 0.1-0.7 msec, while the process of its collapse was accompanied by a strong hydraulic shock. Here the periodic nature of the process is not connected with the feedback mechanisms due to screening of incident radiation. This is confirmed by the fact that the nature of the process was not changed upon switching to pumping through a reflecting prism, when the radiation is fully screened in plasma transition.

Dissipative structure - cavitator

An interpretation of the observed process of heat transformation released in the absorption of optical radiation into the mechanical energy of a hydraulic shock can be given on the basis of the model of a dissipative structure (cavitator) described by the following equation system

$$\frac{dR_b}{dt} = u \tag{13}$$

$$\frac{du}{dt} = \frac{(p - p_o)}{\rho R_b} - \frac{3u^2}{2R_b} - \frac{2\sigma}{\rho R_b^2} \tag{14}$$

$$\frac{dT_v}{dt} = -c_2 \left(\frac{\lambda R_1^2 (T_v - T_l)}{\delta_1} + c_3 p R^2 u \right) \tag{15}$$

$$\frac{dT_l}{dt} = -c_1 \left(\frac{R_2^2 (T_l - T_o)}{\delta 2} + \frac{R_1^2 (T_l - T_v)}{\delta_1} - \frac{W}{4\pi\lambda} \right) \tag{16}$$

where

$$c_1 = \frac{4\pi\lambda}{\rho c v_2}; \qquad c_2 = \frac{4\pi}{\rho c v_1};$$

$$R_1 = \frac{3}{4\pi} \left(v_1 + 4/3\pi R^3 \right)^{1/3}; \qquad R_2 = \frac{3}{4\pi} \left(v_o + 4/3\pi R^3 \right)^{1/3}; \qquad v_2 = v_o - v_1$$

R_b is the bubble radius, T_v is the vapor temperature in the bubble; T_l is the temperature of the heated region of the liquid of volume v_o; v_1 is the volume of liquid layer at the boundary of the bubble and of the heated volume of liquid with a temperature T_v ;δ_1 and δ_2 are constants allowing for the efficiency of heat transfer by Newton law; c_3 is a constant allowing for the heat flux in evaporation; ρ, c and λ are the density, heat capacity and thermal conductivity coefficient of the liquid; σ is the surface tension coefficient; p_o and T_o are the external pressure and temperature; and p is the vapor pressure in the bubble, described by the Clapeyron-Clausius equation [20] for ideal gases.

For $R < R_{cr}$, the radius of critical nucleus , the solution of system (13)-(16) actually comes down to the analysis of only Eq.(16), whose solution is:

$$T_l = T_o + \frac{a}{b}(1 - e^{-bt}) \tag{17}$$

$$a = \frac{W}{c\rho v_o}; \qquad b = \frac{3\lambda}{c\rho\delta_2} \left(\frac{3v_o}{4\pi} \right)^{-1/3}$$

From (17) with known W, p and c and from the maximum period of repetition of the cavitation pulses we can estimate the constant values, $v_o \sim 10 - 11 m^3$ and $\delta_2 \sim 10^{-5} m$.

For $R > R_{cr}$ the system was solved by the Runge-Kutta method on a computer. In [19] was shown a p-v diagram of the self-organizing heat cycle of the cavitator, and the variation of the kinetic energy of the system, from which it was seen that the cavitator consists of a heat engine which "by itself" transforms heat stored in the superheated liquid into the kinetic energy of liquid motion.

Discussion

A comparison of the experimental and calculated data allows us to formulate certain causes and conditions for which a heat cycle is self-organized.

1. In a laser cavitator the part of the liquid volume in which the optical radiation is thermalized serves as the heater, while the remaining liquid serves as the condenser. In the process of the bubble growth the energy is transfered from the heater to the condenser because the size of the bubble becomes larger than the size of the heater.

2. In any circular cycle it is necessary to perform work on the working body in the compression phase due to the energy stored during expansion. In a cavitator this is accomplished without any devices, since the work of expansion goes into an increase in potential energy because in the final stage of the bubble growth the pressure in it decreases below the external pressure.

3. The periodic nature of the process in thermal cavitation is evidently due to the dynamics of heterogeneous nucleation, and is not related with the feedback mechanisms described in [21] since in our experiments no change was observed in the absorption coefficient at the stage of the liquid heating.

PERSPECTIVES

Here we reported the observation of some new types of fluid instability in the event of continuous laser heating, which creates a time-ordered periodic fluid motion. The experiment is interpreted on the basis of simple models of both dissipative structures "circulator" and "cavitator", which are introduced to describe the dynamics of laser heating, heat transfer, and motion of the fluid under the action of gravity and capillary forces. As it seems the proposed researches would open an opportunity to study one of the new aspects of radiation-matter interaction and dynamics of multiphase flows.

This investigation is interesting for the following reasons. On the one hand, liquid is a condensed matter, therefore, a giant light energy per a unit volume can be absorbed. On the other hand, liquid is a fluid matter with a large number of degrees of freedom . These two circumstances do lead to observe a lot of new interesting phenomena. Besides, the investigation in this sphere is interesting for possibility to obtain fresh fundamental knowledge about the nature of liquid on the basis of new laser experiments and techniques.

At last, right now there is opportunity of new applications of instability like "circulator" and "cavitator" in engineering. This are first results in recording of diffraction gratings and images [22],[23]; extraction of binary solution [24]; analytical chemistry[25]; sound wave production in liquids both by CW-laser [26] or solar radiation[26]; body's movement[27], cutting of materials[28], nonlinear optics [29] and conversion of laser beams [30, 31], pumping of liquids [32], investigation of the nature of clusters in solutions[21]. Part of these phenomena has a good opportunity to be used in microgravity conditions[33]

ACKNOWLEDGMENTS

It is a pleasure to thank G.Gouesbet, Y.Enokida, P.Pashinin for fruitful discussions; S.F.Rastopov and S.A.Viznyuk for help in experiments and calculations.

REFERENCES

[1] Ilya Prigogine *" From Being To Becoming" time and complexity in the physical sciences* W.H.Freeman and Germany, 1980.

[2] Nicolis G. and Prigogine I.*"Self-organization in non-equilibrium systems* - N.Y.;Wiley, 1977.

[3] Sukhodolsky A.T. , Light-capillary phenomena *Isvestya Akademii Nauk SSSR Seriya fizicheskaya* vol.50, N6, pp.1095-1102,1986.

[4] Sukhodolsky A.T. Self-organization of heat cycle by lightphoresis *Sov. Phys. - Lebedev Inst. Rep.*; no.8,1988.

[5] L.Landau, E.M.Lifshitz Hydrodynamics *Theoretical physics* vol 6, Moscow, Nauka, 1986 (in Russian)

[6] Gibbs J.W. *The scientific papers.* Vol.1 Dover publication, New York ,1906.

[7] Sukhodolsky A.T. "Selforganization of heatcycle by CW-laser induced mass-transfer of liquid"*Collected articles "Hydrodynamic and heatmasstransfer in space"* Novosibirsk 1988 (Russian)

[8] G.Gouesbet, J.Maquet. Examination of an analogy toward the understanding of thermal lens oscillations, *AIAA.J.Thermo- physics and Heat Transfer* , vol.3, no.1, 27,1989.

[9] C.Roze,G.Gouesbet, J.Maquet, Overstability under simultaneous surface tension, buoyancy and shear effects in a horizontal liquid layer, *American Institute of Aeronautics and Astronautics*, Washington, D.C.,1990.

[10] G.Gouesbet, J.Maquet, C.Roze, R.Darrigo, Surface tension and coupled buoyancy driven instability in a horizontal liquid layer, overstability and exchange of stability, *Physics. Fluids A*,2,6,903,1990.

[11] V.Yu.Bazhenov, M.V.Vasnetsov, M.S.Soskin, V.B.Taranenko, *Pisma v ZhETF* vol 49, 330, 1989.

[12] V.Yu.Bazhenov, M.V.Vasnetsov, M.S.Soskin, V.B.Taranenko, Dynamics of laser-induced bubble and free-surface oscillation in an absorbing liquid, *Appl.Phys.* vol b49,485-489,1989.

[13] S.A.Viznyuk, A.T.Sukhodolsky, Capillary-gravity instability induced by CW-heating of thin layers, *Zh.Tekh.Fiz.*, vol 58,1000,1988 (in Russian),

[14] Takashima M, Surface tension driven instability in a horizontal liquid layer with a deformable free surface. II. - Overstability. *J.Phys.Soc.Jpn.* vol.50, no.8, 2751, 1981.

[15] R.Anthore, P.Flament, G.Gouesbet, M.Rhazi, M.E.Weill, A note on the interaction between a laser beam and some liquid media,*Appl. Optics*, vol.21, no.1,2, 1982.

[16] G.Gouesbet and E.Lefort, Dynamical states and bifurcations of a thermal lens using spectral analysis, *Physical Review A*, vol.37,no.12,4903, 1988.

[17] G.Gouesbet. Simple model for bifurcations ranging up to chaos in thermal lens oscillations and associated phenomena. *Phys.Rev.A* vol 42, no 10, 5928-5945,1990.

[18] G.Gouesbet, P.P.Pashinin, S.F.Rastopov, A.T.Sukhodolsky, Oscillatory instability of liquid drop-lens by CW-laser heating *Soviet Physics - Lebedev Institute Reports*, N11,1991,Moscow

[19] Rastopov S.F.,Sukhodolsky A.T., Self-organization of the heat cycle due to thermal cavitation during continuous laser heating of a liquid *Sov. Phys. Dokl.* vol.32(8), pp.671-673, Aug. 1987

[20] A.V.Lykov, *Theory of Heat Conduction*, Gostekhizdat, Moscow, 1952 (in Russian)

[21] Rastopov S.F., Sukhodolsky A.T., Cluster nucleation in the process of CW-laser-induced thermocavitation *Phys.Lett.* A,149, N4, pp.229-232,1990.

[22] Rastopov S.F.,Sukhodolsky A.T., Use of the laser-induced Marangoni effect for recording of diffraction gratings *Sov. J. Quantum Electron.* vol.17 (8) pp.1091-1092, Aug. 1987

[23] Viznyuk S.A, Sukhodolsky A.T. Using a light-induced Marangoni-effect for recording dynamic diffractive gratings *Lebedev Institute Reports*, N12,1986.

[24] Viznyuk S.A.,Pashinin P.P, Rastopov S.F., Semin V.N., Sukhodolsky A.T. Laser separation of stratifying solutions *JETP Lett.* vol.45, N.12, pp.713-716, June 1987.

[25] Y.Enokida, M.Shiga, A.Suzuki, Laser-induced thermal lens spectroscopy for actinide determination in aqueous and organic solutions (*private communication*).

[26] Rastopov S.F., Sukhodolsky A.T., Sound generation by thermocavitation induced by continuous laser radiation in binary solutions *Acoustic and optical review*, Vol.4,1990.

[27] Rastopov S.F.,Sukhodolsky A.T., Transport of objects along a support by a laser beam Sov. Phys. Tech. Phys. vol.32(9), pp.1098-1099, September 1987

[28] Pashinin P.P.,Rastopov S.F.,Sukhodolsky A.T. Laser processing and cutting of transparent materials *Sov. J. Quantum Electron.* vol.17(4), pp.547-548, April 1987.

[29] Viznyuk S.A.,Sukhodolsky A.T.,Thermocapillary optical nonlinearity in stratified solutions, *Sov.Quant.Elect.* vol.18,no.1,pp.99-103,1991.

[30] Viznyuk S.A., Rastopov S.F., Sukhodolsky A.T., On thermocapillary aberrational transformation of laser beams *Optics Communication*, vol.71, N5, pp.239-243, June 1989.

[31] Viznyuk S.A., Pashinin P.P,, Prokhorov A.M., Pastopov S.F., Sukhodolsky A.T., Phase conjugation by four-wave mixing in a stratifying solution *JETP Lett.* Vol.51, N2, pp.99- 103, Jan. 1990.

[32] Rastopov S.F. Sukhodolsky A.T Light-capillary phenomena *Pis'ma Sov.Phys.-Tech.*,vol.13,No2,pp80-82,1987.

[33] G.Gouesbet, A.T.Sukhodolsky Opportunity of investigation of fluid interface instability and thermocapillary phenomena in space *Proceedings of The International Symposium on Hydromechanics and Heat/Mass Transfer in Microgravity.*

TRANSVERSE AND LONGITUDINAL WAVES AT AN AIR-LIQUID INTERFACE IN THE PRESENCE OF AN ADSORPTION BARRIER

M. Hennenberg [1], X.L. Chu [2], M.G. Velarde [3] and A. Sanfeld [1]

[1] Chimie Physique, Faculté des Sciences, U.L.B.,Campus Plaine, C.P.231, Boulevard du Triomphe, B-1060, Belgium
[2] Chemistry Department, Stanford University, Stanford, CA-94305, U.S.A.
[3] Facultad de Ciencias, U.N.E.D., Apartado 60141, Madrid, E-28071, Spain

INTRODUCTION

Sternling and Scriven [1] showed that the Marangoni elasticity induced by the isothermal transfer of a surface active solute between two adjacent immiscible bulks could result in convective motion .

We will study here the oscillatory case. There are two kinds of surface waves that can be excited or sustained : transverse (denoted TW) and longitudinal (denoted LW) waves. The deformability of the free surface leads to capillary-gravity waves. Inertia enables overshooting of the Laplace overpressure and the surface tends to return to its undeformed shape by damped oscillations unless a big enough supply of energy overcome the viscous effects. The transverse waves will only slightly depend on the mass transfer mechanism. On the contrary, longitudinal surface waves are induced by the fluctuations of surface active substances along the interface as is well known since the pioneering studies by Van den Tempel and Lucassen [2, 3]. In this case, the specific mechanism of the transfer of solute becomes essential since it is the cause leading to the oscillation.

We consider here the competition of the bulk solute diffusion in the liquid phase with a surface adsorption kinetic. Provided local equilibrium between the surface and the bulk can be assumed, there is no mathematical difference between an isothermal solute transfer in the absence of chemical reaction or heat exchange. This is true when adsorption and desorption are infinitely quick processes with respect to bulk diffusion. But the existence of a potential barrier reduces the adsorption speed, so that we study the Marangoni driven oscillations when there is no local equilibrium between the surface and the bulk [4]. A slow adsorption process should play a definite role on the longitudinal oscillations. Furthermore, we will suppose the upper phase to be a mechanically inert gaseous phase which appears only through a desorption process towards it.

In the present development, we will limit ourselves to perturbed quantities of very small amplitude so that a linear perturbation analysis should be sufficient. We will derive from the bulk balance laws and from the surface boundary conditions, the dissipative harmonic oscillator describing the LW and the TW respectively. This approach provides an intuitive explanation of the instability mechanisms and is a guideline for future nonlinear saturation studies We give the instability threshold in a region which we know from previous work [3, 5-7] to be of experimental relevance.

THE LINEAR PERTURBED PROBLEM

The sublayer kinetic and the solute transfer : the reference state

Let us first describe the motionless reference state. We consider an infinitely thick liquid layer which is a binary mixture of a solvent and a surface active solute whose steady flux crosses the deformable surface open to air. The z axis of the reference coordinate system points upwards away from the liquid phase, so that the unperturbed solute distribution is given by :

$$C = C_0 - \beta z \tag{1}$$

where C is the solute concentration in the liquid phase and where β is minus the slope of the unperturbed solute profile. Inside the sublayer region which is a region of negligible width, immediately adjacent to the free liquid-gas surface, the solute is subject to a linear adsorption desorption kinetic which prohibits local equilibrium, so that the bulk diffusion flux which reaches the sublayer, is given by

$$- D\frac{\partial C}{\partial n} = k_a C - k_d^I \Gamma \tag{2}$$

where Γ is the surface excess concentration, D is diffusion coefficient, k_a and $k^I{}_d$ being the desorption and adsorption constants respectively of the sublayer. There is no accumulation of solute along the sublayer. After reaching the interface, the solute will desorb into the gaseous inert phase, thus introducing a second desorption coefficient $k^{II}{}_d$

The balance laws and the surface boundary conditions

We are sketching the transfer mechanism in Fig. 1. We will develop here some aspects of the linear perturbated analysis in terms of dimensionless variables: we scale any length by

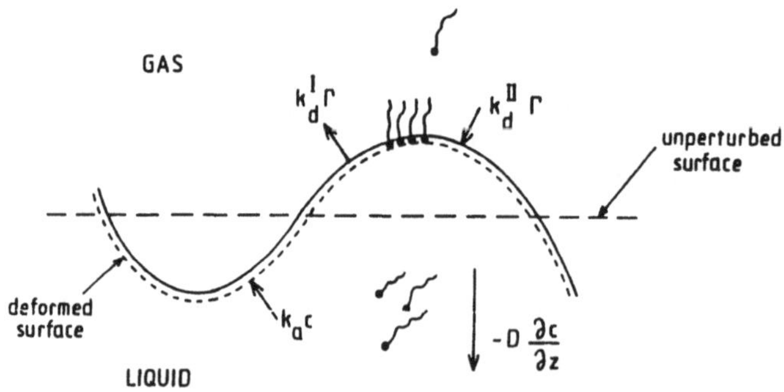

Fig. 1.The model system and the adsorption desorption kinetic.

the capillary length $l = (\sigma/\rho g)1/2$ where σ is the surface tension. The unit of time is l^2/ν where $\nu = \mu / \rho$ is the kinematic viscosity, the ratio of μ the dynamic viscosity and ρ the density. The solute concentration is scaled by $\beta\, l$.

The following equations express the momentum balance for an viscous incompressible fluid. The solute does not affect the density and obeys a convecto-diffusive equation, so that the perturbed Navier-Stokes momentum equation does not contain a term linked to the weight changes.

$$\vec{\nabla}.\vec{v} = 0 \qquad \frac{\partial \vec{v}}{\partial t} = -\nabla P + \Delta \vec{v} \qquad \frac{\partial C}{\partial t} = w + S^{-1}\Delta C \tag{3}$$

The boundary conditions along the perturbed interface $z = \xi$ read :

$$\frac{\partial \xi}{\partial t} = w \tag{4}$$

where w is the vertical component of the velocity. It expresses that the velocity of the surface is the bulk barycentric velocity at the surface : there is no cavitation.

$$\frac{H}{H_z}\frac{M}{S}\vec{\nabla_\Sigma}\Gamma + \vec{\nabla_\Sigma} w + \frac{\partial \vec{u_\Sigma}}{\partial z} = 0 \tag{5}$$

$$\frac{1}{S\,Cr}\Delta_\Sigma \Gamma + \frac{Bo}{S\,Cr}\xi - P + 2\frac{\partial w}{\partial z} = 0 \tag{6}$$

The next two equations (5) and (6) are the tangential and normal momentum balances, along the deformed surface, $u_\Sigma = (u_x,u_y)$ being the horizontal components.of the velocity, P is the pressure. Let us note that the displaced weight due to the deformation of the liquid-gas interface is taken into account in the momentum boundary equation.

$$H\,S\left[\frac{\partial \Gamma}{\partial t} + \vec{\nabla_\Sigma}\,\vec{u_\Sigma} - \frac{1}{S_\Sigma}\Delta_\Sigma \Gamma\right] + \frac{\partial C}{\partial z} + \frac{Bi\,H}{H_z}\Gamma = 0 \tag{7}$$

Equation (7) is the surface mass balance where the surface active concentration of the diffusing solute varies along the surface due to the flux from the bulk or the desorption into the gaseous phase.

$$-Z\frac{\partial C}{\partial z} = \left[C - \xi\right] - \frac{H}{H_z}\Gamma \tag{8}$$

The last Equation (8) is the mass balance in the sublayer. It shows explicitly the competition between the solute bulk diffusion and the adsorption-desorption kinetic along the deformed sublayer. One admits here that the sublayer deformed exactly like the free surface to which it is glued.

We had to introduce a certain number of well known dimensionless parameters : the Schmidt number $S = \nu/D$, the surface Schmidt number $S_\Sigma = \nu/D_\Sigma$ where D_Σ is the surface diffusion coefficient, the capillary number $Cr = \mu D/l\,\sigma$, the Bond number Bo which is equal to 1 and the Marangoni elasticity number M defined by

$$M = -\frac{\partial \sigma}{\partial \Gamma}\frac{\rho\, k_a \beta\, l^2}{k_d^I \nu D} \tag{9}$$

and the surface excess solute $H = \Gamma_0 / \beta l^2$. As we consider only overstable modes, the flux should be directed from the gas towards the liquid phase, as is know to be true from previous work [5,6] Thus β is negative as well as M and H , since for most of the usual surface active substances, the surface tension gradient is negative. Two new dimensionless quantities Bi and Z defined by

$$Bi = \frac{k_d^{II} l^2 H_z}{D} \qquad \text{and} \qquad Z = \frac{D}{k_a l} \tag{10}$$

will also appear now, where $H_z = k_a / k^I_d l$ is the Langmuir adsorption number. Bi accounts for the desorption of the solute into the gaseous phase and it shows the relative importance of the two bulks on the adsorption. Z is the ratio of the characteristic time of the diffusion and the kinetic taking place in the sublayer. The fast adsorption process corresponding to the absence of a potential barrier is obtained by putting Z = 0. Those kinetic characteristic times can be at most comparable so that Z should not be larger than 1. Indeed, for larger values of Z, we have to do with an insoluble film : bulk diffusion is so slow with respect to the surface kinetic that it does not matter anymore.

It is when both Bi = 0 and Z = 0 that we recover the case previously studied by Chu and Velarde [5] which will be guiding our analysis and where one could find a more complete description. When Bi and/or Z are positive quantities, we study the competition between bulk solute diffusion and various surface adsorption kinetics. If Bi Z is larger than 1 the desorption into the liquid is faster than adsorption from the gas. We can guess from the litterature the order of magnitude of all the other dimensionless parameters: Cr varies from 0 for an undeformed surface to 10^{-8} for ordinary deformable surfaces, S is generally around 10^3 , H is - 10^{-10} and the Langmuir adsorption number is between 10^{-5} and 10^{-7}.

DISSIPATIVE HARMONIC OSCILLATOR

We consider the usual Fourier mode decomposition along the plane parallel to the undeformed flat surface, so that each perturbed quantity f(x,y,z,t) can be expressed as :

$$\int_{-\infty}^{+\infty} d\,a \;\exp\left[\lambda\, t + i\,(a_x\, x + a_y\, y\,)\right] f_a(z)$$
$$\text{with} \quad a = \sqrt{a^2{}_x + a^2{}_y} \tag{11}$$

The real part of the time growth constant λ_a determines the stability of the system whereas its imaginary part fixes the transition mode which will be oscillatory if $\mathrm{Im}(\lambda_a) = \omega_a$ is different from zero. This standard procedure reduces the partial differential system of equations to the study of a system of ordinary differential equations depending upon one single mode $f_a(z)$, characterised by its dimensionless wavenumber a. Thus we will drop the subscript index a from now on.

We want to understand what are the different mechanism behind the oscillations, thus to derive from the previous equations (2)-(11) the oscillators, which somehow lie hidden, buried inside those expressions. It means to transform the present problem into equations of the form

$$\frac{\partial^2 g}{\partial t^2} + \psi \frac{\partial g}{\partial t} + \phi\, g = 0 \tag{12}$$

The physical meaning of that last Equation (12) is quite obvious. When Φ is positive and Ψ is equal to zero, we have an ideal oscillator whose frequency ω is the square root of Φ. One has a damped oscillator if Ψ is positive and an unstable one when Ψ is negative. Using the more classical formalism of the compatibility equation, one verify the present approach validity [5-7].

A general derivation is not interesting, nevertheless. We know from previous case [3-7] that the relevant dimensionless frequencies are in a range of finite ω much larger than a^2. This educated guess is an extension of experimental estimates. Nevertheless, the accent should be on the finite value of ω :

$$\varepsilon = \frac{a}{\sqrt{\omega}} \qquad \text{with} \qquad \varepsilon \ll 1 \tag{13}$$

This basic assumption will be used henceforwards. It is equivalent to the two following approximations, used for the marginal case :

$$\sqrt{\lambda} \approx (1+i)\sqrt{\frac{\omega}{2}} \qquad \text{and} \qquad i\frac{\partial g}{\partial t} = -\omega g \quad \forall g$$

Transverse oscillations

Let us neglect the relative surface adsorption Γ_0 so that there is no influence of the surface excess solute number. From Equation (4), one supposes the absence of cavitation and could derive the equation (12) for a harmonic oscillator which will be the deformation of the surface, i.e., $g = \xi$ Such an oscillator corresponds thus to transverse waves. Indeed, if one takes into account the normal component of the momentum balance (3) but writes it along the surface, one obtains :

$$\frac{\partial^2 \xi}{\partial t^2} + \frac{Bo + a^2}{Cr\, S}\frac{\partial \xi}{\partial t} = -2a\frac{\partial w}{\partial z} + \nu \Delta w$$

The mass solute balance (5) is reduced to :

$$[1 + Bi\ Z]\frac{\partial C}{\partial z} + Bi\left(C - \xi\right) = 0$$

so that, after some manipulations, we get an equation of the type of (12) where Ψ is a complex very lengthy expression and where :

$$\phi = \frac{Bo + a^2}{S\, Cr}a$$

This is done without any assumption, whatsoever. If one applies our basic assumption (13), one obtains the ideal TW oscillator corresponding to the neutral state as the law describing waves along an unviscid incompressible fluid, due to overpressure. Let us notice though that this justifies a posteriori the assumption of high frequency since all this domain corresponds now to marginality. We obtain the frequency and the critical Marangoni number as :

$$\frac{Bo + a^2}{S\, Cr} = \frac{\omega^2}{a} \qquad \text{and} \qquad M_{comp} = \frac{M}{(1 + Bi\, Z)} = -\frac{\sqrt{32}\,(S\omega)^{\frac{3}{2}}}{a}$$

The first equation gives the frequency of the transverse oscillations in terms of the usual Kelvin Laplace law [8], and the other relates the frequency ω to an effective surface elasticity M_{comp} which is induced by the desorption since $Bi\, Z = k^{II}_d / k^{I}_d$. The potential barrier does not affect much the transverse wave since this product will be less than 10^{-6}, for all practical purposes. This is the more true since the Langmuir adsorption number does not appear. The minimal value of the surface elasticity able to sustain a transverse oscillation is:

$$M_{comp} = \sqrt{\frac{160}{25}}\left[\frac{S}{Cr}\right]^{\frac{3}{4}}\left[6\sqrt{5}\right]^{-\frac{1}{4}} \quad \text{at} \quad a_c = \sqrt{\frac{1}{5}} \quad \text{and} \quad \omega_c = \sqrt{\frac{6}{\sqrt{125}}\frac{1}{Cr\,S}}$$

Longitudinal oscillations

We consider a very dilute solute on the surface which will be supposed undeformable. Thus, oscillations if they exist, are in the surface : we are looking for longitudinal waves. We will define the LW harmonic oscillator from the solute excess at the surface :

$$g = \frac{H}{H_z}\Gamma = \left[C + Z\frac{\partial C}{\partial z}\right]$$

We assume Cr to tend to zero and the assumption (13) to be valid and put them into the system (3)-(8). We get a single differential equation in the variable g defined above as:

$$\frac{\partial^2 g}{\partial t^2} + \psi\frac{\partial g}{\partial t} + \phi\, g = 0$$

$$\phi = \frac{1}{S\,H_z}\left[-\frac{M\,a^3}{\sqrt{2\,(S\,\omega)^3}} + \frac{\omega\,Bi}{\sqrt{2\,S\,\omega}}\right]\frac{1}{\left(\frac{1}{\sqrt{2\,S\,\omega}} + Z\right)}$$

$$\psi = \frac{1}{S\,H_z}\left[-\frac{M\,a^3}{\sqrt{2\,S^3\omega^5}} + 1 + \frac{Bi}{\left(\frac{1}{\sqrt{2\,S\,\omega}} + Z\right)} + \sqrt{\frac{S\,\omega}{2}}\,H_z + \left(\frac{a}{\omega}\right)^2\frac{M}{\sqrt{S^3}}\right]\frac{1}{\left(\frac{1}{\sqrt{2\,S\,\omega}} + Z\right)}$$

At marginality, one obtains from these two last equations :

$$-M \approx \frac{\left[1 + \omega\left(Z + \sqrt{\frac{2}{S\,\omega}}\right)H_z\,S\right]^3}{\left[\sqrt{2\,\omega}\,Z\,H_z\,S - \frac{Bi}{\sqrt{S}\;\omega} + H_z\sqrt{S}\right]^2} \tag{14}$$

and the critical wavenumber is :

$$a \approx \sqrt{2\,\omega^3}\;Z\,H_z\,S - \frac{Bi}{\sqrt{S}} + \omega\,H_z\sqrt{S} \tag{15}$$

Let us analyse in more details the last two equations which derive from the ideal longitudinal harmonic oscillator.If one has :

$$\omega < \omega_c = \frac{Bi}{H_z\,S} = \frac{k_d^{II}}{\nu\,l^2} \tag{16}$$

the wavenumber is negative. Below ω_c, there is thus no longitudinal wave. But such a wave exists for ω larger than ω_c. This gives us a lower cutoff frequency. For ω larger than ω_c

$$-M \approx \frac{\sqrt{S}}{H_z^2}$$

provided that $1 / \omega$ is greater than $2 S Z^2$. This corresponds to a constant Marangoni elasticity. For even larger frequencies one has now :

$$- M \approx \frac{1}{2 \omega (Z H_z)^2 \sqrt{S}}$$

Due to the potential barrier, the critical longitudinal wave will decrease in that range of ω but this situation is restricted to finite ω, since we will reach a cutoff frequency beyond which we are in a stable region. Indeed as we obtain now the following relation linking the wavenumber and the frequency ω :

$$a \approx \omega^{\frac{3}{2}}$$

the basic hypothesis (13) is no more valid for ω larger than 1. The potential barrier has a definite influence since the Marangoni number should reach a minimum value at a frequency near to the highest cutoff frequency.

NUMERICAL ANALYSIS

The usual procedure is to develop the system of ordinary differential equations which we deduce from the systematic use of Equation (11), applied to the system (3)-(8) The solutions are linear combinations of exponentials with unknown preexponetial parameters. As the boundary conditions (4)-(8) give rise to a linear homogeneous system in these last quantities, the determinant of their coefficients must be equal to zero for our problem to be meaningfull. This leads to a dispersion relationship, which has been studied in great details elsewhere [7] Then, one applies systematically our basic assumption (13), and varies the dimensionless frequency ω, to obtain the critical Marangoni number, all other parameters being fixed.

Cases with and without a potential barrier and desorption into the gas

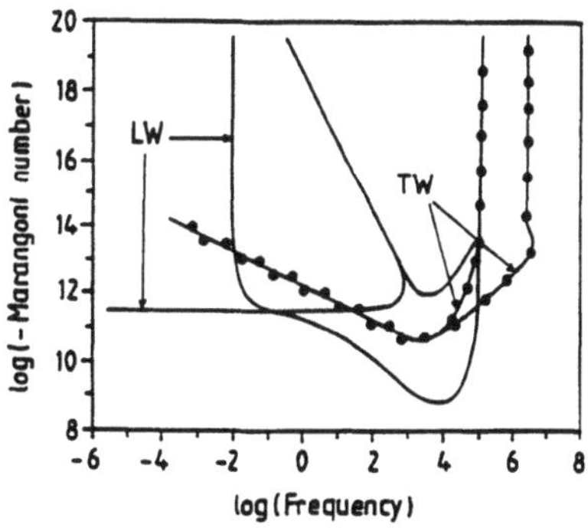

Fig. 2. Comparison between cases with and without adsorption barrier. The LW are solid lines and the TW are dotted.

We compare in Fig. 2 the classical situation where there is no adsorption barrier and no desorption into the gaseous phase (i.e. Z = Bi = 0), to a situation where there exists an adsorption barrier ($Z = 10^{-2}$) and a desorption ($Bi = 10^{-4}$) into the gas phase, all the other parameters being the same ($H = - 10^{-10}$, $H_z = 10^{-5}$, $Cr = 10^{-10}$, $S = S_\Sigma = 10^3$). Each case shows two curves : one of them ,which is denoted the TW curve, is a transverse oscillation normal to the plane of the interface, the other denoted the LW curve is a longitudinal curve, along the liquid gas interface. As one could reasonably expect from our previous analysis, one finds that both TW curves remain pretty much unaffected by the various mechanisms of the solute transfer : this becomes clear when one remembers that they reflect the normal component of the momentum bulk equation. On the contrary the LW changes shape since it is the result of a solute transfer [3-7]. We see that in all cases the very large domain of ω is stable. In a low frequency region appears a cut off frequency below which the LW does not exist for the system with a finite desorption into the gas phase. In the high frequency limit the finite barrier decreases the critical Marangoni number considerably and brings the domain of oscillations of the TW and LW nearer to one another.

Local equilibrium (Z = 0) and desorption from the gas (Bi strictly positive)

We are considering in Fig. 3, a case with no potential barrier and only a desorption to the gas phase. This leads to a cutoff frequency ω_c defined in the previous paragraph, below which no LW exists. Desorption affects the LW oscillation provided that their characteristic times are comparable. For too slow oscillations, desorption is relaxing the variation of surfactant at the surface. For too quick oscillations, the desorption process will not intervene but only the solute bulk diffusion which will give the characteristic frequency of the LW oscillation. At higher frequencies, the viscous dissipation will take over and stabilise everything so that the solute transfer mechanism becomes too weak by itself to induce any oscillation at all.

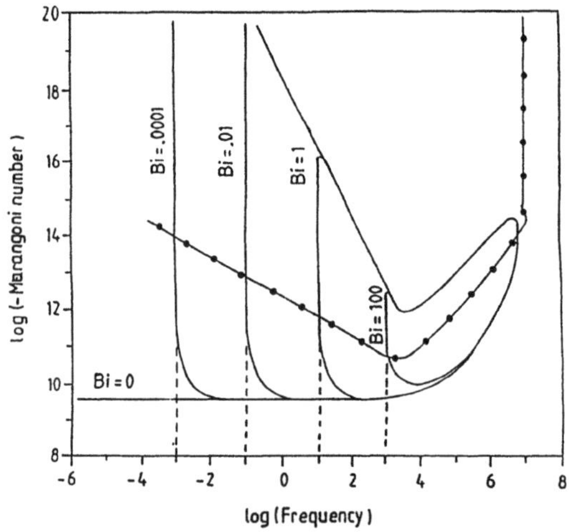

Fig. 3.The LW (solid lines) and the TW (dotted lines) for various Bi (at Z = 0)
$H = -10^{-10}$, $H_Z = 10^{-4}$, $Cr = 10^{-10}$, $S = S_\Sigma = 10^3$

The LW domain of existence for a diffusion coupled to a desorption mechanism, lies between two finite cutoff frequencies. The smallest one is directly function of the gas desorption since Bi must be a non zero quantity. The highest one depends upon the fact that everything will be stabilised by viscous effects.

As normal deformation intervene more and more, the LW behaviour does no more depend on the peculiar mechanism of solute transfer. Indeed the available momentum goes into the overpressure needed to excite the capillary gravity waves. For a large Marangoni number, the surface elasticity will oppose less resistance to a normal deformation. The competing influences between normal and tangential components of the momentum balance give the lead to the normal deformation and will be unable to sustain another excitation.

The potential barrier (Z strictly positive) and no gas desorption (Bi = 0)

To follow the adsorption desorption kinetic described by Equation (2), we will consider in Fig. 4 various values of Z and put Bi = O, thus neglecting the desorption from the gas. Increasing Z reduces the critical Marangoni number for high LW frequencies but does not influence the TW ones.

Slow adsorption enhances the instability. For Z = 0, the solute transfer is a one step process which relaxes only by volume diffusion. This curve is the enveloppe of all the other LW curves in a frequency region where the specific transfer mechanism is of no importance anymore. For Z positive, the transfer of solute becomes a two step process where first the solute reaches the sublayer in a finite time and then crosses it towards the liquid phase through diffusion. The barrier of potential results in an increase of the critical frequency. In the

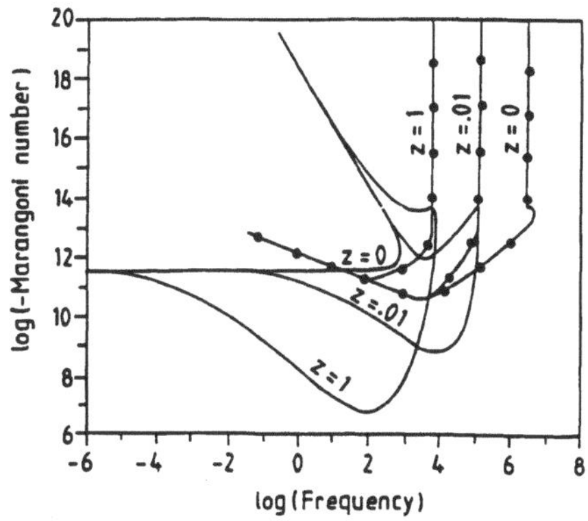

Fig. 4,The LW (solid lines) and the TW (dotted lines) for various Z, (at Bi = 0)
$H = -10^{-10}$, $H_Z = 10^{-5}$, $H_Z = 10^{-6}$, $Cr = 10^{-10}$, $S = S_\Sigma = 10^3$

adsorption dominated case, one observes that M is proportional to $-1/\omega$. Thus at low frequencies the bulk diffusion is responsible for the LW. At high frequencies, the barrier is the leading factor. Increasing Z enhances the destabilizing effect of the potential barrier, in a range of TW frequencies. Thus, less momentum goes into the normal deformation : the surface tends to be flat for only LW sustained waves.

On the contrary, decreasing the resistance to normal deformation will play on the part of the LW curve influenced by the coupling with the TW one and we observe a declining influence of the adsorption barrier.

CONCLUSION

Surfactant transfer accross a deformable interface leads to excitability. It will appear in the form of convective motions or waves according to the direction and relative magnitude of the transfer. An essential factor is the characteristic time for the various phenomena involved in the overall process. The interfacial response is not instantaneous and the various transfer mechanism do not develop on one single time scale. We used a potential barrier and a desorption into the gas phase to account for delays in the actual arrival time of the surfactant from either bulk to the surface. The barrier delinates in the liquid volume a genuine bulk where the surfactant diffuses only and a sublayer glued to the interface where transfer between the layer and the sublayer is far from equilibrium.This should lead to cooperation or to competition between the different phenomena involved.

The previous analysis helps us to uncipher the LW waves and the TW waves. We identified the underlying physico chemical mechanisms for those waves in terms of harmonic oscillators showing that high frequencies were relevant assumptions to describe neutral oscillatory cases. These TW are nearly uninfluenced by the solute transfer. For capillary gravity waves, one should introduce a Marangoni compositional elasticity depending upon the transfer mechanism but the critical frequency is again given by the classical Laplace -Kelvin law. But the LW are very much affected by the specificity of the mechanism. The desorption into the gas has a stabilizing influence at low frequency. Up to $\omega_c = Bi / H_z S$, there is no LW instability since desorption to the gas is a stabilizing factor. For ω larger than this critical value, the LW exists. At first, for a large span of frequencies, it is monitored by the bulk diffusion. Then the adsorption desorption mechanism described by Equation (2) takes over and is mainly responsible for unstability but reduces the absolute value of the critical Marangoni number needed to keep alive the LW excitation. The critical Marangoni number for both waves are reached for frequencies much nearer than in the absence of barrier. Also, as the Marangoni number is increased, the coupling of the LW and the TW becomes more and more important. It follows thus that more and more energy is pumped away from the LW excitations to the TW ones whose relative importance increases and as a consequence, the details of the solute transfer mechanism loose their importance. This will continue until the viscous damping kills away both the LW and the TW and stabilizes thus everything.

ACKNOWLEDGMENTS

This research has been sponsored by the CICYT-GRANT PB 86-651 (SPAIN) and by the "Instituts Internationaux Solvay de Physique et de Chimie" and is part of the ESA-EWG contract n° 9001/90/F/BZ.

REFERENCES

1. C.V. Sternling and L.E. Scriven, Interfacial turbulence : hydrodynamic instability and the Marangoni effect , *A.I.Ch.E. Journal* , 5 : 514 (1959).
2. E.H. Lucassen-Reynders and J. Lucassen, Properties of capillary waves, *Advan. Colloid Interface Sci.*, 2 : 347 (1969).
3. M. Van den Tempel and R.P. Van de Riet, Damping of waves by surface active materials, *J. Chem. Phys.*, 42 : 2769 (1965).
4. R. Defay and G. Petré, Dynamic surface tension, *in* "Surface and Colloid Science", 3 : 27, E. Matijevic Ed.(1971).
5. X.L. Chu and M.G. Velarde, Sustained transverse and longitudinal waves at the open surface of a liquid, *PCH Physicochemical Hydrodynamics*, 10 : 727 (1988).
6. M. Hennenberg, P.M. Bisch, M. Vignes-Adler and A. Sanfeld, Mass transfer, Marangoni effect and instability of interfacial longitudinal waves, *J. of Colloid Interface Sci.*, 74 : 495 (1980).
7. M. Hennenberg, X.L. Chu, A. Sanfeld and M.G. Velarde, Transverse and longitudinal waves at the air liquid interface in the presence of an adsorption barrier, accepted in *J. of Colloid Interface Sci.*.
8. L.D. Landau and E.M. Lifschitz, Capillarity *in* "Fluid Mechanics", Pergamon Press, Oxford (1987).

APPLICATION OF LASER-INDUCED THERMAL LENS OSCILLATION TO CONCENTRATION CONTROL IN SOLVENT EXTRACTION PROCESSES

Youichi Enokida and Atsuyuki Suzuki

The Department of Nuclear Engineering
The University of Tokyo
Tokyo 113, Japan

ABSTRACT

A new method of chemical speciation was proposed by applying the study on the hydrodynamic instability which was initiated by the interaction between a laser beam and organic solution. A concentration of a solute in organic solutions was determined by finding bifurcation point which was observed in laser-induced thermal lens phenomena. The concentration determination was successfully demonstrated for uranyl nitrate extracted with 30 % tri-*n*-butyl phosphate, and the concentration difference equivalent to 1×10^{-3} cm^{-1} was detected. In the experiment, a pulsed dye laser system was operated at a high repetition rate to excite a solute in the organic solution so as to create a local temperature gradient. When the laser beam traveled horizontally just below the free surface of the liquid, the local temperature effect led to a natural convection with a change of refractive index, which was detected by divergence of a probe beam. The time series of the signal obtained through the probe beam exhibited steadiness and periodic behavior (thermal lens oscillation), and the parameters which control the bifurcation to the thermal lens oscillation were exciting beam power, concentration of solute, and distance between the exciting beam and the free surface (d). Since the concentration of solute was found to be related with exciting beam power on the oscillation boundary by using the Lambert-Beer's law, an unknown concentration of the solute was determined by finding an oscillation boundary. A plot of the oscillation boundary on absorbance versus d plain showed that there was a region where fluctuation of the depth of exciting beam was not sensitive to the oscillation boundary. Although we used a pulsed laser system so as to excite a solute, the effect of fluctuation of beam power in time domain was not serious to detect concentration difference of 1×10^{-3} cm^{-1} by applying the procedure discussed in this study. This method can be used for general chemical speciations, and especially it is useful for the detection of excess concentrations of heavy metals, which is important for the process control of the nuclear fuel reprocessing.

Instabilities in Multiphase Flows, Edited by G. Gouesbet
and A. Berlemont, Plenum Press, New York, 1993

INTRODUCTION

The laser-induced thermal lens effect, which was first reported by Gordon et al.,[1] has been developed into a sensitive and quantitative method to determine trace-level chemical species in aqueous and organic solutions. This method, which is called laser-induce thermal lens spectroscopy (LITLS) utilizes a pulsed or continuous-wave laser system to excite chemicals selectively in the solution so as to create a local temperature gradient in the sample solutions. This thermal effect leads to a change of refractive index similar to a thin concave lens which can be detected by divergence of the exciting beam itself or a probe beam. In the past decade, LITLS has been applied successfully for the speciation of the low concentration of actinides.[2-5]

Another well-known interaction between a laser beam and liquid media, which was first reported in 1982,[6] is thermal lens oscillations. Gouesbet et al.[6-20] have intensively and systematically investigated the hydrodynamic instability caused by the thermal lens phenomena from the standpoint of chaos in dissipative dynamical systems. They found several dynamic states and their bifurcations in the experiments[12] as well as theoretical analysis using modern theory of dynamic systems.

While the normal LITLS utilizes the thermal lens phenomena without convective motion (which is illustrated in Figure 1), this paper proposes to utilize the thermal lens phenomena with convective motion for spectroscopy of chemical species, and discusses its importance in the process control in solvent extraction processes for nuclear materials. The fundamental experiments were performed using the laser-based system in order to demonstrate effectiveness of this method for concentration determination of nuclear materials such as uranium and plutonium in 30 % tri-*n*-butyl phosphate (TBP) diluted with *n*-dodecane.

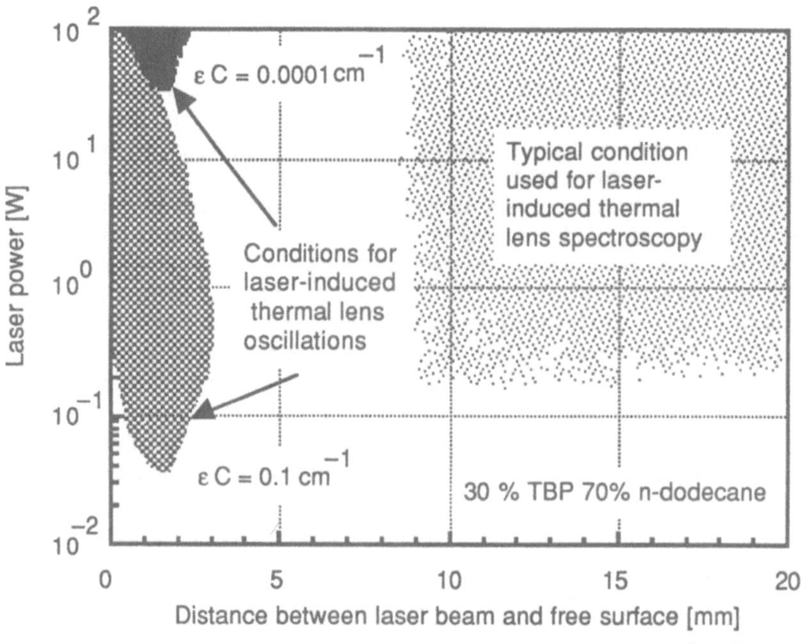

Figure 1. Schematic illustration of dynamical state of laser-induced thermal lens phenomena.

EXPERIMENTS

Experimental Setup

Figure 2 shows the schematic diagram of the experimental system for concentration determination by using thermal lens phenomena. This system used a double-beam configuration, where an unknown solution was placed in the reference configuration relative to the blank cell, so as to reduce the background fluctuation that is brought about by the various non-ideal aspects of the optical and electronic devices.

A Lambda Physik dye laser (Model FL3002E) pumped by a XeCl excimer laser (Model LPX110) was used as a pulsed selective excitation source for a solute in sample solutions. Stilbene 3 (Lambda Physik) in methanol was used as a laser medium of the dye laser for the wavelength range of 412 to 443 nm, and the band width of wavelength was 0.04 cm^{-1}. The maximum pulse energy was approximately 3 mJ/pulse, and the pulse duration was approximately 20 ns. The intensity and pulse repetition rate of the exciting beam was adjusted by the remote controller unit, and the maximum repetition rate was 200 Hz. The laser beam emitted from the dye laser was partially reflected by a beam splitter, and its intensity was measured by a pyroelectric energy detector (Gentec, Model ED 100A). The detection limit for the intensity of the exciting beam was 0.005 mJ with a resolution of 0.001 mJ. The output signal was recorded by a data recorder (TEAC, Model DR-F1). The standard deviation of the pulse intensity was estimated to be approximately 0.15 mJ/pulse for the maximum output. The

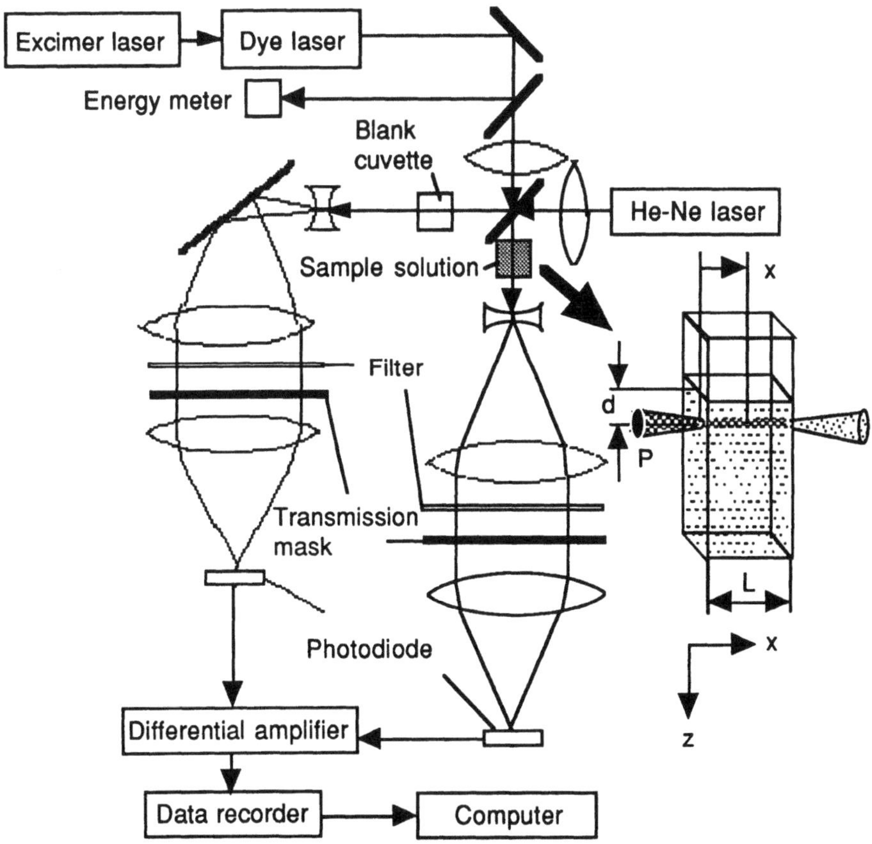

Figure 2. Experimental set-up.

laser system was operated with a high repetition rate (120 Hz) to give the sample solution a quasi-continuous energy input. Rodamine 6G (Lambda Physik) was used as a laser medium of the dye laser for the wavelength range of 569 to 608 nm for the experiments using Oil Red ($C_{24}H_{20}N_4O$), which was used a substitute for uranyl nitrate and provided by Tokyo Chemical Industries.

A probe beam was provided by a He-Ne laser (Japan Laser, Model JLH-P50) which had an average power of 5 mW. If a temperature gradient is formed by the deposited heat, the divergence of the probe beam changes according to the gradient of refractive index that depends on local temperature. A beam splitter was used not only to distribute the exciting and probe beams into the double-beam configuration but also to align the two beams. The exciting and probe beams completely overlapped at the sample cell focusing on the sample through an achromatic lens having a 200-mm focal length. The position of the exciting beam waist was placed approximately at the distance $z=\sqrt{3}\,\pi\omega_p^2/\lambda_e$ beyond the beam waist of probe beam, because this position gives the greatest relative change in the far field beam size by the thermal lens.[21-25] In the actual experimental procedure, position of the exciting beam waist was adjusted so as to maximize the thermal lens signal. After the heat generated by the sample absorption of the exciting beam changed the divergence of the probe beam, the probe beam was propagated beyond the cell and passed through a radially symmetric, parabolic transmission mask (5 cm in diameter), which is originally used for the low concentration measurement (less than 1×10^{-4} M uranyl nitrate) by thermal lens spectroscopy.[26] Since the moving area of probe beam brought about by the thermal lens oscillation is larger than that of normal thermal lens spectroscopy, the position of transmission mask was adjusted to cover the moving of the probe beam center. The degree of the divergence was optically focused and detected by a Si photodiode (Hamamatsu, Model S1722-01). All optical elements were fixed on a stable optical support, and shielded in a black box.

Data Acquisition and Analysis

The difference signal of two photodiodes was electronically amplified (TEAC, Model SA-16U) and stored as digital data by a data recorder (TEAC, Model DR-F1) through an analog-digital converter. The electronic circuits were carefully shielded against electromagnetic noises.

The previous study[27] has shown that the decay curve of the thermal lens signal brought about by a single shot of exciting laser is proportional to $(1 + 2\,t\,/\,\tau)^{-2}$ in the experimental configuration used in the current study, where t and τ are time and the time constant of thermal lens signal, respectively. The τ was measured to be approximately 0.08 s for the 30 % TBP solution. Since the time interval between the exiting laser pulses was about 0.1 τ, the decrease of the thermal lens signal during this interval is not negligible. Therefore, the sampling frequency of the data recorder was set to be 20 Hz to avoid the effect of the illumination frequency of the exciting beam. The total recording time for each experiment was 51.2 s for 1024 sampling points. The starting signal just after the illumination of the exciting beam (in the first 1024-sampling period) showed the transient behavior probably arising from the initiation of hydrodynamic motion. The measurement was started 3 min after the initiation of the exciting beam illumination to avoid collecting the early transient signals. Since the stored data in the data recorder gives the information on the thermal lens oscillation, the data were fed to a fast Fourier transform (FFT) analyzing program on a personal computer (NEC, Model PC9801VM2).

Sample Solutions

The organic samples were prepared by diluting 0.5 M stock solution of uranyl nitrate in 30 % TBP (70% *n*-dodecane) equilibriated with 2.7 M HNO_3. The stock solution was originally prepared by a batch-wise solvent extraction between 30 % TBP and aqueous nitric acid solution of uranyl nitrate. The chemical form of uranyl nitrate in the organic solution was expected to be $UO_2(NO_3)_2{\cdot}2TBP$.[28] The nitric acid concentrations of the organic samples were approximately 0.4 M, as determined by titration.

Figure 3 shows a portion of absorption spectrum of uranium nitrate extracted by 30 % TBP observed by absorption spectroscopy using a spectrophotometer (Shimaz, UV-3101) in the wavelength region of 412 to 443 nm. This spectrum illustrates the fine structure of the principal absorption band. The positions of these peaks are similar to those observed in the aqueous nitric acid solution, and can be interpreted in terms of transition between several energy band series, as reviewed in Ref. 29. The maximum absorption coefficient is given at 414 nm (molar absorption coefficient (ε) is 11.2 cm^{-1} M^{-1}). Laser intensity was, however, strongest at 420 to 440 nm, such that all experiments were performed at 425 nm. The molar absorption coefficient at this wavelength is 10.1 cm–1 M–1 as a result of enhancement effect caused by organic solvent, because 6.5 cm^{-1} M^{-1} is observed in the aqueous solution. This enhancement effect in organic solution leads to the probability of the lower detection limit for concentration measurement.

The sample solution was sealed in a spectroscopy quartz cell in the experiment of concentration determination, and vertical location was adjusted with an accuracy equal to 0.01 mm. The effective width and height of the cell were 10 mm and 45 mm, respectively, and the beam path length (l) was 10 mm. The outer wall of the cell was cooled by air-ventillation. All measurements were performed at room temperature (21±1 °C).

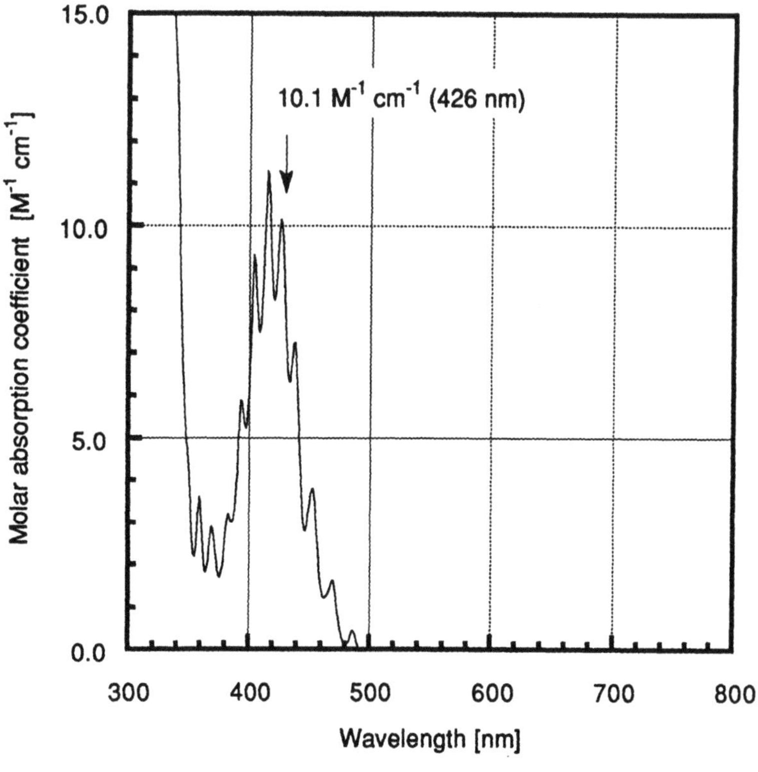

Figure 3. Absorption spectrum of uranyl nitrate complexed with TBP showing fine structure.

THERMAL LENS OSCILLATIONS IN 30 % TBP SOLUTION

Two dynamic states, which corresponded to steady Rayleigh convection and periodic oscillation, were observed for the organic solutions containing uranyl nitrate in the current study. Figure 4 illustrates typical signals in the time domain obtained in the experiments in which the pulsed dye laser was operated at the repetition rate of approximately 120 Hz. In these results, the difference of signal behavior is clearly arising from the change of d.

Gouesbet et al. clarified the phenomenological mechanisms that cause the oscillation as follows.[12] When a exciting laser beam is passed through an absorbing sample, heat is generated and deposited into the sample by beam absorption followed by non-radiative transition. A thermal plume, which is generated by continuous absorption of beam in the sample cell, is driven to the free surface in a convection-dominated regime. When the thermal plume reaches the free surface, it is disrupted by the Marangoni effect that drives the hot liquid toward the lateral boundaries of the cell. The liquid in the cell is then moved, driving a cold liquid at the exciting beam location, thus ending the process cycle. When a thermal plume is generated continuously, a new cycle is ready to begin. Oscillatory behavior would be linked to a continuous disruption and creation of thermal plumes. Two time scales are associated with the above scenario; Rayleigh time corresponding to the ascending of the thermal plume, and Marangoni time corresponding to its disruption at the free surface. The Marangoni time is usually much smaller than the Rayleigh time. It was also pointed out that exciting beam energy (P) and a distance (d) between the laser beam position and the free surface are control parameters for the different dynamic states.

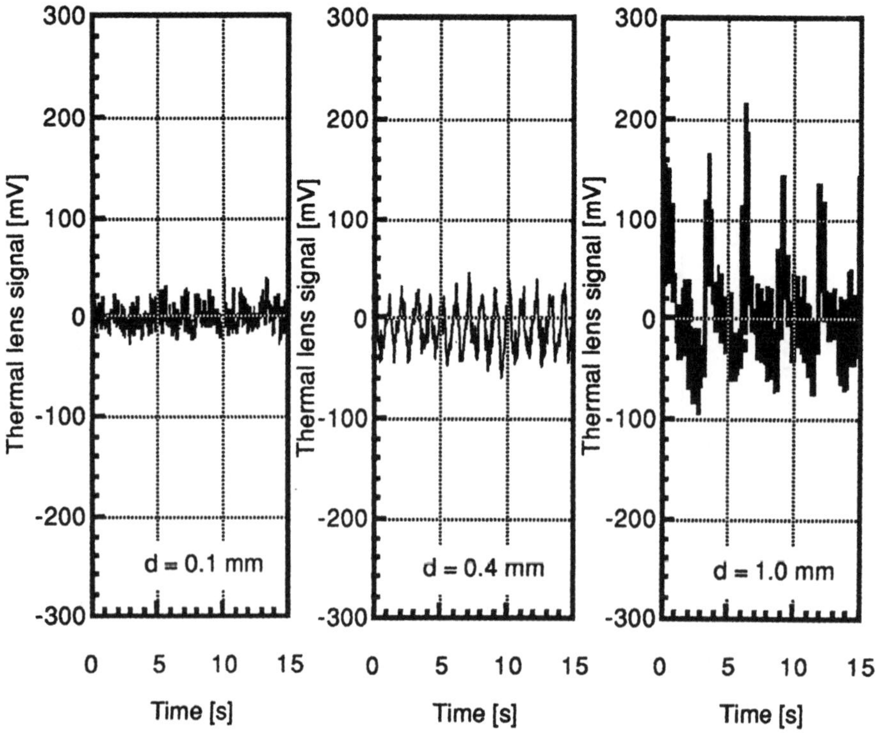

Figure 4. Thermal lens signal in the time domain (P = 0.1 W, C = 0.04 M).

Under the dynamic states observed in dissipative systems, the minor noise brought by various causes can be suppressed. Therefore, spectrum analysis on the time series signal gives us information about the major oscillation dominating the system. This feature of the dissipative system is expected to eliminate the disturbance from the noise for concentration determination.

The results of Fast Fourier Transfer (FFT) analysis are shown in Figure 5 for three conditions corresponding to those in Figure 4. The two states, steady and periodic convection states, are well distinguished by FFT analysis, and it is suggested that the bifurcation point between the two states may be determined by the plot of peak height of spectrum density versus d. The steep decrease of peak height of the spectrum density illustrated in Figure 6(a) shows feasibility of FFT analysis to determine the bifurcation point. The depth of the bifurcation point for C = 0.04 M is calculated to be 0.284 mm by linear extrapolation based on the two nearest points under P = 0.1 W irradiation.

The exciting laser power was one of the control parameters (as well as d) for the bifurcation of laser-induced thermal lens phenomena in the previous study.[12] Since we are aiming at concentration determination of a solute in organic solution, a slight modification is useful to take account of the absorption effect by the solute. When the Lambert-Beer's law holds, the energy absorbed in the organic solution containing the solute at x (position in the axial direction of laser beam illustrated in Figure 2) is proportional to $\varepsilon\, C\, P\, 10^{-\varepsilon C x}$, where C is molar concentration of the solute in the organic solution.

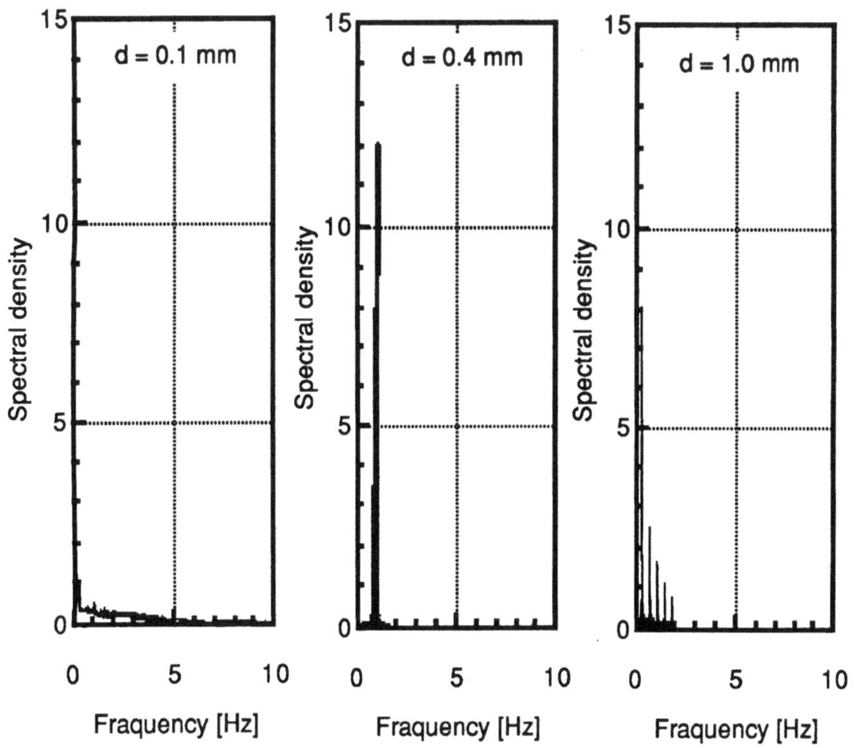

Figure 5. Results of FFT analysis on time series data.

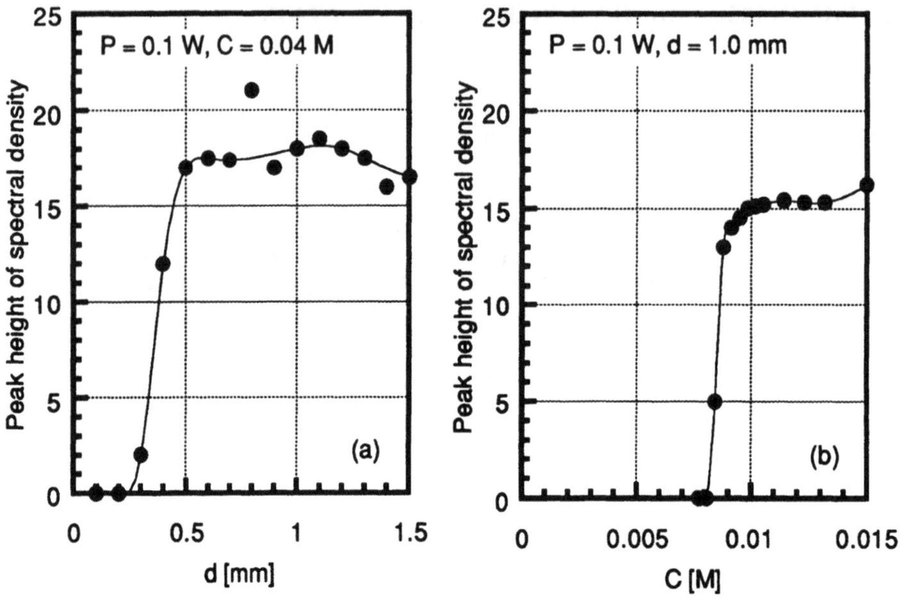

Figure 6. Dependency of the peak height of FFT power spectrum on a distance between laser beam and free surface (d), and solute concentration (C). (P = 0.1 W, C = 0.7 M).

By applying this relationship, we can introduce another control parameter C as well as P. The boundaries between the different dynamic states under the constant intensity of the exciting beam can be related with the solute concentration. The bifurcation points for various solute concentrations were also determined by the FFT spectrum analysis. In Figure 6(b), steep increase of peak height of power density by FFT analysis was observed and it was feasible to determine the boundary concentration of uranyl nitrate under constant power and depth of the exciting beam.

When the position of the exciting beam was fixed at d = 1 mm, the concentrations of uranyl nitrate on the boundary between the two states were experimentally determined for several values of exciting beam power. The concentration of uranyl nitrate was adjusted by precise dilution of the stock solution, and the error in uranyl nitrate concentration was estimated to be ± 1×10^{-4} M. Figure 7(a) illustrates such concentrations on the boundary. Since the exciting laser beam was focused at the center of the light path in the sample cell, the local temperature rise was expected to be highest at x = L/2 in the axial direction of the laser beam. The absorption rates of laser energy at x = L/2 by assuming the Lambert-Beer's law were calculated for the boundary concentrations, and plotted against 1/P as in Figure 7(b). The linear relation between 1/P versus absorption energy rate was obtained over the concentration range of 0.004 to 0.012 M uranyl nitrate in 30 % TBP. This relation confirms that exciting beam absorption in the sample solution obeys the Lambert-Beer's law for the bifurcation points. No experiments were performed at the low concentration region less than 0.004 M U(VI) because of the insufficient intensity of the exciting beam. Further study is necessary to apply this method to the determination of the lower concentration by using an exciting beam of high intensity or some other assisting method. Since it has been found that the electrically heated hot wire can induce the similar thermal effect as laser-induced thermal lens oscillation in the organic solution[14], we can use the hot wire system as a heat source to set the system just under the

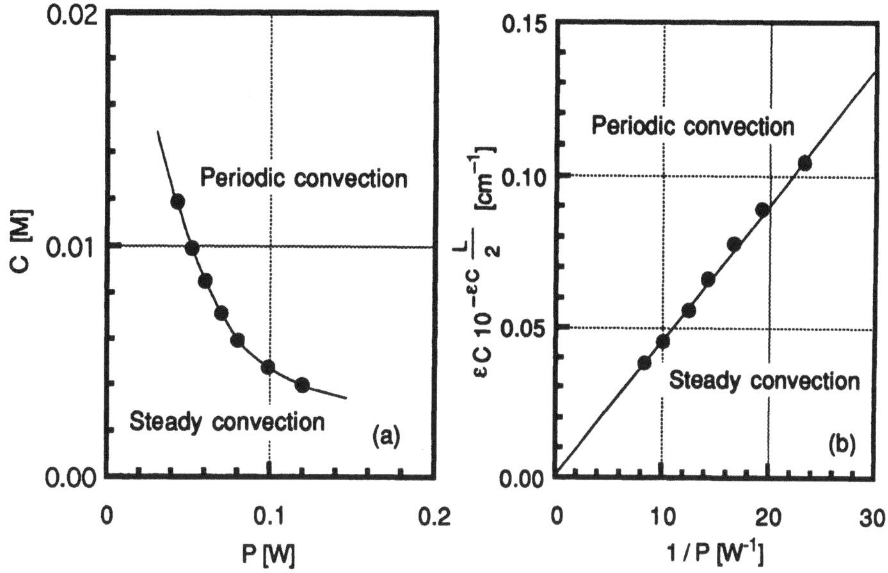

Figure 7. The boundary of different dynamic states in the plain (C, P) and Lambert-Beer's law on the boundary.

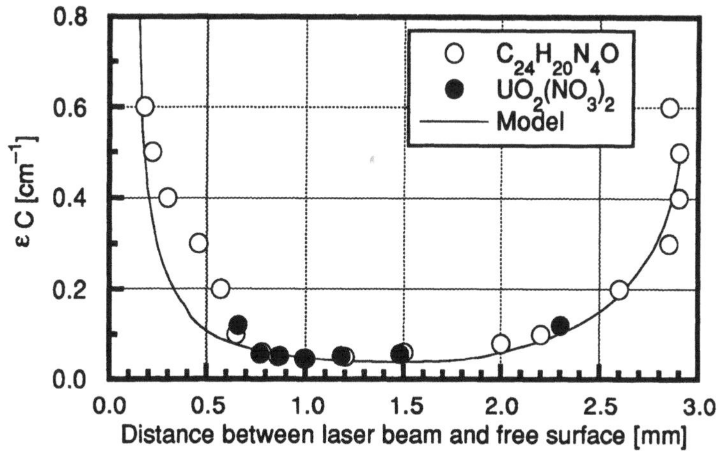

Figure 8. The chart for boundary of different dynamical states (P = 0.1 W).

bifurcation point. By using this hybrid system, we can expect spectral selectivity by tunable laser even without the high intensity of exciting laser beam.

The parameter plane (ε C, d) was systematically investigated for the bifurcation condition between states of steady convection and periodic behavior in the 30 % TBP solution that contained 0.4 M HNO_3 as well as beam absorbing solute. Most of the points plotted in Figure 8 were determined by using Oil Red as a substitute for uranyl nitrate with 0.1 W exposure of the exciting beam. The data obtained at d = 1.0 mm for the two different solutes were identical, and the concentration determination method proposed in this study can be used not only for uranyl nitrate in 30 % TBP but also for other liquid system where thermal lens oscillation occurs. The boundary curve in Figure 8 also shows that the sensitivity of position was lowest near d = 1 mm. Since the position control of the exciting beam with high accuracy seems to be relatively difficult for plant application, it is reasonable to measure concentration on the selected region where d is not sensitive.

A theoretical study[19] has developed a simple model to explain the thermal lens oscillation behavior. We have modified the model by taking account of Lambert-Beer 's law, and found that the modified model agrees well with the experimental observation as illustrated in Figure 8. The zone of thermal lens oscillation illustrated in Figure 1 was calculated by using this model.

The accuracy of the concentration determination based on the thermal lens oscillation was checked by determination of the slight change of the concentration of uranyl nitrate in 30 % TBP. The exciting beam of approximately 0.05 W was illuminated at the position of d=1 mm for the sample solution, and a concentration difference equivalent to 1×10^{-3} cm^{-1} was detected by using 1.19 and 1.20×10^{-2} M uranyl nitrate solutions. By normal absorption spectroscopy, such a slight concentration difference is difficult to distinguish because fluctuations arising from light source, optics and or electronic circuits directly affect the accuracy of concentration measurements. For example, the error of concentration determination for 1×10^{-2} M is estimated to be 6 % of the concentration by assuming 1 % fluctuation of light source for normal absorption spectroscopy. Since the fluctuation of the exciting beam used in the current study was at least 5 % as a standard deviation, the non-ideal effect brought about by light source fluctuation was partially suppressed as expected for the dynamics of nonlinear system.

CONCENTRATION CONTROL IN SOLVENT EXTRACTION PROCESSES

The experimental results described above are still preliminary in nature, but the techniques discussed in this article can be applied to determine concentration of actinides which are found in the reprocessing of nuclear fuels by the PUREX[28] process. For example, our method of concentration determination is directly useful for detection of the excess concentrations of heavy metals which must be detected under upset process conditions.[30] Recently, reduction in the number of solvent extraction cycles in the PUREX process is proposed[31] not only as a mean to

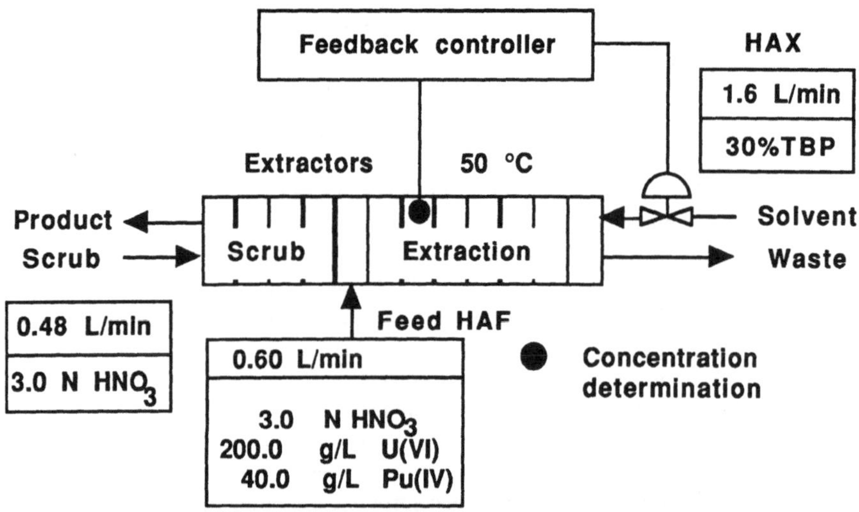

Figure 9. Typical chemical flow sheet of nuclear fuel reprocessing by the PUREX process.

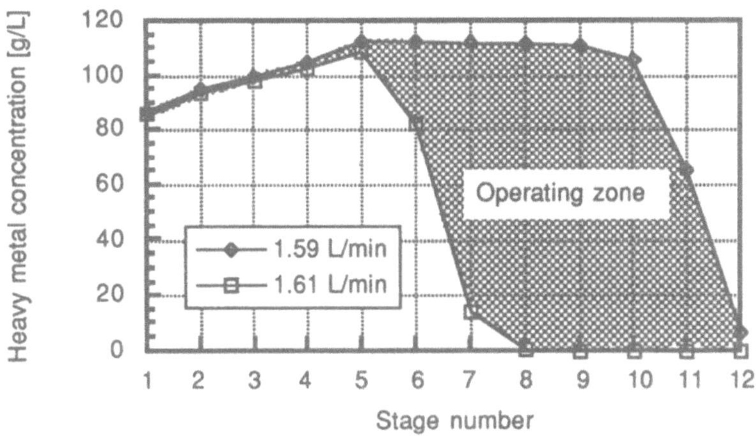

Figure 10. Effect of solvent flow rate control on heavy-metal loading of 30 % TBP.

reduce the total number of hardware units in a reprocessing facility to improve the nuclear safety, but also to minimize the volume of waste produced by the plant. However, both high decontamination factors (DF) and high product yield of heavy metals from fission products are required in the first solvent extraction process. High heavy metal loading of the organic solvent (30% TBP) is effective in achieving a high DF. Based on calculations made using the SOLE computer program[32] for a typical chemical flow sheet shown in Figure 9, the sensitivity of solvent flow rate upon product yield of heavy metal was high as illustrated in Figure 10. This result means that the organic solvent flow rate must be controlled within a very narrow range of values to maintain the improved decontamination level, while avoiding loss of heavy metal to the waste stream. This process control is feasible by applying a feedback control when we place a concentration monitor at 7th stage, for which the method discussed in this article is promising.

Acknowledgements

The authors acknowledge helpful comments by Prof. G. Gouesbet of Institute National des Sciences Appliquees de Rouen in France during the preparation of this paper. They also thank Mr. T. Kakudo for his contribution to some experiments in this work.

REFERENCES

1. J. P. Gordon, R. C. Leite, R. S. Moore, S. P. Porte, and J. R. Whinerry, "Long-Transient Effect in Lasers with Inserted Liquid Samples," App. Phys., 36: 3 (1965).
2. J. V. Beitz and J. P. Hessler , "Oxidation State Specific Detection of Transuranic Ions in Solution", Nucl. Technol., 51: 169 (1980).
3. T. Berthoud, P. Mauchien, N. Omenetto, and G. Rossi., "Determination of Low Levels of Uranium(VI) in Water Solutions by Means of the Laser-Induced Thermal Lensing Effect", Anal. Chim. Acta, 153: 265 (1983).
4. J. I. Kim, Chemical behavior of transuranic elements in natural aqueous systems, in: "Handbook on the Physics and Chemistry of the Actinides, Volume 4," North-Holland Publishing, Amsterdam (1986).
5. Y. Enokida, M. Shiga, and A. Suzuki, "Determination of U(VI) over Wide Concentration Ranges in Aqueous Nitric Acid and 30 % Tri-*n*-Butyl Phosphate by Laser-Induced Thermal Lens Spectroscopy," to be published in Radiochim. Acta.
6. R. Anthore, P. Flament, G. Gouesbet, M. Rhazi, and M. E. Weill, "The Interaction between a Laser Beam and some Liquid Media," Appl. Opt., 21: 2 (1982).
7. M. E. Weill, M. Rhazi, and G. Gouesbet, "Oscillations d'une surface libre chaufée sous l'interface à l'aide fil," Acad. Sci., 294: 567 (1982).

8. G. Gouesbet, M. Rhazi, and M. E. Weill, "New Heartbeat Phenomenon and Concept of 2-D Optical Turbulence", Appl. Opt., 22: 304 (1983).
9. G. Gouesbet, M. E. Weill, "Complexities and Entropies of Periodic Series with Application to the Transition to Turbulence in the Logistic Map," Phys. Rev. A, 30: 1442 (1984).
10. G. Gouesbet, M. E. Weill, and E. Lefort, "Convective and Free Surface Instabilities Provoked by Heating below an Interface", AIAA Journal, 24: 1324 (1986).
11. G. Gouesbet and E. Lefort, "Thermal Lens Oscillations at Low Laser Powers," App. Opt., 26: 2940 (1987).
12. G. Gouesbet and E. Lefort, "Dynamical States and Bifurcations of a Thermal Lens Using Spectral Analysis", Phys. Rev. A, 37: 4903 (1988).
13. G. Gouesbet, "Une revue sur les travaux concernant une instabilité interfaciale menant au coeur d'anneaux," Ann. Phys. 13: 103 (1988).
14. G. Gouesbet and J. Maquet, "Examination of an Analogy toward the Understanding of Thermal Lens Oscillations," AIAA J. Thermophys. Heat Trans., 3: 27 (1989).
15. G. Gouesbet and J. Maquet, "A Simple Model to Understand Overstability in Thermal Lensing," Int. Comm. Heat Mass Trans., 16: 133 (1989).
16. G. Gouesbet, "Oscillatory Instabilities Produced by Laser or Hot-Wire Heating Below an Interface," JSME Int. J., Series II, 32: 301 (1989).
17. G. Gouesbet ,"Dynamical States and Bifurcations in a New Thermo-Dynamic Nonlinear System: Optical Heartbeats and Associated Phenomena, Entoropie, 153/154: 47 (1990).
18. G. Gouesbet, J. Maquet, C. Roze, and R. Darrigo, "Surface-Tension and Coupled Buoyancy-Driven Instability in a Horizontal Liquid Layer. Overstability and Exchange of Stability," Phys. Fluids, 2: 903 (1990).
19. G. Gouesbet ,"Simple Model for Bifurcations Ranging up to Chaos in Thermal Lens Oscillations and Associate Phenomena," Phys. Rev. A, 42: 5928 (1990).
20. G. Gouesbet, "Reconstruction of the Vector Fields of Continuous Dynamical Systems from Numerical Scalar Time Series," Phys. Rev. A, 43: 5321 (1991).
21. S. J. Sheldon, L. V. Knight, and J. M. Thore, "Laser-Induced Thermal Lens Effect: A New Theoretical Model", App. Opt., 21: 1663 (1982).
22. C. A. Carter and J. M. Harris, "Comparison of Models Describing the Thermal Lens Effect", App. Opt., 23: 476 (1984).
23. R. O. Carman and P. L. Kelley, "New Thermooptical Measurement Method and a Comparison with Other Methods", App. Phys. Lett., 12: 241 (1968).
24. C. Hu and J. R. Whinnery, "New Thermooptical Measurement Method and a Comparison with Other Methods", App. Opt., 12: 72 (1973).
25. K. Mori, T. Imasaka, and N. Ishibashil, " Thermal Lens Spectrophotometry Based on Pulsed Laser Excitation", Analytical Chemistry, 54: 2034 (1982).
26. K. L. Jansen and J. M. Harris, "Thermal Lens Measurements by Optical Computation of the Laser Beam Spot Size", Analytical Chemistry, 57: 1698 (1985).
27. Y. Enokida, M. Shiga, and A. Suzuki, "Laser-Induced Thermal Lens Spectroscopy for Actinide determination in Aqueous and Organic Solutions", Proceedings of the Third International Conference on Nuclear Fuel reprocessing and Waste Management, 2: 875 (1991).
28. M. Benedict, T. H. Pigford, and H. W. Levi, "Nuclear Chemical Engineering," MacGraw-Hill, New York (1981).
29. J. J. Katz, G. T. Seaborg, and L. R. Morss, "Actinide Elements, Volume 1," Chapman and Hall, New York (1986).
30. K. Ebert and R. V. Ammon, "Safety of the Nuclear Fuel Cycle, Volume 4," VCH, Weinheim (1989).
31. H. Schmieder and G. Petrich, "IMPUREX: A Concept for an IMproved PUREX Process, Radiochim. Acta, 48: 181 (1989).
32. Y. Enokida, R. Ohmori, T. Usami, and A. Suzuki, "The SOLE Computer Program for PUREX Solvent Extrcation Processes with Idealized Mixer-Settlers, in preparation.

THEORETICAL AND EXPERIMENTAL INVESTIGATION OF PERIODIC INTERFACIAL WAVES BETWEEN TWO VISCOUS FLUID LAYERS

François Charru and Jean Fabre

Institut de Mécanique des Fluides de Toulouse
Unité Associée au CNRS
Avenue du Professeur Camille Soula
31400, Toulouse, France

INTRODUCTION

The stability of the interface between two fluids is addressed in various two-phase flow situations, related namely to the flow pattern transitions, the transport of highly viscous fluids by film lubrication or emulsification, the determination of bubble and droplet sizes in turbulent flow, the formation of spray... At low Reynolds number the contrast of the two viscosities at the interface may play a complex role in competition with surface tension and gravity. Among the different flow configurations which may exist, the case of Couette-Poiseuille flow in channels has received a special attention in the recent years. Nevertheless few experimental results are available at the present time. The theoretical formulation of the problem is similar for both Couette and Poiseuille flow, while their experimental design leads to very different technical difficulties.

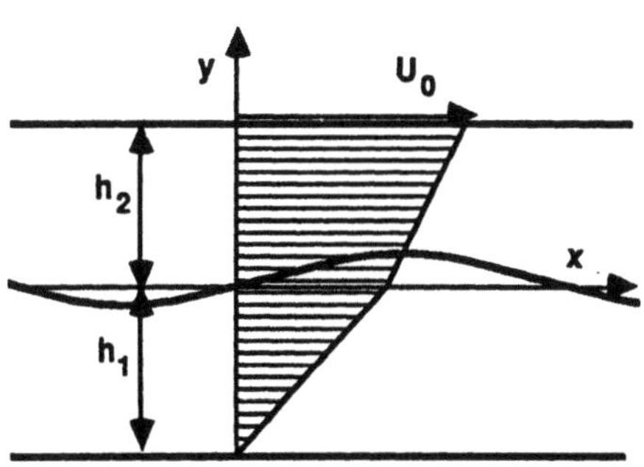

Figure 1. Couette flow of two fluid layers.

Instabilities in Multiphase Flows, Edited by G. Gouesbet
and A. Berlemont, Plenum Press, New York, 1993

In this study we focus on the stability of Couette flow illustrated in figure 1. For some fluid properties and under some flow conditions, disturbances $\eta(x,t)$ may appear at the interface. The behaviour of these perturbations is analyzed in the present paper with both theoretical and experimental points of view.

EXPERIMENTAL FRAMEWORK

Very few experimental results on the stability of the interface between two fluids flowing in a channel have been reported in the literature. To our knowledge, only some experiments of Poiseuille flow in rectangular channels have been carried out (Charles and Lilleleht, 1965; Kao and Park, 1972), the goal being to investigate the space evolution of the waves. However the instability generally occurs near the inlet where the flow is never developed.

Unfortunately, there does not exist a simple method to realize an experiment for studying the stability of a Couette or Poiseuille plane flow of two fluid layers. The choice which was retained here, privileges the absence of flow development. The flow geometry pictured in figure 2, consists of an annulus of rectangular cross section set in horizontal position, in which two fluids are stratified. It can be viewed as a curved rectangular channel with neither inlet nor outlet, satisfying the aforementioned requirement. Couette flow can be studied by rotating the upper wall around the axis. In the present study, the channel height is h=20 mm, its width L=40 mm and the annulus mean diameter D=400 mm. The maximum velocity of the upper plate is about U_0=0.5 m/s corresponding to an angular velocity ω_0=2.5 rd/s. Thus the ratio of the centrifugal force to the gravity force $U_0\omega_0/g$ remains small enough since it is nearly equal to 0.1: the effect of curvature might be disregarded. More critical is the aspect ratio L/h=2: this leads to 3D effects which must be taken into account in the interpretation of the results. It must be kept in mind that in such a geometry, only the waves such that $\lambda=\pi D/n$ can exist, where n is the number of waves around the periphery.

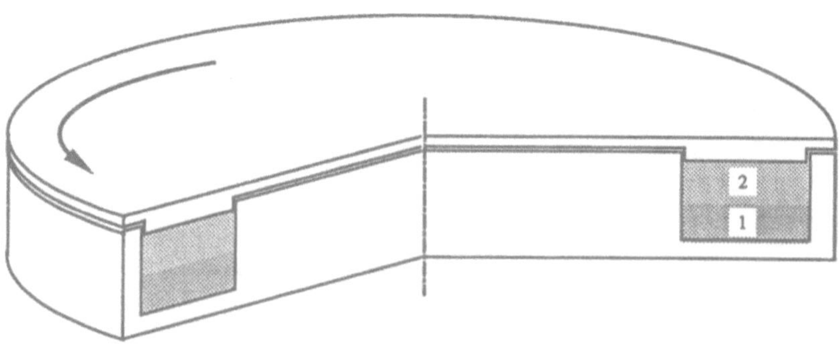

Figure 2. Principle of the Couette flow experiment in curve channel.

Both the channel and the rotating cover plate was lathed in plexiglass. The upper plate is entrained by a servo-motor through a speed reducer. The angular velocity of the shaft can be continuously varied up to 200 revolutions per second, with the mean of an electronic control system driven from the analog output of a microcomputer.

The waves appearing at the interface were identified by using two conductive probes consisting each of two wires of 0.3 mm diameter distant of 5 mm, fixed vertically at the bottom of the channel and emerging from the conductive lower layer. The two probes were distant from each other of an azimuthal angle of 88 degrees. The probe system is supplied with a 10 kHz oscillator, the differential output voltage being demodulated to provide a D.C. signal proportional to the interface height. The probe signals were recorded on the hard disk of a microcomputer Mac II, through a Mac Adios acquisition system. The signal processing, involving power spectra and cross-correlations have been made with the Mathematica software.

It is indeed expected that there exists a critical angular velocity above which interface instability occurs: thus the velocity is increased in order to determine this velocity as well as the wave characteristics. To avoid however the influence of the flow history which affects the reproducibility of the results, a computerized procedure was used: the angular velocity of the upper plate was increased by steps consisting of a transient time of about 1 mn with increasing (or decreasing) velocity followed by about 10 mn during which the velocity is kept constant and probe signals are recorded.

Two immiscible fluids have been used. The heavier is a mixture of 42-58% in mass of KCl aqueous solution and glycerin, having a density ρ_1=1.152 kg/m^3; the lighter is a mineral oil with a density ρ_2=0.845 kg/m^3. Their viscosity are μ_1=12 10^{-3} Po and μ_2=31 10^{-3} Po at 20°C. In order to avoid local concentration gradients leading to undesirable interfacial behaviour, the fluids are prepared one month before so that thermodynamic equilibrium can be obtained.

THEORETICAL FRAMEWORK

Up to now, the stability of two-phase Couette flow has only received a fragmentary theoretical answer, since this nonlinear problem involves a complex competition between gravity, surface tension, inertia and viscosity of each phase. The general solution must depend on six dimensionless parameters : the ratios d, m, r of the film thicknesses, the viscosities, and the densities, plus a Reynolds number, a Froude number and a Weber number, defined as follows

$$d=\frac{h_2}{h_1} \quad m=\frac{\mu_2}{\mu_1} \quad r=\frac{\rho_2}{\rho_1} \quad Re=\frac{\rho_1 h_1 U_I}{\mu_1} \quad Fr=\frac{U_I^2}{(1-r)gh_1} \quad We=\frac{\rho_1 h_1 U_I^2}{\sigma} \qquad (1)$$

where U_I is the velocity of the undisturbed interface, g the gravity and σ the surface tension.

The linear stability analysis of two-phase Couette flow, bounded by two plate, leads to the following results :

— In the limit of long wave approximation ($kh_2 \ll 1$), a stable flow can become unstable by increasing the viscosity of one of the two fluids; in addition this instability persists at arbitrarily small Reynolds number, the growth rate being proportional to $(kh_2)^2$ (Yih, 1967).

— For thin lower layer, satisfying $h_1/h_2<(kh_2)^{1/3}$, the Yih's analysis fails (Hooper, 1985). When the thin layer is the less viscous, the interface is stable. When the thin layer is the more viscous, then the interface is unstable and the growth rate is proportional to $(kh_2)^{4/3}$.

— For wavelengths that are short compared to the total thickness but not too short in comparison with the diffusion scale, Renardy (1985) has shown that unstable situations are missed out by the long wavelength asymptotic analysis. In other words, there exists some situations in which long waves are stables whereas unstable modes appear for k^{-1} neither short nor long compared to the thickness h.

The results of nonlinear stability analysis of Couette flow have been obtained from two different methods:

— Starting from the fact that short and O(h) wavelengths are expected to be damped by high enough surface tension, Hooper and Grimshaw (1985) used a multiple scale perturbation method for long waves of small amplitude. The interface displacement then follows a Kuramoto-Sivashinsky equation (K-S). The linear stability analysis reveals the existence of a critical wavenumber k_c above which the waves are stabilized by surface tension. Investigating the time evolution of an initial periodic wave having few unstable harmonics, Hooper and Grimshaw observed two different behaviours when the wavenumber k is decreased: for $k<k_c<2k$, a periodic steady solution is found resulting from the stabilization of the unstable first harmonic by the stable second harmonic; for smaller k, a bouncy state results from the successive control of the unstable first harmonic and unstable second harmonic.

— Using the bifurcation theory, Renardy (1989) derived a Landau equation near marginal stability for the complex amplitude of periodic waves. Supercritical waves are found, even when the upper fluid is the heavier.

A similar result than Hooper and Grimshaw was obtained later on by Charru and Fabre (1990). It leads to a more general equation which reduces to the K-S equation when small terms are ignored. Let us recall the method, similar to this used by Benney (1966), Lin (1974) and Nakaya (1975) for the free falling films.

As we focus upon the long wave instability, the wavelength λ is chosen so as to satisfy $\mu=h_1/\lambda<<1$. The physical quantities are thus expected to vary significantly over h_1 in the y-direction, and over λ in the x-direction. The governing equations are put in dimensionless form by scaling the y- and x-coordinates by h_1 and λ respectively, the time, streamfunctions and pressure, by λ/U_I, h_1U_I and $\rho_1U_I^2$ respectively. The streamfunction of each phase k is decomposed into the streamfunction $\overline{\Psi}_k$ of the unperturbed flow and the perturbation ψ_k. The general solution is obtained from the Navier-Stokes equation written for each phase with the no slip condition at the walls and the interface and, in addition, the conditions for no mass transfer and momentum balance at the interface.

The solution of the unperturbed steady flow leads to the classical linear velocity profile in each phase:

$$\overline{\Psi}_{k,y} = k_k\, y + 1 \qquad (2a,b)$$

where $k_1=1$ and $k_2=1/m$

The solution for the perturbation can be obtained by solving the following partial differential equation for each phase k, in which the small parameter μ appears:

$$\begin{aligned}\psi_{k,yyyy} &= \mu\chi_k \mathrm{Re}\left[(\psi_{k,yyt} + (\overline{\Psi}_{k,y} + \psi_{k,y})\psi_{k,xyy} - \psi_{k,x}(\overline{\Psi}_{k,yyy} + \psi_{k,yyy})\right] - 2\mu^2\psi_{k,xxyy} \\ &+ \mu^3\chi_k \mathrm{Re}\left[(\psi_{k,xxt} + (\overline{\Psi}_{k,y} + \psi_{k,y})\psi_{k,xxx} - \psi_{k,x}\psi_{k,xxy}\right] - \mu^4\psi_{k,xxxx} \qquad (3a,b)\end{aligned}$$

where $\chi_1=1$, and $\chi_2=r/m$. Moreover ψ_k must satisfy the aforementioned boundary conditions at the lower wall ($y=-1$), the upper wall ($y=d$) and the interface ($y=\eta$):

$$\psi_{1,y} = 0 \;;\; \psi_{1,x} = 0 \qquad \text{at } y=-1 \quad (4a,b)$$

$$\psi_{2,y} = 0 \;;\; \psi_{2,x} = 0 \qquad \text{at } y=d \quad (4c,d)$$

$$\psi_{1,x} = \psi_{2,x} \;;\; \overline{\Psi}_{1,y} + \psi_{1,y} = \overline{\Psi}_{2,y} + \psi_{2,y} \qquad \text{at } y=\eta \quad (4e,f)$$

$$\begin{aligned}&\Delta_m\left[(\overline{\Psi}_{,yy} + \psi_{,yy} - \mu^2\psi_{,xx})(1-\mu^2\eta_{,x}^2) - 4\mu^2\psi_{,xy}\,\eta_{,x}\right] = 0 \qquad \text{at } y=\eta \quad (4g) \\ &-\frac{1}{\mu\,\mathrm{Re}}\Delta_m[\psi_{,yyy}] + \Delta_r\left[\psi_{,yt} + (\overline{\Psi}_{,y} + \psi_{,y})\,\psi_{,xy} - \psi_{,x}(\overline{\Psi}_{,yy} + \psi_{,yy})\right]\end{aligned}$$

$$-\frac{\mu}{Re}\Delta_m\Big[\,2[\psi_{,xxy}+\eta_{,x}\psi_{,xyy}+\eta_{,x}[\psi_{,xyy}+\eta_{,x}(\overline{\Psi}_{,yyy}+\psi_{,yyy})]+\eta_{,xx}(\overline{\Psi}_{,yy}+\psi_{,yy})\Big]$$

$$+\psi_{,xxy}-\tfrac{1}{2}\eta_{,x}^2\psi_{,yyy}-\eta_{,x}\psi_{,xyy}\Big]=\frac{\mu^2}{We}\eta_{,xxx}-\frac{1}{Fr}\eta_{,x}+O(\mu^2)\quad \text{for } y=\eta \quad (4h)$$

$$\eta_{,t}+\big[\overline{\Psi}_1(\eta)+\psi_1(\eta)\big]_{,x}=0 \qquad \text{at } y=\eta \quad (4i)$$

where for simplicity we put $\Delta_m[\psi]=\psi_1-m\psi_2$ and $\Delta_r[\psi]=\psi_1-r\psi_2$. Eq. (4h) comes in fact from a derivative along the interface of the momentum jump condition in which the pressure gradients have been eliminated by using the Navier-Stokes equations. The resulting equation is developed in power series of μ, and truncated at μ^2-order. Nevertheless, surface tension is assumed to be high enough for its stabilizing effect to be retained, leading to the condition $We_0\equiv We/\mu^2=O(1)$.

The solution for ψ_k is obtained as expansions in powers of the small parameter μ

$$\psi(x,y,t)=\sum_{i=0}^{\infty}\mu^i\,\psi^{(i)}(x,y,t) \qquad (5)$$

the quantity μRe being assumed small compared with 1. At each order, the $\psi_k^{(i)}$ are solution of fourth-order differential equations, homogeneous at the zeroth-order, non-homogeneous at the upper orders, and they have to satisfy the boundary conditions (4a–h). The streamfunctions are thus determined as functions of y and of the unknown interface displacement $\eta(x,t)$ which has to satisfy the kinematic condition (4i). The solutions at the zeroth and first order are given elsewhere (see Charru, 1991). They are of the form:

$$\psi_k^{(0)}=\sum_{j=0,3}a_{kj}\,y^j \qquad (6a)$$

$$\psi_k^{(1)}=\sum_{j=0,7}\Big[Re\,b_{kj}\,\eta_{,x}+\frac{Re}{We_0}c_{kj}\,\eta_{,xxx}\Big]\,y^j \qquad (6b)$$

where the a_{kj}, b_{kj}, c_{kj} are forbiddingly complicate functions of η, having the form of rational fractions involving powers of η up to 16. It must be noted in passing, that from the zeroth order solution applied to the interface condition (4i), one obtains the velocity c_0 of the linear waves as a function of the thickness and viscosity ratios:

$$c_0=1+\frac{2\,d^2\,(1+d)\,(1-m)}{(d^2-m)^2+4\,m\,d(d+1)^2} \qquad (7)$$

The interface displacement has thus to satisfy the partial differential equation (4i) which becomes

$$\eta_{,t}+\big[\overline{\Psi}_1(\eta)+\psi_1^{(0)}(\eta)+\mu\,\psi_1^{(1)}(\eta)\big]_{,x}=0 \qquad \text{at } y=\eta \quad (8)$$

The complexity of the foregoing equation is reduced if the analysis is limited to small amplitude waves. Eqs. (6) are thus developed in Taylor series for $\eta(x,t)$. On retaining h_1 as the only length scale the small parameter μ disappears. However the same notation will be kept for x.

The interface equation valid up to $O(\eta^3)$ and $O(\mu^2)$ becomes

$$\eta_{,t}+\left\{c_0\eta+V\eta^2+\left[\mathrm{Re}\,(R+T\eta)\,\eta_{,x}+\frac{\mathrm{Re}}{\mathrm{We}}\,(S+U\eta)\,\eta_{,xxx}\right]\right\}_{,x}=0 \qquad (9)$$

where the coefficient V, R, T, S, U depend on the physical properties and flow condition as follows:

$$R = R_0(m, d, r) - S(m,d)\,Fr^{-1} \quad ; \qquad T = T_0(m, d, r) - U(m,d)\,Fr^{-1}$$

$$S = S(m,d) \quad ; \qquad U = U(m,d) \quad ; \qquad V = V(m,d)$$

They have been calculated with the software *Mathematica* by Charru (1991). On considering that $\eta=O(\varepsilon)$ and $\partial/\partial x=O(\mu)$, and assuming that all coefficients are of same order, the dominant terms of the above equation give the K-S equation for $\varepsilon=\mu$. The linear coefficients c_0 and R, corresponding to the velocity and the growth rate of infinitesimal disturbances, agree with Yih's results. The other coefficients V and S have been previously obtained by Hooper & Grimshaw. Nevertheless their expressions for the three coefficients R, V, S contain some errors which have been put into light by handling the properties of symmetry of the problem (Charru, 1991).

LINEAR STABILITY ANALYSIS

What We See in Experiments

The experiments of linear stability consisted in determining whether a critical angular velocity of the upper plate above which interfacial waves are formed exists. As pointed out by Hooper, the flow is destabilized when the thickness of the less viscous fluid layer increases. Careful experiments during which the velocity was increased gently by successive steps, were undertaken for different thickness ratios. The identification of the critical velocity at which the waves develop resulted from visual observation and from analysis of the time traces of interface displacement.

At low enough velocity U_0 of the upper plate, the interface is flat. When U_0 is increased above a given critical value U_{0c}, a periodic disturbance of small amplitude (of a few tenths of mm) appears. This disturbance disappears when U_0 is decreased below U_{0c}. The phenomenon defines unambiguously the critical velocity U_{0c} since it is reproducible. In addition, the perturbation may be clearly identified as shown in the two upper time traces of figure 3, for which U_0 differs from less than 2%. This periodic disturbance will be called "supercritical wave" for obvious reason discussed further.

The power spectrum of time traces allows to determine the wave frequency f. The time lag Δt between the signals recorded from each of the two probes is determined by cross-correlation. The wave velocity $c=\Delta x/\Delta t$ is obtained from the distance Δx between the two probes. The wavelength $\lambda=c/f$ is then deduced: it was observed to be about 4 cm.

In figure 4, the critical velocity has been plotted versus the thickness ratio $d=h_2/h_1$. As expected U_{0c} increases when d increases, confirming experimentally the thin layer effect. For the couple of fluids utilized in the present experiments, the critical velocity behaves linearly with the thickness ratio. In figure 5, the wave velocity near criticality, normalized by the theoretical interfacial velocity of the unperturbed flow $U_I=U_0/(1+d/m)$, is plotted versus the thickness ratio. It is shown that the wave velocity decreases with increasing thickness ratio, with typical values lying between 0.4 and 0.8 times the theoretical interfacial velocity of plane flow.

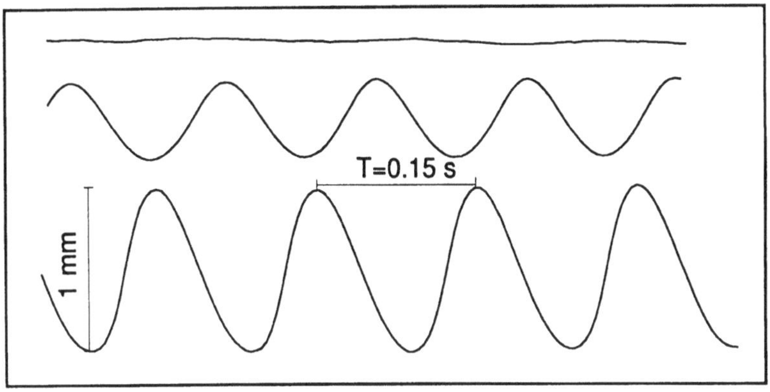

Figure 3. Time traces of interface vertical displacement for h_1=16 mm, h_2=4 mm and, from top to bottom, U_{0c}=0.40, 0.41 and 0.50 m/s.

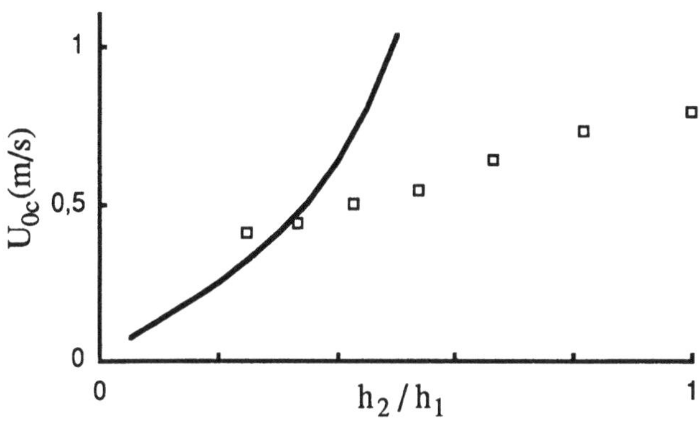

Figure 4. Critical velocity versus thickness ratio (experiments: □, theory: —)

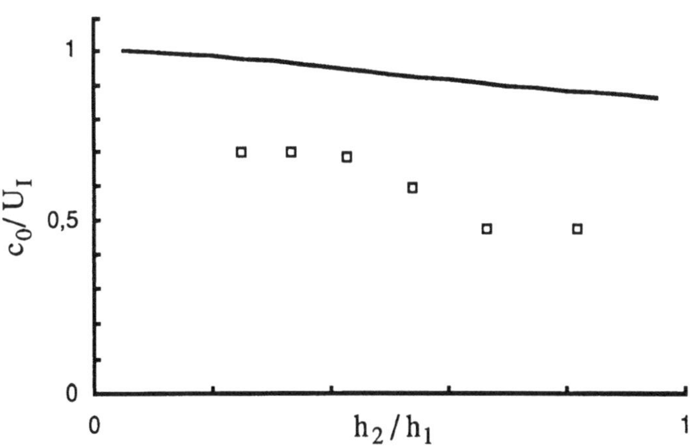

Figure 5. Wave velocity near criticality versus thickness ratio (experiments: □, theory: —)

Beside the supercritical wave, a periodic disturbance whose wavelength is equal to the perimeter πD, was sometimes observed. This disturbance which may appear at an unidentifiable velocity lower than U_{0c} has the following features: it has a very small growth rate (0,01 mm/mn), it does not appear systematically and its presence does not influence the occurrence of the supercritical wave. As the phenomenon was hard to reproduce, no conclusion can be deduced.

What the Theory Tells Us

The linear stability results are recovered from eq. (9) by imposing an harmonic disturbance of the form $\eta = \exp\{i(kx-\omega t)\}$, with real wavenumber $k=2\pi/\lambda$ and complex frequency $\omega=\omega_r+i\omega_i$, with $\omega_r=2\pi f$. The dispersion relation gives :

$$\omega_r = c_0 k \quad \text{and} \quad \omega_i = \mathrm{Re}\, k^2 \left[R - \frac{S}{We} k^2\right] \tag{10}$$

The first equation gives the velocity c_0 of infinitesimal disturbances which are not dispersive at this order. The second equation concerns the growth rate. On the one hand, the flow appears to be marginally stable against $k=0$ disturbances. On the other hand, and keeping in mind that S is always positive, the flow is stable if $R<0$, and if $R>0$ there exists a critical wavenumber k_c :

$$k_c = \left(\frac{We\, R}{S}\right)^{1/2} \tag{11}$$

above which surface tension stabilizes the flow. The linear stability diagram of figure 5 plotted in (WeR/S, k) coordinates is thus composed of two branches corresponding to $k=0$ and $k=k_c$: in between the flow is unstable.

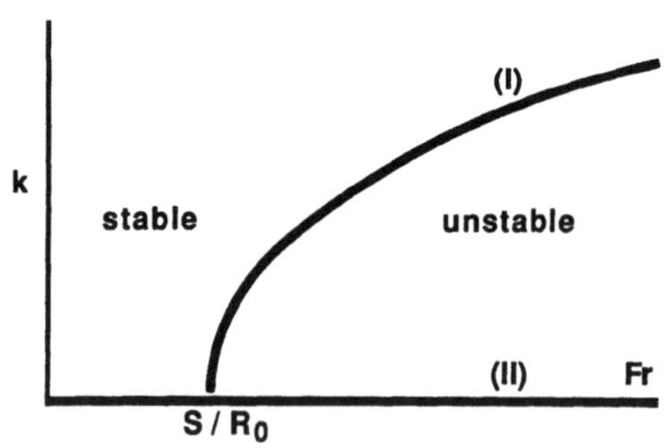

Figure 6. Stability diagram.

Note that if the Bond number defined by $Bo = (1-r)\,\rho_1 g\, h_1^2 \sigma^{-1}$ is introduced, the growth rate can be rewritten :

$$\omega_i = \mathrm{Re}\, k^2 \left[R_0 Fr - S\,(1 + Bo^{-1} k^2)\right] \tag{12}$$

This shows that for given fluids and layer thicknesses such that $R_0>0$, there exists a critical Froude number, $Fr_c=S\, R_0^{-1}$, and thus a critical interface velocity, above which the flow is unstable. In fact each of the three other physical parameters m, d and r can play the role of the bifurcation parameter. In contrast to the case of free falling films there exists no critical Reynolds number.

The theoretical solution for the plane Couette flow has been plotted in figures 4 and 5 for comparison with the actual experimental case. If the orders of magnitude and the tendencies are comparable, the experimental results do not quantitatively agree with the theory.

On the one hand the critical velocity increases with the thickness ratio less in experiments than in theory: surprisingly both agree well for a thickness ratio about equal to 0.5 for which 3D effects are expected to be the weakest in the thicker layer. This might suggests that due to the lateral confinement, the flow is far from plane.

On the other hand, the wave velocity near criticality is about 0.7 times less in experiments tan in theory. Again the 3D effects might be important since the theoretical velocity at the interface which was used for scaling c_0 is probably greater than the actual one.

NONLINEAR STABILITY ANALYSIS

What happens beyond criticality in experiments

When the angular velocity of the plate is increased above the critical velocity, the role of nonlinear effects in limiting the wave amplitude appears clearly in the experimental results (figure 3). Also clear is the fact that the amplitude increases with the velocity U_0. In the same time, the wave departs from sinusoidal shape. This has been confirmed and quantified from spectrum analysis.

In figure 7, the variance of interface fluctuation σ_h is plotted versus the velocity of the upper plate for h_1=16 mm and h_2=4 mm. Close to the bifurcation, σ_h varies linearly with U_0. In this neighborhood the waves remain nearly sinusoidal so that the wave amplitude A is proportional to $\sqrt{\sigma_h}$. The dependance of the wave amplitude near criticality may be expressed by:

$$A \propto \sqrt{U_0 - U_{0c}} \qquad (13)$$

Thus the wave amplitude evolves in the same manner as in Hopf bifurcation. This is a remarkable result which agree with the predictions of the bifurcation theory.

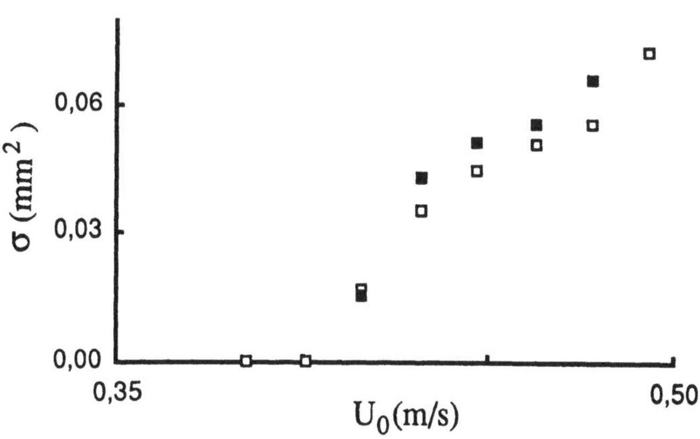

Figure 7. Evolution of the variance of interfacial height with respect to the plate velocity for h_1=16 mm and h_2=4 mm

About the Nonlinear Wave Behavior

The experimental results suggest to study the solutions of eq. (9) corresponding to particular waves travelling with a constant velocity c by keeping their shape. Let us focus upon weakly nonlinear waves near criticality by introducing the small bifurcation parameter: $\varepsilon = c - c_0$. In addition we introduce a diffusion length $k_d^{-1} = \sqrt{ReR/V}$. In a moving frame defined by $\xi = x - ct$, these travelling waves are solution of:

$$-\varepsilon\eta + \eta^2 + k_d^{-2}(1 + \eta T/R)\,\eta_{,\xi} + (k_c k_d)^{-2}(1 + \eta U/S)\,\eta_{,\xi\xi\xi} = 0 \qquad (14)$$

which may be usefully transformed into a set of first order differential equations:

$$\begin{bmatrix} \eta \\ \eta_{,\xi} \\ \eta_{,\xi\xi} \end{bmatrix}_{,\xi} = \begin{bmatrix} 0 & 1 & 0 \\ 0 & 0 & 1 \\ \varepsilon(k_c k_d)^2 & -k_c^2 & 0 \end{bmatrix} \begin{bmatrix} \eta \\ \eta_{,\xi} \\ \eta_{,\xi\xi} \end{bmatrix} - \begin{bmatrix} 0 \\ 0 \\ 1 \end{bmatrix} k_c^2 \left\{ k_d^2 \eta^2 + (T/R - U/S)\eta\eta_{,\xi} + \mathbf{O}(3) \right\} \qquad (15)$$

The components of the unknown vector are the coordinates of the phase space. This O.D.E. has two stationary solutions given by the two fixed points $\mathbf{P_0}(0,0,0)$ and $\mathbf{P_1}(\varepsilon,0,0)$ corresponding to $\eta_0 = 0$ and $\eta_1 = \varepsilon$. The properties of the trajectories near $\mathbf{P_0}$ and $\mathbf{P_1}$ depend on the nature of the eigenvalues α of the linearized problem (Pumir et al, 1983).

For $\mathbf{P_0}$, α is solution of

$$\alpha^3 + k_c^2\,\alpha - (k_c k_d)^2\,\varepsilon = 0 \qquad (16)$$

For $\varepsilon > 0$, one of the two roots, say α_1, which is real and positive, defines a one-dimensional unstable manifold, whereas the two other roots α_2 and α_3 which are complex-conjugated with negative real parts, define a two-dimensional stable manifold (figure 8). For $\varepsilon < 0$ the stable manifold is one-dimensional whereas the unstable one is two-dimensional.

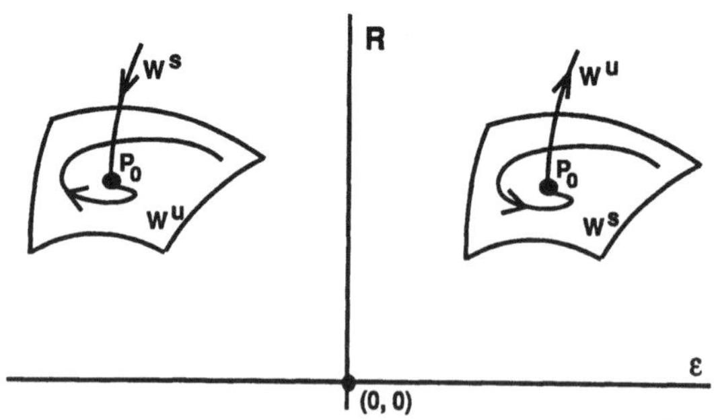

Figure 8. Local pictures of the phase space around the fixed point P_0 in each region of the parameter space. W^u and W^s are the unstable and stable manifolds respectively. Axes are bifurcation lines.

For $\mathbf{P_1}$, it can be found that α satisfies:

$$\alpha^3 + k_c^2\,\{1 + \varepsilon\,(T/R - U/S)\}\,\alpha + (k_c k_d)^2\,\varepsilon = 0 \qquad (17)$$

The conclusions of stability analysis for $\mathbf{P_1}$ are opposite to the previous case, yielding a one-dimensional stable manifold and a two-dimensional unstable manifold for $\varepsilon > 0$, and vice versa for $\varepsilon < 0$.

For $\varepsilon=0$ the sign of the real eigenvalue changes (transcritical bifurcation) as well as the real part of the complex eigenvalues (Hopf bifurcation) and the two fixed points coalesce. The center manifold is then three-dimensional like the space itself. The simultaneous occurrence of two bifurcations leads to a complex behavior of the system dynamics. Nevertheless, the Hopf bifurcation suggests the existence of periodic waves in the vicinity of $\varepsilon=0$. Their existence may be proved from the study of the normal form of eqs. (15). Its calculation has been made by Charru (1991) and will not be detailed here: it follows the method used by Chang (1986 & 1987) for the stability of free falling films. In cylindrical coordinates (r, θ, z) this normal form is:

$$k_d^{-2}\, z_{,\xi} = \frac{\varepsilon^2}{4} - z^2 - 2r^2 \tag{18a}$$

$$k_d^{-2}\, r_{,\xi} = zr \tag{18b}$$

$$k_c^{-1}\, \theta_{,\xi} = 1 + \frac{T/R - U/S}{2}\left[z + \frac{\varepsilon}{2}\right] \tag{18c}$$

The physical variables are related to the cylindrical coordinates by the following expressions:

$$\eta \approx z + 2\, r\cos\theta + \varepsilon/2 \quad ; \quad \eta_{,\xi} \approx -2\, k_c\, r \sin\theta \quad ; \quad \eta_{,\xi\xi} \approx -2\, k_c^2\, r\cos\theta \tag{19}$$

The first two differential equations are θ-independant : this means that the phase portrait is symmetrical with respect to z-axis. Thus the solutions may be studied within the meridian plane of figure 9 (Poincaré map). Two heteroclinic trajectories are lying between the two fixed points P_0 and P_1 corresponding to shock solutions. In addition, the map displays the point P_2 located at $(|\varepsilon|/2\sqrt{2},\, 0)$ which corresponds to a periodic orbit in the phase space. This point is surrounded by a family of closed curves that are the intersections between the meridian plane and tori centered on the z-axis. A linear stability analysis shows that P_2 is a center linked to the conjugated eigenvalues $\pm i\, |\varepsilon|\, k_d^2/\sqrt{2}$. Each of these tori imbeds doubly periodic orbits whose characteristic length scales correspond to the rotations around P_2 and the z-axis.

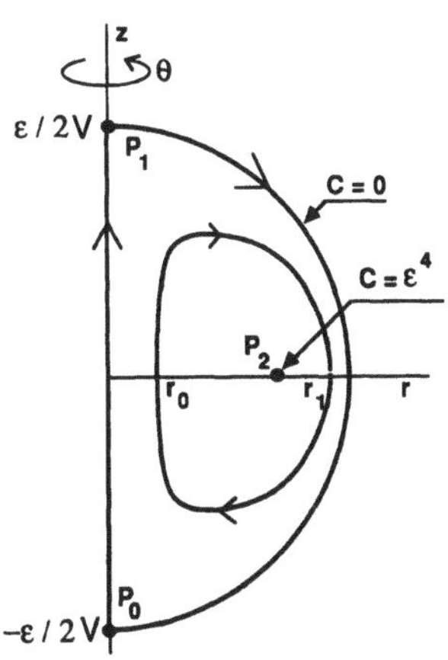

Figure 9. Poincaré map of the phase portrait of the normal form.

The first two equations (18a,b) admit a first integral giving the equation of the closed curves in the (r,z)-plane :

$$64\,r^2\left(\frac{\varepsilon^2}{4} - z^2 - r^2\right) = C \tag{20}$$

where C is a constant specific to each trajectory, varying from 0 (heteroclinic orbits) to ε^4 (periodic orbit). In between there exists a non-enumerable infinite set of orbits on tori representing doubly periodic waves.

The periodic wave, corresponding to the orbit $\mathbf{P}_2$ may be found by using eqs. (19)

$$\eta \approx \frac{\varepsilon}{2}\left(1 + \sqrt{2}\cos k_1\xi\right) \tag{21}$$

where its wavenumber k_1 is expressed by:

$$k_1 = k_c\left(1 + \frac{\varepsilon}{4}(T/R - U/S)\right) \tag{22}$$

Eq. (21) shows that periodic waves exist near criticality. Eq.(22) gives a wavenumber-amplitude equation or a nonlinear dispersion relationship through eq. (21). It appears that this periodic wave bifurcates with wavenumber k_c from the upper marginal stability branch (figure 6).

CONCLUSION

The linear and nonlinear stability of the interface between two fluid layers in laminar Couette flow has been investigated. Both experimental and theoretical results are reported. On the one hand, an original experimental set-up has been designed for studying the interfacial periodic waves. On the other hand, the interface equation has been derived and the existence of nonlinear periodic waves has been demonstrated. The comparison between experiment and theory leads to the following conclusions.

Concerning the linear stability, it has been shown that experimental results qualitatively agree with the asymptotic analysis for long waves. Concerning the nonlinear stability, the existence of a Hopf bifurcation has been found experimentally. From the theory, this bifurcation appears together with a transcritical bifurcation. The discrepancies might be explained from the small aspect ratio of the channel. In addition the flow conditions are not in the validity domain of the perturbation method, for the wavelength and the Reynolds number. However the experiments gave very original and reproducible results which would be confronted to numerical stability calculations. These comparisons will shed light on the troublesome role of viscosity.

REFERENCES

Benney, D.J., Long waves on liquid films, *J. Math. & Phys.*, 45:150 (1966).

Chang, H.C., Evolution of nonlinear waves on vertically falling films - a normal form analysis, *Chem. Engng. Sci.*, 42:515 (1987).

Chang, H.C., Travelling waves on fluid interfaces : Normal form analysis of the Kuramoto-Sivashinsky equation, *Phys. Fluids*, 29:3142 (1986).

Charles, M.E. and Lilleleht, L.U., , An experimental investigation of stability and interfacial waves in a co-current flow of two liquids, *J. Fluid Mech.* 22:217 (1965).

Charru, F. and Fabre, J., Ondes longues non linéaires à l'interface de deux fluides visqueux, *C. R. Acad. Sci. Paris*, 310 II:1711 (1990).

Charru, F., *Stabilité de l'interface entre deux fluides visqueux*, Thèse, INP Toulouse, (1991).

Hooper, A.P. and Grimshaw, R., Nonlinear instability at the interface between two viscous fluids, *Phys. Fluids*, 28:37 (1985).

Hooper, A.P., Long-wave instability at the interface between two viscous fluids : Thin layer effects, *Phys. Fluids*, 28:1613 (1985).

Kao, M.E., and Park, C., Experimental investigation of the stability of channel flow, *J. Fluid Mech.* 52:401 (1972).

Lin, S.P., Finite amplitude side-band stability of a viscous fluid, *J. Fluid Mech.*, 63:417 (1974).

Nakaya, C., Long waves on a thin fluid layer flowing down an inclined plane, *Phys. Fluids*, 18:1407 (1975).

Pumir, A., Pomeau, Y. and Manneville, P., On solitary waves running down an inclined plane, *J. Fluid Mech.*, 135:27 (1983).

Renardy, Y., Instability at the interface between two shearing fluids in a channel, *Phys. Fluids*, 28:3441 (1985).

Renardy, Y., Weakly nonlinear behavior of periodic disturbances in two layer Couette-Poiseuille flow, *Phys. Fluids A*, 1:1666 (1989).

Yih, C.S., Instability due to viscous statification, *J. Fluid Mech.* 27:337 (1967).

INSTABILITIES IN EVAPORATING LIQUID FILMS

S. W. Joo[1], S. H. Davis[1], and S. G. Bankoff[2]

[1] Engineering Sciences and Applied Mathematics
[2] Chemical Engineering
Northwestern University
Evanston, IL 60091

INTRODUCTION

A viscous liquid layer placed on a heated plate can experience a variety of hydrodynamic instabilities due to the effects of local density and surface-tension variations and, for volatile layers, evaporation. If the plate is inclined and the liquid drains down, another family of instabilities can exist (even for isothermal layers) due to the effects of the mean flow. In this study, we consider extremely thin volatile layers, where the instabilities due to the evaporation and the mean flow are particularly important, and study the dynamics of the flow in the presence of these instabilities.

In a horizontal heated layer, the mean flow is absent, but a vertical temperature gradient is present. The liquid particles thus can vaporize at the interface between the liquid and the ambient gas until the temperature of the interface reaches a saturation value. In general, the density of the vapor particles is much smaller than that of the liquid particles, so that there exists a substantial velocity jump upon evaporation. This velocity jump induces extra normal stress at the interface, so called the vapor recoil, and is responsible for the evaporative instability.

This evaporative instability has been studied, among others, by Bankoff (1971), Spindler et al. (1978), and Burelbach et al. (1988). Burelbach et al. considered two-dimensional layers in the presence of evaporative mass loss, vapor recoil, thermocapillarity, and van der Waals attractions. They used an *one-sided model* of evaporation, where the dynamics of the vapor are decoupled from those of the liquid, and studied the development of dryout via the long-wave evolution equation, which approximate the flow properly when the disturbance wavelength is much larger than the film thickness.

In an inclined layer, the mean flow is present, and can induce instabilities when the flow rate (or the film thickness) is large enough to overcome the hydrostatic and capillary stabilization. The interface then becomes corrugated and various wave motions set in downstream. In most cases, small-amplitude waves appear at some downstream location, called the wave-inception line, and develop into finite-amplitude two-dimensional waves. This surface-wave instability is identified by Yih (1955, 1963) and Benjamin (1957), and has been studied intensively, as summarized by Lin (1983) and Lin & Wang (1985). Further downstream, the two-dimensional waves tend to become three-dimensional and erratic, as observed most recently by Lacy et al. (1991) and explained by Joo & Davis (1992).

When both the evaporation and the mean flow are present, the aforementioned instabilities interact with each other and can greatly influence the linear and nonlinear flow development. Joo et al. (1991) studied such cases for two-dimensional layers by deriving a long-wave evolution equation and performing a linear stability analysis and spectral computations. They show the existence of a stability window, analogous to that for layers with mean flow and thermocapillarity (Kelly et al. 1986), where the instability disappears due to the hydrostatic pressure. They also noticed that the surface-wave instability does not accompany substantial local thinning of the layer while there is a tendency toward wavebreaking, and thus can either retard the rupture time or cause the rewetting of the dry region after the rupture occurs.

In this study we extend the work of Joo et al. (1991) to three-dimensional layers, with particular attention to the effects of coupled instabilities on the nonlinear flow development. For nonvolatile layers with thermocapillarity, it is known that longitudinal patterns can develop and lead to the formation of rivulets (Joo et al. 1992). In volatile layers, similar pattern formations can be expected in the absence of the thermocapillarity, with the vapor recoil playing the role of the thermocapillarity.

LONG-WAVE THEORY

We consider a Newtonian liquid of constant density ρ, viscosity μ, thermal conductivity k, and thermal diffusivity κ flowing downwards, due to gravity, on an inclined plate heated at a constant temperature T_H. The layer is unbounded in the streamwise and spanwise (or cross-stream) directions, but bounded above by its interface with the vapor phase of far-field temperature T_S ($< T_H$). The liquid is volatile, and vaporizes into vapor particles at the interface. The heat conducted across the liquid layer from the heated plate is used to vaporize the liquid particles and impart kinetic energy to the vapor particles. These heat losses determine the temperature T_I of the interface, which varies in time and space due to the interfacial motion.

The mean thickness of the layer is initially d_0, but decreases in time due to the evaporation. The streamwise component of the velocity reaches its maximum at the interface, and is $gd_0^2 \sin\beta/\nu$ for an undisturbed layer, where g is the gravitational acceleration, β is the angle of inclination, and $\nu = \mu/\rho$. The Reynolds number based on this initial maximum velocity becomes $G \sin\beta$, where

$$G = \frac{gd_0^3}{\nu} \tag{1}$$

is a parameter measuring the film thickness.

The volatility of the liquid is described by the evaporation number

$$E = \frac{k(T_H - T_S)}{\mu \mathcal{L}}, \tag{2}$$

where $\mathcal{L}$ is the latent heat. If $E \to 0$ (or ∞), the evaporation is extremely slow (or fast), and the rate of evaporative mass loss approaches zero (or ∞). The vapor recoil then is measured by E^2/D, where

$$D = \frac{3}{2}\frac{\rho^V}{\rho} \tag{3}$$

is the density ratio of the vapor to the liquid. In the one-sided evaporation model used in this study, the vapor and the liquid phases are coupled only through the vapor recoil. Burelbach et al. (1988) give a detailed explanation of the model.

The characteristic length of the interfacial disturbances that appear is, in general, much larger than the thickness of the liquid layer. A short disturbance would be rapidly suppressed by the action of the interfacial tension. Thus long-wave asymptotics of the Benney type allow description of the flow development. The full system of the Navier-Stokes and energy equations then can be reduced into a single evolution equation for the local layer thickness, $h(x, y, t)$:

$$h_t + \frac{E}{h+K} + Gh^2 h_x \sin\beta + \frac{2G^2}{15}\left(h^6 h_x\right)_x \sin^2\beta + E\frac{5G}{24}\left(\frac{h^4}{h+K}\right)_x \sin\beta$$

$$+\nabla\cdot\left\{\left[\frac{E^2}{D}\left(\frac{h}{h+K}\right)^3 - \frac{G}{3}h^3\cos\beta\right]\nabla h + Sh^3\nabla\left(\nabla^2 h\right)\right\}$$

$$+EP\left(\frac{h}{h+K}\right)^3\left[\frac{E}{3(h+K)} + \frac{G}{120}(7h - 15K)hh_x\sin\beta\right] = 0, \tag{4}$$

where x and y are, respectively, streamwise and spanwise coordinates, and ∇ is the gradient operator (∂_x, ∂_y). The parameters K,

$$S = \frac{\sigma d_0}{3\rho\mu^2}, \tag{5}$$

and

$$P = \frac{\nu}{\kappa} \tag{6}$$

measure, respectively, degree of thermal nonequilibrium at the interface (see Burelbach et al. 1988), interfacial tension σ, and thermal diffusivity (Prandtl number). Here, the parameters are $O(1)$ except $E = O(\epsilon)$, $D = O(\epsilon^2)$, and $S = O(\epsilon^{-2})$, where ϵ is a small number measuring the ratio of film thickness to a typical wavelength of the disturbance.

One state of the flow of particular interest is represented by a flat layer of infinite lateral extent that drains downward and disappears in a finite time. The velocity component in the streamwise direction has a parabolic profile, while those in the other directions are identically zero. The heat transfer in the liquid layer is purely conductive, and thus the temperature decreases linearly from the bottom plate to the interface. The temperature along the interface is constant, and is determined by the instantaneous film thickness, degree of interfacial thermal nonequilibrium K, evaporation number E, and the Prandtl number P.

The layer thickness is independent of x and y, can be obtained by solving the evolution equation (4) after setting $\partial_x = \partial_y = 0$:

$$h_0(t) = \Omega - K + \frac{EP}{3\Omega}\left[\frac{K(2K^2+6K+3)}{1+K} + \frac{3K^2}{2}\ln\frac{\Omega}{1+K} + \frac{K^3}{\Omega} - 3K\Omega - Et\right], \quad (7)$$

where $\Omega(t) = \sqrt{(1+K)^2 - 2Et}$ and h_0 indicates a basic state. From (7) one can obtain the disappearance time t_D, when the liquid layer has completely evaporated. As given by Burelbach et al. (1988),

$$t_D = \frac{1+2K}{2E} \quad (8)$$

if $P = 0$. In the limit $E \to 0$, (7) recovers the basic state $h_0 = 1$ for isothermal or nonvolatile layers.

The linear stability of the basic state (7) can be analyzed by considering a small-amplitude simple harmonic disturbance of wavenumber vector $\vec{k}$ superimposed on the flat state:

$$h = h_0(t) + \left[H(t)e^{i\vec{k}\vec{x}} + \text{c.c.}\right], \quad (9)$$

where the disturbance amplitude $2H(t) << 1$, $\vec{k} = (k\cos\theta, k\sin\theta)$, and $\vec{x} = (x, y)$. Here, the oblique angle θ is zero for purely two-dimensional waves (no spanwise variation) and $\pi/2$ for purely longitudinal rolls (no streamwise variation). If we substitute the above expression into (4) and linearize in the small amplitude, we obtain

$$\frac{\dot{H}}{H} = \Gamma(t) - ikc(t) + \Gamma_{ev}(t), \quad (10)$$

where

$$\Gamma = k^2\left[\frac{E^2}{D}\left(\frac{h_0}{h_0+K}\right)^3 + \frac{2G^2}{15}h_0^6\cos^2\theta\sin^2\beta - \frac{G}{3}h_0^3\cos\beta - k^2\bar{S}h_0^3\right] \quad (11)$$

is the effective growth rate,

$$c = G\left[1 + \frac{5E}{24}\left\{\frac{4h_0}{h_0+K} - \left(\frac{h_0}{h_0+K}\right)^2 + \frac{Ph_0^2(7h_0-15K)}{25(h_0+K)^3}\right\}\right]h_0^2\cos\theta\sin\beta \quad (12)$$

is the linearized phase speed, and

$$\Gamma_{ev} = \frac{E}{(h_0+K)^2} + E^2P\frac{h_0^2(h_0-3K)}{3(h_0+K)^5}. \quad (13)$$

The stability is determined by the sense of Γ; Γ_{ev} measures the initial disturbance amplitude relative to the decaying thickness of the basic state and gives an algebraic variation in H. If $\Gamma > 0$ (or < 0), the initial small disturbance will grow (or decay) in time, and so the flat basic state is unstable (or stable).

Fig. 1 shows a stability diagram in the (E^2/D, G)-plane. The solid curve represents neutral stability, where $\Gamma = 0$. On the horizontal axis ($E^2/D = 0$), the layer is isothermal, and the surface-wave instability appears for sufficiently large film thickness ($G > G_c$). This critical thickness (or G) approaches zero as β increases toward $\pi/2$, so that for vertical layers the stable region disappears. For a nonvertical layer, there is a stability window when the intensity of heating $T_H - T_S$ is small enough ($E^2/D < (E^2/D)_c$). If one fixes the intensity of heating (or E^2/E) and keeps increasing the layer thickness (or G) from zero, the layer becomes unstable due to the vapor recoil, stable due to the hydrostatic effects, and then unstable again due to the mean flow. If $E^2/D > (E^2/D)_c$, the stability window disappears, resulting in complete destabilization of the flow for nonvertical layers.

NONLINEAR EVOLUTION

The nonlinear development of the instability can be studied by posing an initial-value problem in a spatially-periodic domain and integrating the evolution equation (4) numerically. As an initial condition, we use a small-amplitude three-dimensional disturbance, harmonic in both streamwise and spanwise directions:

$$h(x, y, 0) = 1 - \delta_1 \cos(k_1 x) - \delta_2 \cos(k_2 y), \tag{14}$$

where $\delta_1 = \delta_2 = 0.1$, and k_1 and k_2 are, respectively, the streamwise and spanwise wavenumbers. The computational domain then is one spatial period, $0 \le x \le 2\pi/k_1$ and $0 \le y \le 2\pi/k_2$, of the initial disturbance. The growth of subharmonics, which can occur in laboratory situations, is not allowed.

In order to update h temporally, the Hamming modified predictor-corrector method is used, while the spatial derivatives are computed using the Fourier spectral method. The rupture time t_R (smaller than the disappearance time t_D) is defined as the time required for any part of the layer thins below the absolute error bound 10^{-12} of the computation. The computation is terminated as soon as the rupture occurs or more than 64×64 collation points are required to suppress the aliasing error, beyond which the evolution equation (4) no longer describes the flow adequately.

EVAPORATIVE INSTABILITY

We first consider horizontal volatile layers heated from below by setting $\beta = 0$. The dominant effects are then mass loss, hydrostatic pressure, vapor recoil, and surface tension.

As shown above, mass loss is not a deciding factor for instability. Surface tension determines the range of unstable wavenumbers. Therefore, for sufficiently long disturbances, i.e., small k_1 or k_2 in (14), the instability appears if the vapor recoil overcomes the hydrostatic effects ($E^2/D > G/3$). An initial small disturbance grows, with the troughs thinning toward the bottom plate and the crests growing upward.

Fig. 1 shows a typical unstable evolution induced by the vapor recoil. Instantaneous free-surface configurations are shown for one spatial period when $k_1 = k_2 = 0.05$, $E = 0$ (small mass-loss limit), $E^2/D = 5$, $G = 1$, and $S = 100$. The initial configuration (14) is shown in Fig. 1a. The layer is thinnest at the center and thickest at the corners. Therefore, as time progresses, the center thins while the corners grow, as shown in Figs. 1b, 1c, and 1d.

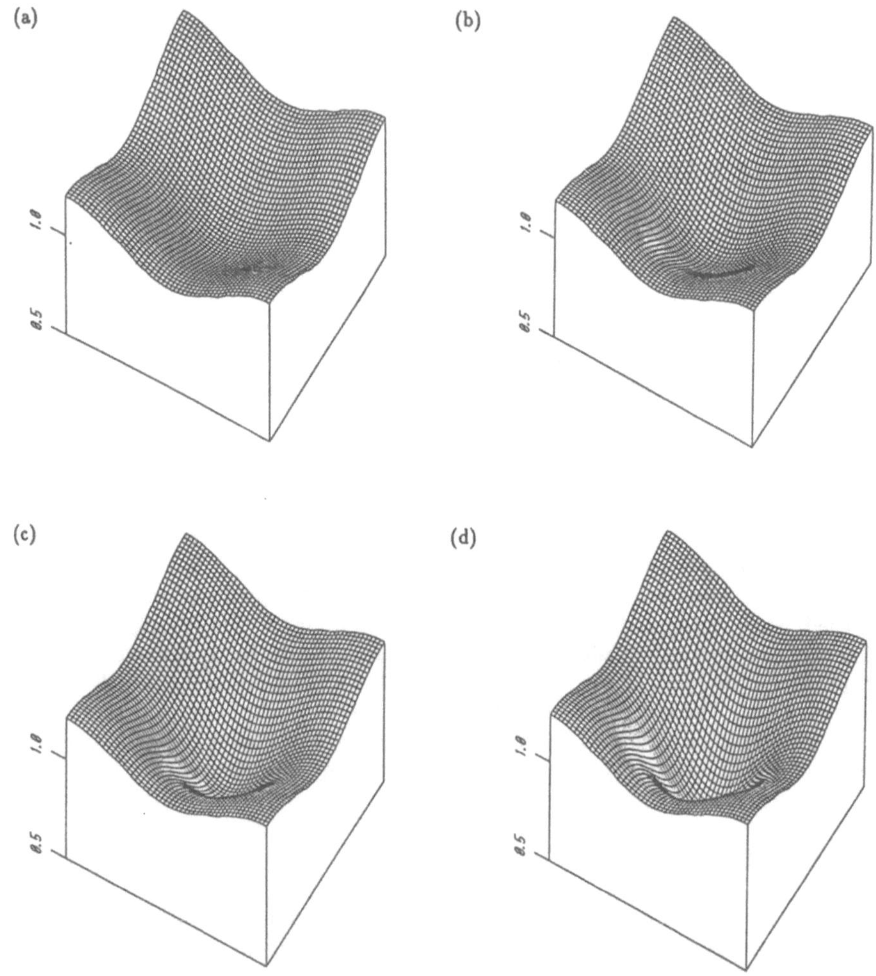

Figure 1. Free-surface configurations for a horizontal layer. $E = 0$, $E^2/D = 5$, $G = 1$, $S = 100$, and $k_1 = k_2 = 0.05$. (a) t=0; (b) t=100; (c) t=200; (d) t=300.

The center thins toward the bottom by displacing fluid particles away from it, inducing pressure gradients. The pressure gradients increase as the thinning develops further due to the bottom effect. Therefore, after substantial thinning, the center flattens due to the high pressure and can eventually grow back upward. Similar phenomena have been identified in thin layers subject to Rayleigh-Taylor instability (Yiantsios & Higgins 1989) and thermocapillary instability (Joo et al. 1991). In two-dimensional layers, the free surface takes the form of two fingers, and thus a two-point rupture can occur. In three-dimensional layers, the region around the center can be isolated from the bulk due to the three-dimensional fingering (Joo et al. 1992).

When the mass loss is taken into account ($E \neq 0$), the nonlinear development of the instability can be quite different. For instance, when the thinning occurs, the fluid particles in the center region are displaced not only laterally but also across the free surface due to the evaporation. Sufficiently large mass loss, therefore, can prevent the fingering. An extreme example would occur when E is very large. The mass loss is so rapid that no instability development can be seen within the small disappearance time t_D.

COUPLED INSTABILITIES

We now consider volatile layers on a vertical plane by setting $\beta = \pi/2$ in (4). The effects of the mean flow (surface-wave instability) are coupled with the evaporative instability. The hydrostatic stabilization does not exist.

The disturbance wave travels downstream, and also grows due to the instabilities. The surface-wave instability does not induce significant thinning of the troughs, unlike the evaporative instability, but steepens the wave owing to the thickness-dependent local phase speed; a thicker region travels faster than a thinner region. This steepening in turn give rise to changes in evaporation. Joo et al. (1991) have shown an example of this interaction for layers with thermocapillary instability, where the finger closer to the steepening front grows much faster toward the bottom than the other.

In three-dimensional layers, spanwise flow is allowed, so that the liquid particles can travel around adverse regions, such as dry patches and extremely shallow (thin) places. As the layer thins locally, the flow tends to detour the thinned region, possibly forming longitudinal rolls where most of the downstream fluid motion occurs.

Fig. 2 shows an example of unstable evolution, where longitudinal rolls are formed. The parameters are kept the same as in Fig. 1, i.e., $k_1 = k_2 = 0.05$, $E = 0$, $E^2/D = 5$, $G = 1$, and $S = 100$, except now $\beta = \pi/2$. The initial state of the disturbance is as shown in Fig. 1a.

As the layer flows downstream, surface wave develops while the effects of vapor recoil induce local thinning. Initially, the evaporation effect is minimal, and the flow development is similar to that for an isothermal layer, as shown in Fig. 2a. The free surface still shows a strongly three-dimensional pattern, as in Fig. 2b, where the contours of constant thickness are plotted. As the layer thins further (locally), the flow begins to concentrate along the edges, where the thickness is larger and has less resistance to flow. The surface then gradually becomes longitudinal in pattern. This transition from three-dimensional to longitudinal pattern is shown in Fig. 2c through 2f.

Similar evolutions toward longitudinal patterns exist for layers with thermocapillary instability (Joo et al. 1992). It is interesting to note that neither the surface-wave or

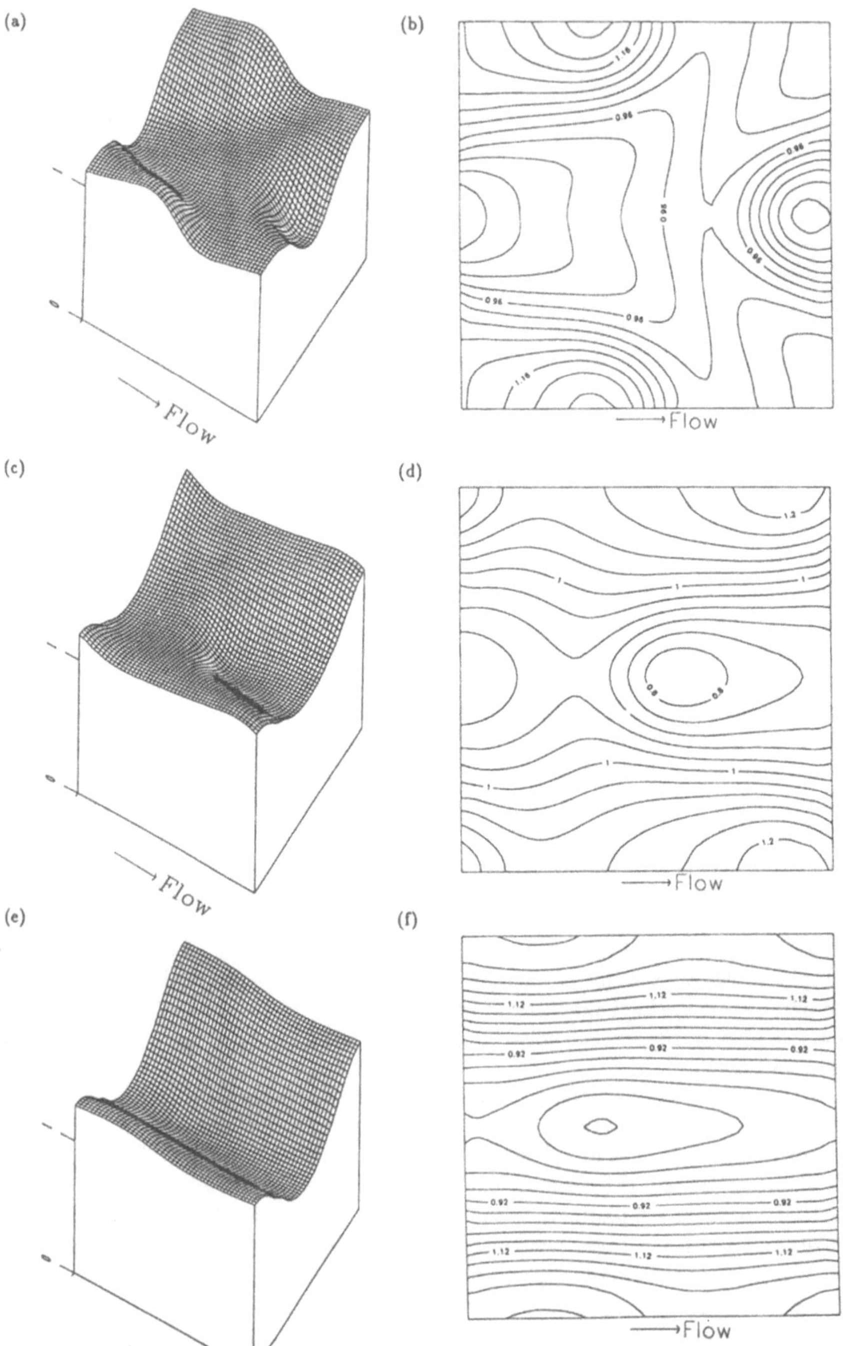
(a)
Flow
(b)
1.16
0.96
Flow
(c)
Flow
(d)
1.2
1
0.8
Flow
(e)
Flow
(f)
1.12
0.92
Flow

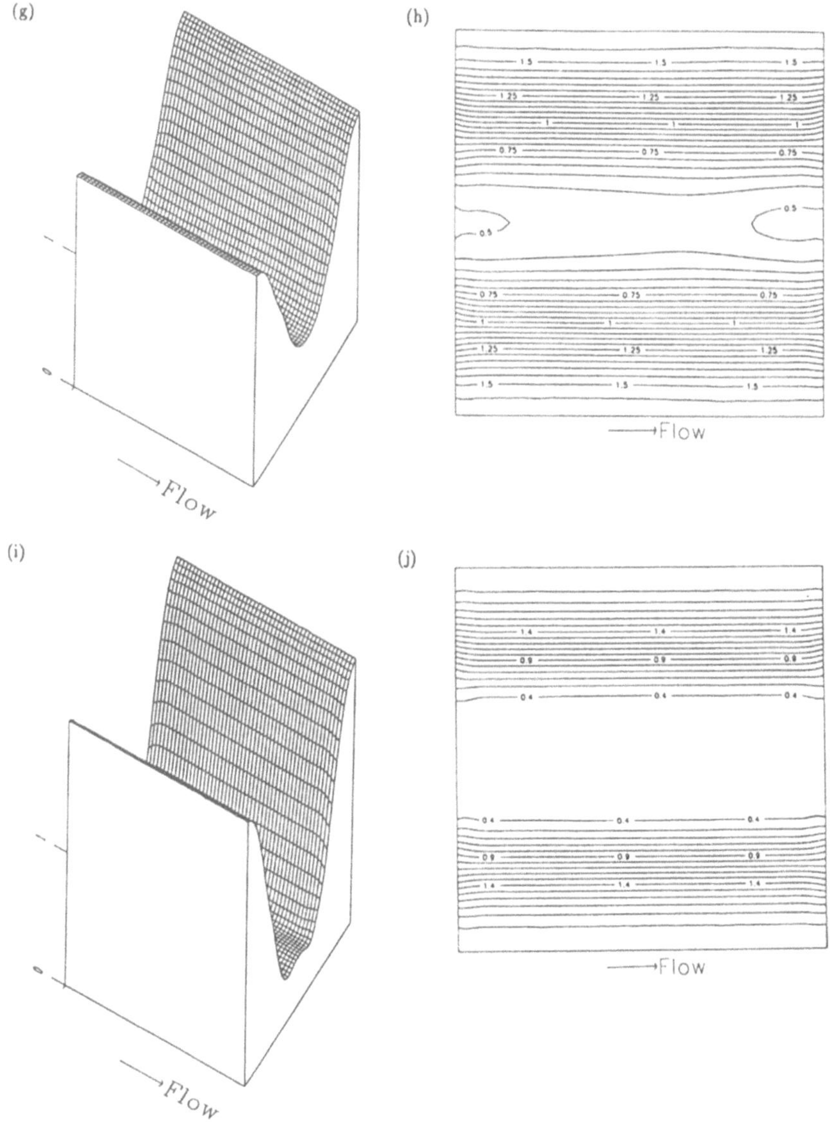

Figure 2. Free-surface configurations and contours of constant thickness for a vertical layer. $E = 0$, $E^2/D = 5$, $G = 1$, $S = 100$, and $k_1 = k_2 = 0.05$. (a)&(b) t=25; (c)&(d) t=50; (e)&(f) t=100; (g)&(h) t=200; (i)&(j) t=300.

the evaporative (or thermocapillary) instability by themselves can generate longitudinal patterns. The pattern is developed through the nonlinear interaction between the two different modes.

After the longitudinal pattern is fully developed, the subsequent evolution is mainly in the spanwise direction. Along the flow direction, there is virtually no variation. The spanwise mode keeps growing due to the evaporative instability. The troughs (now aligned in the flow direction) thin further toward the bottom while the crests (also along the flow direction) grow upward. Figs. 2g and 2h show this stage.

As the troughs thin substantially, they experience high pressure and flatten, as in the horizontal layers. Then, fingering occurs and new longitudinal rolls, which are smaller and between the main rolls, appear and develop. This is shown in Figs. 2i and 2j. After the layer ruptures, these rolls will be separated by dry regions, resulting in a series of rivulets. Due to the fingering, two (or more, depending upon secondary flow development not pursued here) rivulets can be expected from one initial spanwise disturbance period for this particular case.

As in the horizontal layers, rapid mass loss can suppress the nonlinear flow developments discussed above. If the disappearance time t_D is too small for the instability to develop, the layer will simply dry out before longitudinal rolls appear.

CONCLUDING REMARKS

Thin volatile liquid layers flowing on a heated plate are studied via a long-wave evolution equation of the Benney (1966) type. The effects of mass loss, wave propagation, mean flow, hydrostatic pressure, vapor recoil, and surface tension are modelled in the equation.

Linear stability analysis shows that the mean flow and the vapor recoil induce surface-wave and evaporative instabilities, respectively. The mass loss does not contribute to instability. Due to hydrostatic stabilization, there exists a stability window, below and above which the flow is unstable due to, respectively, vapor recoil and mean flow.

The nonlinear development of the instability is studied by posing an initial-value problem in a periodic domain and integrating the evolution numerically. In horizontal layers, interesting flow developments occur due to the vapor recoil in conjunction with the pressure gradient induced through thinning. The location of rupture can be different from the initial location of the trough, and in many cases a small drop isolated from the bulk can form after rupture. For non-horizontal layers, the mean flow and interact with the evaporative instability, and create longitudinal rolls. This is particularly interesting because neither of the two instabilities by itself can generate longitudinal patterns.

This study shows that the evolution of volatile layers with vapor recoil is very similar to that of non-volatile layers with thermocapillarity we studied earlier, except that here we have an additional parameter (mass loss), which makes the layer disappear in a finite time. If this disappearance time is extremely small, the layer will dry out before the instabilities develop.

ACKNOWLEDGEMENT

This work was supported by U.S. Department of Energy, Division of Basic Energy Sciences, through Grant no. DE FG02-86ER13641.

REFERENCES

Bankoff, S. G. 1971 Stability of liquid flow down a heated inclined plane. Int. J. Heat Mass Transfer, **14**, 377-385.

Benjamin, T. B. 1957 Wave formation in laminar flow down an inclined plane. J. Fluid. Mech., **2**, 554-574.

Benney, D. J. 1966 Long waves on liquid films. J. Math. & Phys., **45**, 150-155.

Burelbach, J. P., Bankoff, S. G. & Davis S. H. 1988 Nonlinear stability of evaporating/condensing liquid films. J. Fluid Mech., **195**, 463-494.

Joo, S. W., Davis, S. H., and Bankoff, S. G. 1991 Long-wave instabilities of heated falling films: two dimensional theory of uniform layers. J. Fluid Mech., **230**, 117-146.

Joo, S. W. and Davis, S. H. 1992a Instabilities of thee-dimensional viscous falling films. To appear in J. Fluid Mech.

Joo, S. W., Davis, S. H., and Bankoff, S. G. 1992 A mechanism for rivulet formation in heated falling films. Pending publication.

Kelly, R. E., Davis, S. H. & Goussis D. A. 1986 On the instability of heated film flow with variable surface tension. Proc. 8th Int. Heat Transfer Conf., **4**, 1937-1942.

Lacy, C. E., Sheintuch, M., and Dukler, A. E. 1991 The flow of thin, wavy vertical films: problems in characterization by methods of deterministic chaos. Submitted to AIChE Journal.

Lin, S.-P. 1974 Finite amplitude side-band stability of a viscous film. J. Fluid. Mech., **74**, 417-429.

Lin, S.-P. & Wang, C.-Y. 1985 Modeling wavy film flows. Encyclopedia of Fluid Mechanics, **1**, 931-951.

Spindler, B., Solesio, J. N. & Delhaye, J. M. 1978 On the equations describing the instabilities of liquid films with interfacial phase change. Two-Phase Momentum, Heat and Mass Transfer in Chemical Process and Energy Engineering Systems (ed. F. Durst, G. V. Tsiklauri & N. H. Afgan), vol. 1, pp. 339-344, Hemisphere Press.

Yiantsios, S. G. & Higgins B. G. 1989 Rayleigh-Taylor instability in thin viscous films. Phys. Fluids, **1**, 1484-1501.

Yih, C.-S. 1955 Stability of parallel laminar flow with a free surface. Proc. 2nd U.S. Congr. Appl. Mech., Amer. Soc. Mech. Engrs., 623-628.

Yih, C.-S. 1963 Stability of liquid flow down an inclined plane. Phys. Fluids, **6**, 321-334.

KELVIN - HELMHOLTZ TWO-PHASE FLOW INSTABILITY

Emile Danho* and Michel Bouthier**

*Department of Mathematics, University of Abidjan, 22 B.P. 582 Abidjan 22, Côte d'Ivoire
**C.N.R.S., Paris 6 University, Tour 66, 4 Pl. Jussieu, 75232 Paris Cedex 05 , France
personal address : 19 rue Dedouvre, 94250 Gentilly, France

1- INTRODUCTION

Kelvin-Helmholtz stability waves are of great importance in the field of fluid mechanics and for quite a while now special attention has been focused on studying them in detail. Indeed two layered fluid flows occur in many physical situations, and even now, numerous works continue to approach the matter (BONTOZOGLOU & Al [2] [3], CHIMONAS [4], HORTON & Al [12], KULIKOSKII & Al [13], COURTIS & Al [6]).

Recently, much interest has arisen concerning the stability of flows involving clear fluid regions contiguous to fluid layers containing particles as dispersed phase. The earliest works on this subject seem to be the ones of DANHO [7] and ACRIVOS & AL [19], [20]. ACRIVOS & Al considered newtonian viscous fluid flow in a channel and they took into account particles contained in a lower fluid layer by means of viscosity coefficient and volume mass depending on the particles concentration; the dynamic and kinematic quantities are then supposed to vary rapidly but continuously through the frontier between the two layers.

In our work we deal with ideal fluids filling the two superposed half spaces $z < 0$ and $z > 0$ (fig. 1-a). They move at different velocities U_1 and U_2 parallel to the x-direction. The linear stability of this Kelvin-Helmholtz system is investigated when the region $z>0$ contains particles as dispersed phase and when gravity is neglected. There are no capillary forces and particles are assumed to be free to leave one embedding fluid, to run into the other and also to reverse (fig. 1-5).

Now, the governing equations (see below) appear to be non-conservative. In order to obtain the jump conditions, we interpret the equations via the formalism of COLOMBEAU's generalized functions [5]. Simutaneously, we are led to split the frontier

Instabilities in Multiphase Flows, Edited by G. Gouesbet
and A. Berlemont, Plenum Press, New York, 1993

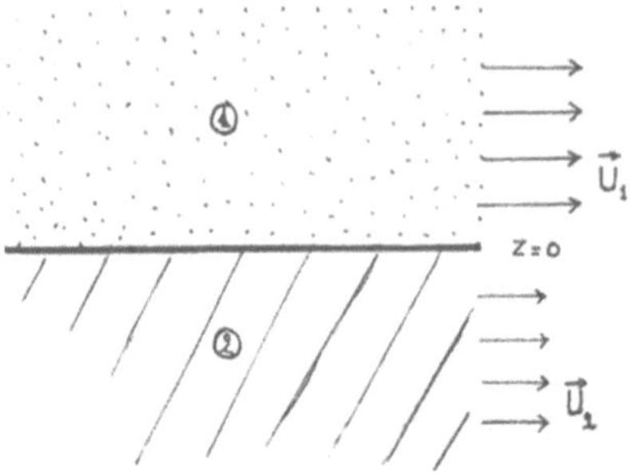

Fig.1-a. Basic stationary flow

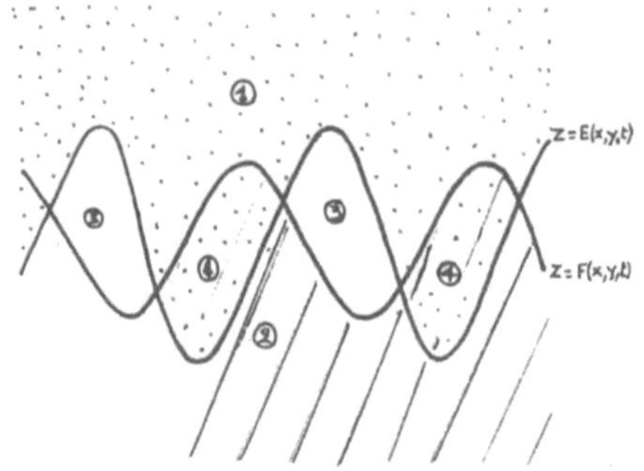

Fig.1-b. Splitting of frontiers related to instability waves

into two surfaces of discontinuity; one of these latter (located at z = E(x,y,t)) stands for the boundary of the half space occupied by particles, whereas the other (located at z = F(x,y,t)) is the frontier between the two layers of embedding fluids. Here it is pointed out that the stability problem is thoroughly solvable when DREW-SEGEL's equations [8] hold in describing continuous dispersed phase media and MARBLE's equations [14] hold in inferring the jump conditions. The eigenvalue relation then proves to be a third degree polynomial equation. With help of numerical calculations, wave parameters are discussed below for the first time.

2- CONSTITUTIVE EQUATIONS FOR CONTINUOUS PARTICULATE MEDIA

The governing equations can be stated with taking care of writing the convective terms in divergence form; according to MARBLE [14] and DREW-SEGEL [8] [9] , they are :

mass conservation of embedding fluid,

$$\frac{\partial}{\partial t}(1-\alpha)\rho + \mathrm{div}(1-\alpha)\rho\mathbf{U} = 0, \tag{1-a}$$

momentum equation for embedding fluid,

$$\frac{\partial}{\partial t}(1-\alpha)\rho\mathbf{U} + \mathrm{div}(1-\alpha)\rho\mathbf{U}\otimes\mathbf{U} = (1-\chi\alpha)\nabla P + \alpha m(\mathbf{G}-\mathbf{U}), \tag{1-b}$$

particles mass conservation,

$$\frac{\partial}{\partial t}\alpha\rho_p + \mathrm{div}\ \alpha\rho_p\mathbf{G} = 0, \tag{1-c}$$

particles momentum equation,

$$\frac{\partial}{\partial t}\alpha\rho_p\mathbf{G} + \mathrm{div}\ \alpha\rho_p\mathbf{G}\otimes\mathbf{G} = -\chi\alpha\nabla P + \alpha m(\mathbf{U} - \mathbf{G}); \tag{1-d}$$

The embedding fluid, as well as the particles, are assumed to be incompressible, therefore :

$$\frac{\partial \rho}{\partial t} + U\, \nabla\rho = 0,$$

or equivalently :

$$\frac{\partial}{\partial t}(1\text{-}\alpha) + \mathrm{div}(1\text{-}\alpha)\mathbf{U} = 0,$$

where ρ and **U** stand respectively for volume mass and fluid velocity; α, ρ_p, **G** designate respectively volume fraction, volume mass and velocity related to particles; the scalar m denotes a drag coefficient, and P is the averaged pressure. A pressure coupling parameter χ has been introduced for conciseness, it is defined by :

$$\chi = \begin{cases} 0 & \text{when dealing with SAFFMAN [18] or MARBLE [14]'s constitutive equations,} \\ 1 & \text{when dealing with DREW-SEGEL's one.} \end{cases}$$

In clear fluid regions, the classical equations for ideal fluids hold true : they can be obtained merely by setting $\alpha = 0$.

3- JUMP CONDITIONS

In order to determine jump conditions, the divergence terms appearing in the system (1) can be treated by means of classical shock procedure; the drag forces can be disregarded since their expression involves no derivatives; there remains the pressure term $\alpha\nabla P$ which amounts to the product of a discontinuous function by a derivative of discontinuous function; such a product can be interpreted but with help of COLOMBEAU's generalized functions theory. Within this theory, $\alpha\nabla P$ turns to be equal to :

$$(\alpha\nabla P) + \frac{\alpha_+ + \alpha_-}{2}\, [[\, P\,]]\, \mathbf{n}\, \delta_\Sigma,$$

where ($\alpha\nabla P$) stands for the function $\alpha\nabla P$ defined everywhere except on Σ ; and δ_Σ is the Dirac distribution associated to Σ ; here $[[\, P\,]] = P_+ - P_-$; the index + or - are referring to each side of the surface of discontinuity Σ .

Then, from the system (1), one can infer the following jump conditions :

$$[[\, (1-\alpha)\, (U_n - W_\Sigma)\,]] = 0 \tag{2-a}$$

$$[[\, (1-\alpha)\, \rho\, (U_n - W_\Sigma)\,]] = 0 \tag{2-b}$$

$$[[\, \alpha\rho_p\, (G_n - W_\Sigma)\,]] = 0 \tag{2-c}$$

$$[[\alpha\rho_p\, (G_n - W_\Sigma)\, \mathbf{G}\,]] + \chi\hat{\alpha}[[\, P\,]]\, \mathbf{n} = 0 \tag{2-d}$$

$$[[\, (1-\alpha)\, \rho\, (U_n - W_\Sigma)\, \mathbf{U}\,]] + (1-\chi\hat{\alpha})\, [[\, P\,]]\, \mathbf{n} = 0 \tag{2-e}$$

Here the index n refers to n-ormal velocities, and W_Σ is the velocity of Σ .

When analysing these jump conditions on z=F(x,y,t) and z=E(x,y,t) , we find that the pressure P is continuous through z=F(x,y,t). The frontier z=E(x,y,t) is found to be a shock wave and P is discontinuous through it as consequence of the α-jump ($\alpha_+ \neq 0$ and $\alpha_- = 0$).

4- LINEARIZATION, MODAL WAVES, AND EIGEN VALUE PROBLEM

According to the classical method, a small perturbation will be superposed to the stationary flow. The dependent functions are thus supposed to take the forms :

$$U(x,y,z,t) = Ui + (\tilde{u}(x,y,z,t), \tilde{v}(x,y,z,t), \tilde{w}(x,y,z,t)),$$

$$G(x,y,z,t) = Ui + (\tilde{q}(x,y,z,t), \tilde{r}(x,y,z,t), \tilde{s}(x,y,z,t)),$$

$$\alpha(x,y,z,t) = \alpha + \tilde{\theta}(x,y,z,t), \qquad P(x,y,z,t) = P + \tilde{p}(x,y,z,t).$$

Now, the equation resulting from the jump conditions for the two surfaces of discontinuity can be added in order to form a single system imposed at z=o ; this one is :

$$w_1(x,y,o,t) = (\frac{\partial}{\partial t} + U_1\frac{\partial}{\partial x})\left[F - \frac{\alpha}{1-\alpha}\sup(0, E-F)\right], \tag{3-a}$$

$$s(x,y,o,t) = (\frac{\partial}{\partial t} + U_1\frac{\partial}{\partial x})E, \tag{3-b}$$

$$(\frac{\partial}{\partial t} + U_1\frac{\partial}{\partial x})w_1 - \frac{1}{1-\alpha}(\frac{\partial}{\partial t} + U_1\frac{\partial}{\partial x})w_2 = -\frac{\alpha}{1-\alpha}(\frac{\partial}{\partial t} + U_1\frac{\partial}{\partial x})(\frac{\partial}{\partial t} + U_2\frac{\partial}{\partial x})E; \tag{3-c}$$

$$p_1(x,y,o,t) = p_2(x,y,o,t). \tag{3-d}$$

The terme sup(0, E - F) which appears in (3-a) is due to frontier crossing. For it, every perturbation quantity can be choosen as being sinusoïdal whereas F(x,y,z,t) is to be sinusoïdal

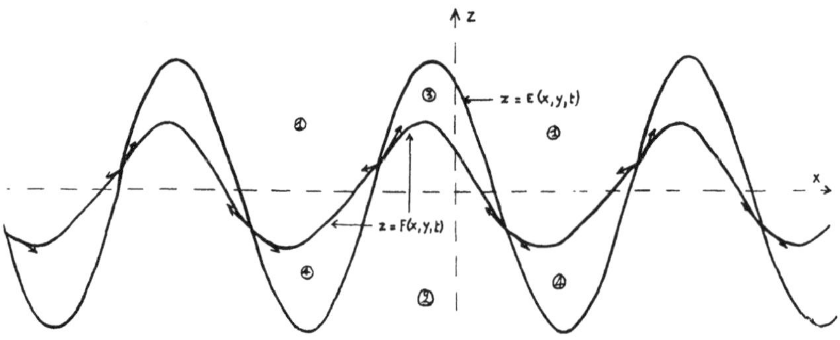

Fig.2. The interface relative to sinusoïdal waves : E(x,y,t) represents the sinusoïdal boundary of the particulate domain; F(x,y,t) represents the frontier of the embedding fluid. This latter one is sinusoïdal, segment by segment.

segment by segment only (see fig 2). We shall thus transform every dependent function A, except F, according to :

$\tilde{A}$ (x,y,z,t) = A (z) exp i (kx+ly+ ωx) ,

where k, l, ω stand for wave numbers and frequency .

All this provides the following eigenvalue problem :

in the region (1) (where z > 0) ,

$$i\rho_1 (kU_1+\omega)\begin{pmatrix}u_1\\v_1\\w_1\end{pmatrix}+\begin{pmatrix}ikp_1\\ilp_1\\p'_1\end{pmatrix}-\frac{\alpha}{1-\alpha}m\begin{pmatrix}q-u_1\\r-v_1\\s-w_1\end{pmatrix}=\begin{pmatrix}0\\0\\0\end{pmatrix} \quad \text{(4-a)}$$

$$i(ku_1+lv_1)+w_1'+\alpha(1-\alpha)^{-1}(kq+ilr+s')=0 \quad \text{(4-b)}$$

$$i\rho_r (kU_1+\omega)\begin{pmatrix}q\\r\\s\end{pmatrix}+\begin{pmatrix}ikp_1\\ilp_1\\p_1'\end{pmatrix}-m\begin{pmatrix}u_1-q\\v_1-r\\w_1-s\end{pmatrix}=\begin{pmatrix}0\\0\\0\end{pmatrix} \quad \text{(4-c)}$$

in the région (2) (where z < 0) ,

$$i\rho_2 (kU_2+\omega)\begin{pmatrix}u_2\\v_2\\w_2\end{pmatrix}+\begin{pmatrix}ikp_2\\ilp_2\\p_2'\end{pmatrix}=\begin{pmatrix}0\\0\\0\end{pmatrix} \quad \text{(5-a)}$$

$$i(ku_2+lv_2)+w'_2=0 \quad \text{(5-b)}$$

at z = 0 ,

$$(kU_1+\omega)E=s \quad \text{(6-a)}$$

$$p_1=p_2 \quad \text{(6-b)}$$

$$(1-\alpha)(kU_2+\omega)w_1-(kU_1+\omega)w_2+\alpha(kU_2+\omega)s=0\,. \quad \text{(6-c)}$$

Futhermore, we can suppose the perturbations to be attenuated at infinity.

5- THE EIGENVALUE EQUATION

This eigenvalue problem leads to a spectrum comprising a continuous spectrum and a proper spectrum. The continuous spectrum results from the boundless domain of the flow, and the proper spectrum, for which the previous system admits a non null solution, is defined by :

$$\rho_1\left[\bar{\alpha}\rho_1+\alpha\rho_r-i\bar{\alpha}\rho_r\tau k(c-U_1)^2\right](c-U_1)^2+\rho_2\left[\rho 1-i\bar{\alpha}(\alpha\rho_1+\bar{\alpha}\rho_r)\tau k(c-U_1)\right](c-U_2)^2=0$$

$$\bar{\alpha}=1-\alpha$$

where $c=-\omega k^{-1}$ is the wave velocity and the parameter $\tau=\rho_1 m^{-1}$ is homogeneous to a relaxation time. This eigenvalue equation is a cubic equation with respect to the unknown c and, hence, there exist three proper eigenvalues. Among the three roots of (7), two ones form a conjugate pair approaching the two classical Kelvin-Helmholtz modes :

$$c = \frac{\rho_1 U_1 + \rho_2 U_2 \pm i(U_2 - U_1)\sqrt{\rho_2 \rho_1}}{\rho_1 + \rho_2} \qquad \text{(when } \alpha \ll 1\text{)}.$$

The third root of (7) is revealed as corresponding to a new mode; it is always stable, and, it tends toward the value $c = U_1 - i\rho_1 \rho_p^{-1} (k\tau)^{-1}$ (when $\alpha \ll 1$). Just as for clear fluids, the most unstable waves are the short waves. When α either vanishes or equals 1, as also when k either vanishes or becomes infinite, the conjugate modes satisfy the well known relation of Kelvin -Helmholtz :

$$(c-U_1)^2 + a(c-U_2)^2 = 0,$$

where $a > 0$ is depending upon ρ_1, ρ_2, ρ_p and α.

We can prove that, in the complex plane (c_r, c_i), the points verifying (8) are located on the circle of radius $|U_2 - U_1|/2$ centered on the x axis at $x = (U_1+U_2)/2$, say the Kelvin-Helmholtz's circle. On account of this occurence, the stability of Kelvin -Helmholtz two phase flow will be studied with regarding this circle.

6- NUMERICAL RESULTS

The stability curves can be drawn in the c- plane. There is always an unstable mode within the pair of conjugate eigenvalues. When the wave number k is varied from $-\infty$ to $+\infty$, this mode describes a closed curve (fig.3,4,5). The set of these closed curves for different values of a are stituated in the neighbourhood of Kelvin -Helmholtz's circle. This circle separates eigen modes related to positive values of k (solid curves) from the ones related to negative values of k (dashed curves). Three cases must be distinguished :

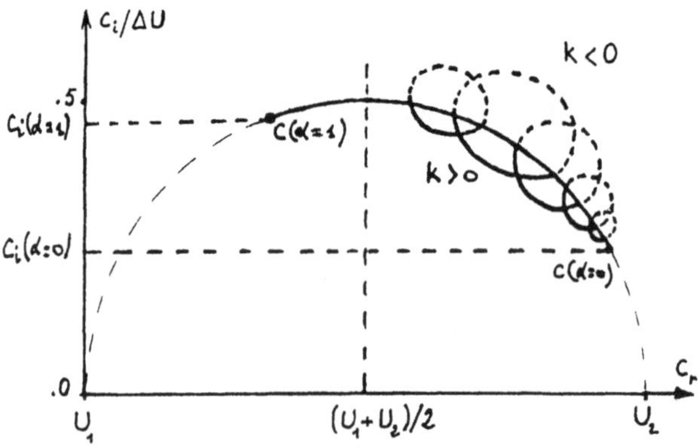

Fig.3.- Sketch relative to flow destabilization

1) If $c_i(\alpha=0) < c_i(\alpha=1)$ (fig.3), then for any value of the concentration α, the flow will be destabilized on the whole since all the stability curves will be above the straight line $c_i = c_i(\alpha=0)$.

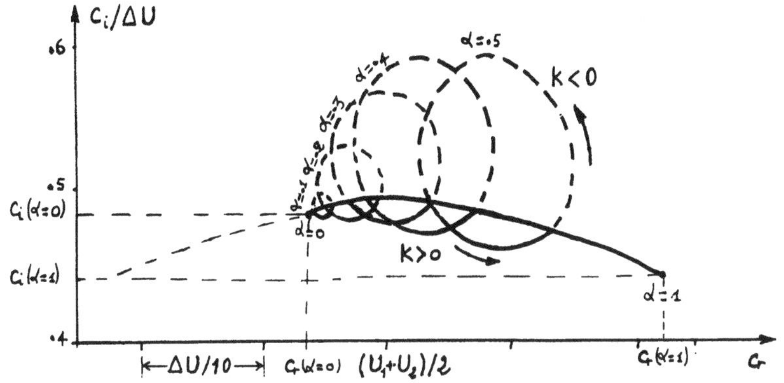

Fig.4. Stability curves when the flow is destabilized, and stabilized next.

2) When $c_i(\alpha=0) > c_i(\alpha=1)$, as also when $c_r = (U_1+U_2)/2$ pertains to $(c_r(\alpha=0),c_r(\alpha=1))$ (fig.4), there exist two values α_m and α_M of concentration such that :

a) for $\alpha < \alpha_m$, the flow is destabilized owing to particle presence.

b) for $\alpha_m \leq \alpha \leq \alpha_M$, there exist particular values $k_0(\alpha)$ and $k_1(\alpha)$ such that for wave numbers $k \in [k_0(\alpha), k_1(\alpha)]$ the flow will be less unstable, whereas for wave numbers k such that $0 < k < k_0(\alpha)$ or $k_1(\alpha) < k < -\infty$ the partcles will reinforce the instability of the flow.

c) for $\alpha > \alpha_M$, the flow is less unstable.

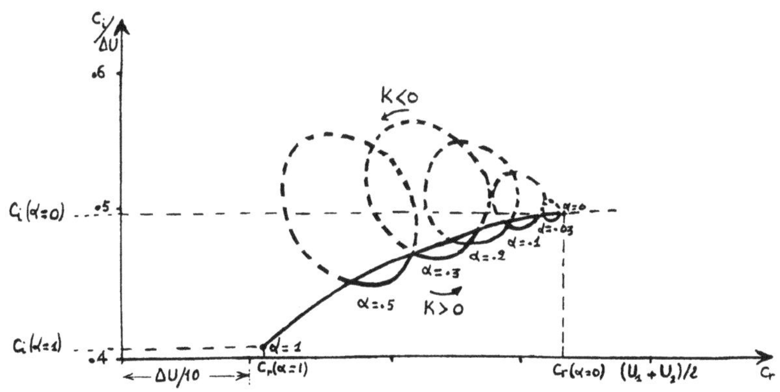

Fig.5. Stability curves when the flow is stabilized

3) When $c_i(\alpha=0) > c_i(\alpha=1)$ and $c_r = (U_1+U_2)/2$ does not pertain to $[c_r(\alpha=0), c_r(\alpha=1)]$, the flow is less unstable.

As we can easily calculate the values of c when $\alpha = 0$ or 1, the stability conditions we discussed previously can be expressed via a simple criteria depending on the mass parameters ρ_1, ρ_2 and ρ_p , as follows :

1) if ρ_1 lies between ρ_2 and $\sqrt{\rho_1\rho_p}$, there is stabilization,

2) if ρ_2 lies between ρ_1 and $\sqrt{\rho_1\rho_p}$, there is destabilization, and stabilization next,

3) if $\sqrt{\rho_1\rho_p}$ lies between ρ_1, ρ_2, there is destabilization.

7- CONCLUSION AND COMMENTS

For Kelvin-Helmholtz two-phase flow (fig.1), we have performed a study of stability based upon the constitutive equations of both MARBLE [14] and DREW-SEGEL [8]. No capillary forces are taken into account, and therefore the particles are allowed to leave an embedding fluid and to run into the other one. In our model, DREW-SEGEL's equations are used to describe the continuous flow regions whereas jump conditions are deduced from MARBLE's ones. It turns out that the interface dividing the uniform flows comprises two surfaces of discontinuity; the boundary of the particulate region stands for a sinusoïdal wave, whereas the fluid frontier is fitted with sinusoïds segment by segment only. The former surface of discontinuity stands for a shock wave. The solution of the eigenvalue problem related to this model can be thoroughly carried out by means of algebra. We have found out a continuous spectrum and three proper modes. One proper mode alone is unstable; it is standing close to the classical Kelvin-Helmholtz unstable mode. Therefore, in spite of the presence of particles, the flow is always unstable whatever the volume ratio α. It is shown that for ρ_1 lying between ρ_2 and $\sqrt{\rho_1\rho_p}$ the flow is less unstable, whereas it is destabilized when $\sqrt{\rho_1\rho_p}$ lies between ρ_1 and ρ_2. If ρ_2 is situated between ρ_1 and $\sqrt{\rho_1\rho_p}$ the flow is first destabilized and then less unstable when α increased from 0 to 1.

BIBLIOGRAPHY

1. G.K. BATCHELOR, A new theory of instability of a uniform fluidized bed - J.F.M. (1988) vol. 193, pp. 75-110.
2. V.BONTOZOGLOU , T.J.HANRATTY, Effects of finite depth and current velocity on large amplitude Kelvin-Helmholtz waves - J.F.M. (1988) vol.196, pp. 187-204.
3. V.BONTOZOGLOU, Weakly non linear Kelvin-Helmholtz waves between fluids on finite depth - Intl J. Mult. Fl., (1991) vol.17 n°4, pp. 509-518.
4. G.CHIMONAS,The combined Rayleigh, Kelvin-Helmholtz problem - Phys. Fl., (1986) vol. 29 (7), pp. 2061-2066.
5. J.F. COLOMBEAU, Elementary Introduction to new Generalized Functions - North Holland, Math Studies 113 Amsterdam (1985).
6. N.COURTIS, J.M. DELAHAYE , R. NAKACH - Two-phase flow modelling : the closure issue for a two-layer flow. Intl J.Mult. Fl., (1989) vol 15, n°6, pp. 977-983 .
7. E.DANHO - Etude de la stabilité des ecoulements diphasiques cisaillés et stratifiés : Détermination des conditions aux discontinuités pour les systèmes non conservatifs à l'aide des distributions non linéaires - Ph. D. Paris 6 University Paris (1989).
8. D.A.DREW & L.A.SEGEL, Averaged equations for two-phase flows-Studies in Applied Math, (1971) vol n°3, pp. 205-231 .
9. D.A.DREW, G.S. ARNOLD , R.T.LAHEY. Relation of microstucture to constitutive equations - Two phase flows and waves - The IMA vol. in Math. and its Appl. vol 26 , pp. 45-56 - edited by D.D. JOSEPH and D.G. SCHAEFFER - Springer-Verlag Berlin (1990).

10. D.A.DREW, LAHEY, The virtual mass and lift force on a sphere in rotating and straining inviscid flow - Intl J.Mult. Fl., (1987) vol 13, n°1, pp.113-121.
11. R.ESTRADA , R.P. KANWAL, Distributional analysis for discontinuous fields - J. of Math. Anal. and Appl. (1985) vol. 105, n°2, February 1 .
12. W.HORTON, T.TAJIMA & T.KAMIMURA, Kelvin-Helmholtz instability and vortices in magnetized plasma - Phys. Fl., (1987) vol 30 (11), pp. 3485-3495.
13. A.G.KULIKOVSKII & I.S. SHIKINA, On the asymptotic behaviour of localized perturbations in the presence of Kelvin-Helmholtz instability - Fluid Dynamics, (1985) n°2, pp 186-193.
14. F. E.MARBLE, Dynamics of Dusty Gases - Ann. Rev. Fl. Mech., (1970) vol. 2, pp. 397-446.
15. A.PROSPERETTI , J.V.SATRAPE, Stability of two-phase flow models - in Two phase flows and waves , edited by D.D.JOSEPH & D.G.SCHAEFFER, IMA volumes in maths and its Appl., Springer-Verlag Berlin (1990), pp.98-118.
16. J.MOREAU, Jump functions of a real interval to a banach space - Annales Faculté des Sciences de Toulouse (1989), vol. 97.sp.
17. H.GOUIN, Variational theory of mixtures in continum mechanics - Eur. J. Mech., B/Fluids, (1990) vol. 9, n°5, pp.469-491.
18. P.G.SAFFMAN, On the stability of laminar flow of a dusty gas - J.F.M., (1962) vol.13, pp.120-128.
19. U.SCHFLINGER, A. ACRIVOS & K.ZHANG, Viscous resuspension of a sediment within a laminar and stratified flow - Int. J.Mult.fl., (1990) vol. 16, n°4, pp. 567-578.
20.K.ZHANG, A. ACRIVOS & U.SCHAFLINGER, Stability in a two dimensional Hagen-Poiseuille resuspension flow - Intl J.Mult.Fl., (1992) vol 18, n°1, pp. 51-63.
21. Yu. V. EGOROV, A contribution to the theory of generalized functions - Russian Math. Surveys (1990) 45 : 5 pp.1-49.
22. D.D.JOSEPH, T.S.LUNDGREN, RJACKSON & D.A.SEVILLE, Ensemble averaged and mixture theory equations for incompressible fluid - particle suspensions - Int J.Mult. Fl., (1990) vol 16, n°1,pp.35-42.
23. D.D.JOSEPH & D.G. SCHAEFFER, Two phase flows and waves - The IMA vol. math. appl., Springer Verlag, Berlin (1990).
24. Ch.B.NAMURATOV & A.S.SOLOVEV, Effect of suspended particles on the stability of plane Poiseuille flow - Fl.Dynamics (1986) n°1, pp.38-44.

46. D. DREW, LAHEY, The virtual mass and lift force on a sphere in rotating and straining [illegible] 1987 [illegible]

[illegible]

INSTABILITIES, WAVES AND SLUGS IN PIPELINE TWO-PHASE FLOWS

Malcolm Watson

BP Research
Chertsey Road
Sunbury-on-Thames
Middlesex, TW16, 7LN
United Kingdom

SUMMARY

This paper is concerned with multi-phase fluid flow models that can be represented by systems of one-dimensional hyperbolic equations, with particular application to pipeline two-phase flows. It is shown how the governing equations can be rewritten in a form which enables analogies to be drawn with the flow of liquid down open, inclined channels. For such flows, smooth periodic solutions do not satisfy the basic equations but a train of piecewise smooth waves joined together by shocks can be constructed. These waves evolve non-linearly to their asymptotic form following a physical instability occurring in the flow. For pipeline two-phase flows, the top of the pipe puts a constraint on the maximum amplitude of the waves but also encourages the growth of the instability. When the wave bridges the pipe a liquid slug is formed, which is often an unwelcome problem in production facilities, especially offshore.

These ideas can be extended to other systems of multi-phase flows. For example, a third phase can be added to pipeline flows and the governing equations can be rearranged into a form enabling similar solutions to be obtained. Dispersed two-phase flows and granular flows can also be represented by analogous sets of equations with similar solutions for appropriate values of the physical parameters. These types of solutions, viz piecewise smooth waves joined together by shocks are typical for systems of hyperbolic equations with source terms indicating that instabilities and growth of waves in a wide variety of multi-phase systems are closely related phenomena.

INRODUCTION

Understanding pipeline multi-phase flows is of increasing importance to the oil and gas industry, especially with the development of small offshore reservoirs. Depending on the respective flowrates, many flow regimes are possible. If both the oil and gas flowrates are small the gas flows smoothly over the stratified oil layer underneath. As the flowrates increase an instability may occur on the interface between the oil and gas leading to the formation of large amplitude waves which may bridge the pipe and form slug flow. Current methods for the prediction of flow regimes are based on either incomplete mathematical models or on experimental data, typically obtained from small diameter pipelines with water and air often the operating fluids.

This paper describes solutions of the pipeline one-dimensional equations for gravitationally separated two-phase flows. These equations can be manipulated into a form which enables analogies to be made with liquid flows down open inclined channels. It is believed that the same ideas can be applied to many other multi-phase systems with corresponding similar solutions.

BASIC EQUATIONS

We consider one-dimensional fluid flow models that can be represented by hyperbolic systems of the form

$$\frac{\partial \mathbf{U}}{\partial t} + \frac{\partial \mathbf{F}}{\partial x} = \mathbf{G}, \tag{1}$$

where $\mathbf{U}$ is the vector of dependent variables, $\mathbf{F(U)}$ contains the flux terms and $\mathbf{G(U)}$ consists of the sources. In particular we look at systems such that non-trivial equilibrium flows $\mathbf{U_0}$ exist satisfying

$$\mathbf{G(U_0)} = \mathbf{0}. \tag{2}$$

We then examine the stability of these equilibrium flows to small disturbances. For appropriate values of the physical parameters the flow may be unstable and will evolve non-linearly to large amplitude waves. These ideas may be explained more easily by looking at specific examples.

LIQUID FLOW DOWN AN OPEN INCLINED CHANNEL

Consider liquid flow down an open, inclined channel with friction, as described by Dressler (1949). The one-dimensional, shallow water equations for the conservation of mass and momentum may be written in the dimensionless form

$$\frac{\partial h}{\partial t} + \frac{\partial}{\partial x}(hu) = 0, \tag{3}$$

$$\frac{\partial}{\partial t}(hu) + \frac{\partial}{\partial x}\left(hu^2 + \frac{1}{2}h^2\right) = \lambda h - u^2, \tag{4}$$

where the liquid depth is h and the fluid velocity is u. The positive parameter λ measures the relative importance of the gravitational and frictional terms.

The equilibrium solutions, given by a zero subscript, satisfy

$$\lambda h_0 - u_0{}^2 = 0. \tag{5}$$

Hence if we look for small perturbations from the equilibrium solutions by setting

$$h = h_0 + h', \; u = u_0 + u', \tag{6}$$

in equations (3) and (4), linearising in the perturbation quantities and looking for normal mode solutions with x and t dependence of the form $\exp ik(x-ct)$ the resulting dispersion equation is:

$$(c-u_0)^2 + 2i\frac{u_0}{kh_0}(c-u_0) - \frac{i\lambda}{k} - h_0 = 0. \tag{7}$$

The stability boundary is found by setting the imaginary part of c to be zero and by using (5) the critical value λ_c is given by

$$\lambda_c = 4, \tag{8}$$

with instability for $\lambda > \lambda_c$. At the onset of instability all wavelengths are unstable.

An alternative approach is to follow the ideas of Dressler (1949) and look for travelling wave solutions of the full non-linear equations. If the moving coordinate

$$X = x - ct, \tag{9}$$

where c is the wavespeed, is introduced, then equations (3) and (4) become

$$\frac{d}{dX}h(u-c) = 0 \tag{10}$$

$$\frac{dh}{dX} = \frac{\lambda h - u^2}{h - (u-c)^2}. \tag{11}$$

Dressler showed that it was not possible to construct periodic smooth solutions of (10) and (11). However, it is possible to construct a train of periodic, piecewise smooth waves joined together by shocks. In order to construct these waves, Dressler showed that there must be a critical liquid height, h_* and velocity u_*, say, such that

$$h_* - (u_*-c)^2 = 0. \tag{12}$$

Equation (12) determines the wavespeed c in terms of the critical quantities. Using (12), (11) becomes (removing the root at $h = h_*$)

$$\frac{dh}{dX} = \frac{\lambda h^2 - (1+2\lambda^{1/2})h_* h + h_*^2}{h^2 + h_* h + h_*^2}. \tag{13}$$

There is another liquid height, h_{min} say, such that $dh/dX = 0$ and this is given by

$$\frac{h_{min}}{h_*} = \frac{(1+2\lambda^{1/2}) + (1+4\lambda^{1/2})^{1/2}}{2\lambda}. \tag{14}$$

Equation (13) determines the smooth part of the wave but it does not admit periodic solutions. In order to construct a train of waves they must be joined together by means of shocks where the liquid height jumps from a level h_+ to h_- where

$$h_{min} < h_- < h_* < h_+, \tag{15}$$

and $h_\pm$ (and also $u_\pm$) satisfy the usual jump conditions

$$-c[h] + [hu] = 0 \tag{16}$$

$$-c[hu] + [hu^2 + \frac{1}{2}h^2] = 0. \tag{17}$$

The construction of the waves is indicated in Figure 1.

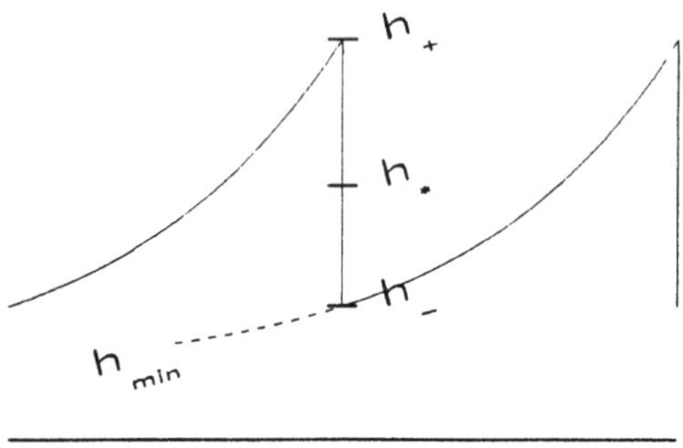

Figure 1. Construction of roll waves.

In order for these waves to exist the inequalities in (15) must be satisfied. The limiting case is given when the inequalities are replaced by equalities and this gives a critical value of $\lambda_c = 4$. This is the same as obtained above from the stability analysis and it is what we would expect. It is also interesting to note that for a given value of $\lambda > \lambda_c$ a whole range of wavelengths is possible and described parametrically by varying h_- between h_* and h_{min} with the longest waves having the largest amplitudes when $h_- = h_{min}$.

Another approach is to investigate equations (3) and (4) directly by numerical techniques. This has been done by Watson (1990) who solved (3) and (4) using the two-step Lax-Wendroff method with the anti-diffusion of Boris and Book to remove spurious oscillations at shocks (see for example Sod, 1985). It would be expected that for values of $\lambda > \lambda_c$ small disturbances to a uniform flow will grow non-linearly to the roll waves described by Dressler and also above. For subcritical values $\lambda < \lambda_c$ small disturbances will decay. Figure 2 shows the growth of small sinusoidal disturbances into the non-linear roll waves for a supercritical case with $\lambda = 5$. The dotted line shows the undisturbed equilibrium value for the liquid height. The paper by Watson gives more results which also demonstrate the dependence of the final roll wave solutions on the frequency of the disturbances and the independence of the roll waves on the amplitude of the disturbances.

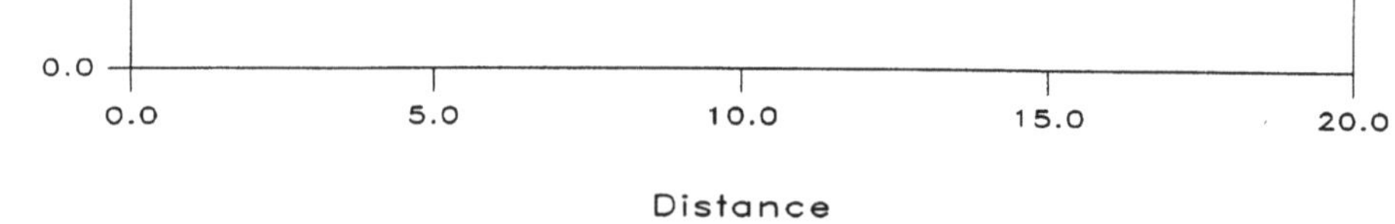

Figure 2. Growth of small disturbance into roll waves.

PIPELINE TWO-PHASE FLOWS

We wish to show that similar ideas may be applied to pipeline two-phase flows. We consider incompressible, gravitationally separated, two-phase flow of liquid and gas in a circular pipe of diameter d inclined at an angle θ to the horizontal, as shown in Figures 3 and 4.

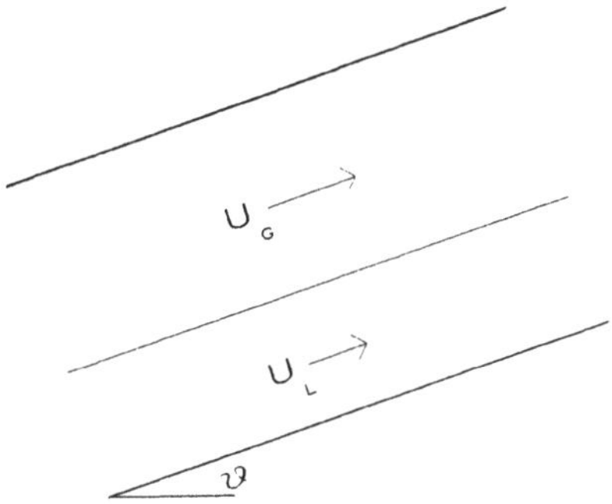

Figure 3. Side view of circular pipe inclined at angle θ to the horizontal.

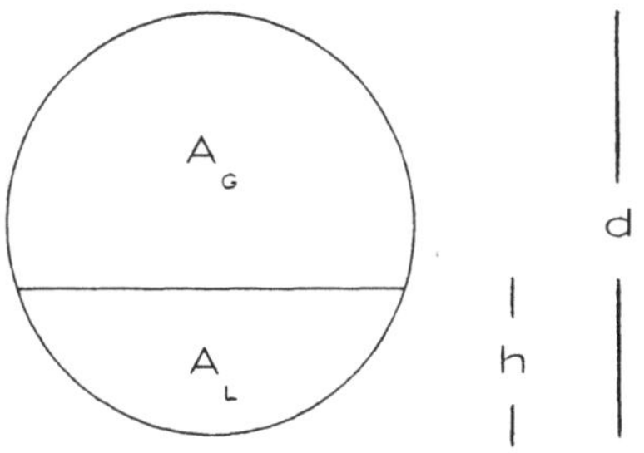

Figure 4. Cross-section of circular pipe.

If the flow properties are averaged over the cross-section then the mass and momentum conservation equations for each phase are

$$\frac{\partial}{\partial t}(\rho_L A_L) + \frac{\partial}{\partial x}(\rho_L A_L U_L) = 0, \tag{18}$$

$$\frac{\partial}{\partial t}(\rho_L A_L U_L) + \frac{\partial}{\partial x}(\rho_L A_L U_L^2) = -A_L \frac{\partial p_I}{\partial x} - \rho_L g \cos\theta \, A_L \frac{\partial h}{\partial x}, \quad - \rho_L g \sin\theta \, A_L + \tau_I S_I - \tau_L S_L, \tag{19}$$

$$\frac{\partial}{\partial t}(\rho_G A_G) + \frac{\partial}{\partial x}(\rho_G A_G U_G) = 0, \tag{20}$$

$$\frac{\partial}{\partial t}(\rho_G A_G U_G) + \frac{\partial}{\partial x}(\rho_G A_G U_G^2) = -A_G \frac{\partial p_I}{\partial x} - \rho_G g \cos\theta \, A_G \frac{\partial h}{\partial x}, \quad - \rho_G g \sin\theta \, A_G - \tau_I S_I - \tau_G S_G, \tag{21}$$

where S_L and S_G are appropriate wetted perimeters, S_I is the interfacial chord and τ_I, τ_L τ_G are the usual interfacial, liquid/wall and gas/wall shear stress respectively. For single phase channel flow (18) and (19) simplify to (3) and (4) for the liquid flow in an open inclined channel.

The set (18)-(21) may be rewritten in conservation form (Watson, 1990) by forming a total mass conservation equation by adding together (18) and (20) to give

$$\frac{\partial}{\partial t}(\rho_L A_L + \rho_G A_G) + \frac{\partial}{\partial x}(\rho_L A_L U_L + \rho_G A_G U_G) = 0. \tag{22}$$

The interfacial pressure p_I may be eliminated from the two momentum equations (19) and (21) to give

$$\frac{\partial}{\partial t}(\rho_L U_L - \rho_G U_G) + \frac{\partial}{\partial x}\left(\frac{1}{2}\rho_L U_L^2 - \frac{1}{2}\rho_G U_G^2 + (\rho_L-\rho_G)g\cos\theta\, h\right)$$

$$= -(\rho_L-\rho_G)g\sin\theta + \left(\frac{1}{A_L} + \frac{1}{A_G}\right)\tau_I S_I$$

$$+ \frac{1}{A_G}\tau_G S_G - \frac{1}{A_L}\tau_L S_L. \tag{23}$$

As shown by Whitham (1974) it is possible in general to construct a whole hierarchy of conservation equations. For an alternative version of the difference of momentum equation for two-phase flows see Watson (1990).

The two hyperbolic equations (22) and (23) must be supplemented with two algebraic equations. One is obtained from the geometric constraint that the areas occupied by the liquid and gas phases must fill the pipe, so that

$$A_L + A_G = A. \tag{24}$$

Since the phases have been assumed to be incompressible, (18) and (20) may be used to give

$$A_L U_L + A_G U_G = Q(t), \tag{25}$$

where $Q(t)$ is a known function of time dependent on the inlet flowrates.

The equilibrium flow satisfying the RHS of (23) is zero forms the basis of the work by Taitel and Dukler (1976). For low flowrates the interface between the liquid and the gas will be quite smooth and their approach works quite well. For increased flowrates waves will appear on the interface and the neglect of the LHS of (23) is invalid. In order to take account of waves a common practice is to modify the interfacial friction term but this is not really satisfactory. The approach in this paper is to retain all terms in (23) and not make any further approximations.

The two-phase flow equations have been put in a similar form to the equations governing liquid flow in an open channel, equations (3) and (4) above. It might be expected, therefore, that similar solutions will exist. In particular we could expect that periodic, large amplitude, piecewise smooth, travelling waves joined together by shocks analogous to Dressler's roll waves will satisfy equations (22)-(25). Such roll wave solutions of the pipeline two-phase flow equations have been investigated by Miya et al (1971), Cercignani and Battara (1983) and Watson (1989). It is also possible to examine the stability of the flow to infinitesimal disturbances. This approach has been taken by Ferschneider et al (1985) and Wu et al (1987). It was shown above for liquid flow in open channels that the criterion for the onset of the roll waves is exactly the same as the classical stability boundary. This is also the case for the pipeline two-phase flow equations (Watson,

1989) since the limiting case of zero amplitude Dressler type roll waves coincides with the stability analyses of Ferschneider et al (1985) and Wu et al (1987). Again, on the stability boundary all wavelengths are unstable and in the non-linear regime the largest amplitude waves have the longest wavelengths.

As for the liquid flows in channels it is also possible to solve the two-phase equations (22)-(25) numerically using finite difference techniques. This has been done by Watson (1990) again using the two-step Lax-Wendroff method with Boris and Book anti-diffusion. Figure 5 show a numerical simulation demonstrating the growth of waves on the interface between the liquid and the gas. The results are for a horizontal pipe with air and water being the operating fluids. A small sinusoidal disturbance of amplitude 0.1% has been imposed on the inlet flowrates and the results show that the waves grow very quickly to their asymptotic form. The hold-up is the fraction of the cross-sectional area occupied by the liquid (A_L/A). More results may be found in Watson (1990). As for the liquid flows in open channels, the final asymptotic forms are independent of the amplitude of the initial disturbance but the wavelength is dependent on the frequency.

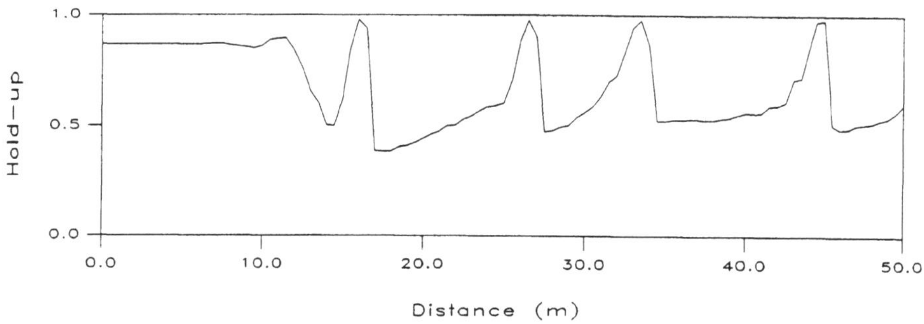

Figure 5. Growth of small disturbance into large amplitude waves for two-phase flows.

It is not necessary to impose a disturbance in the flowrates in order to generate the waves. Figures 6 and 7 show results for two-phase flow into a dip, although the pipe has been straightened out for presentation purposes.

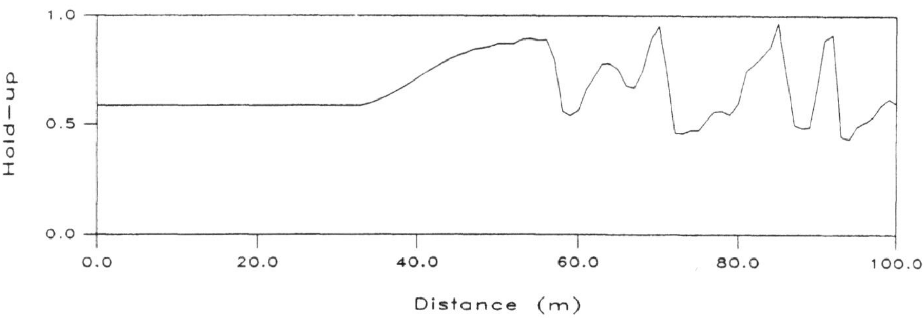

Figure 6. Large amplitude waves generated by dip in pipeline.

The pipe is 100 m long comprising two sections of pipe of length 50 m with inclinations of -1° and +1°. In general for the same flowrates, the equilibrium flows in downward sloping pipes are more stable than those in upward sloping pipes. The inlet flowrates are perfectly steady but it can be clearly seen that large amplitude waves are generated by the dip and exist in the upward sloping part of the pipe. Figure 7 is for the same flowrates as in Figure 6 but with a smaller radius of curvature in the pipe at the dip. The resulting asymptotic form has waves of smaller amplitude and shorter wavelength. The disturbance caused by the dip is enough to lead to the formation of large amplitude waves even though the inlet flowrates are perfectly steady.

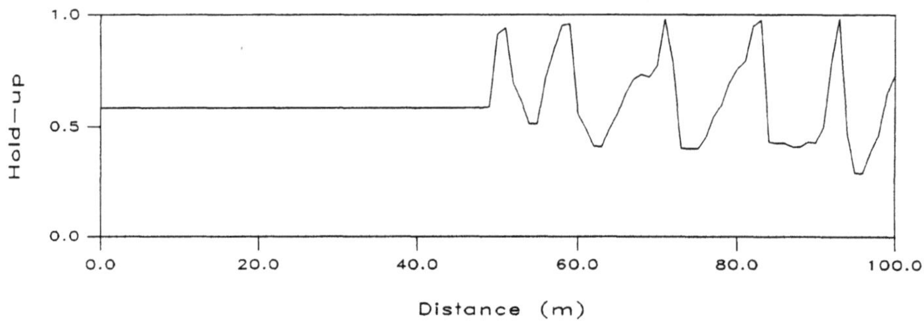

Figure 7. Large amplitude waves generated by dip in pipeline. The radius of curvature at the dip is smaller than for the results shown in Figure 6.

OTHER FLOWS

Pipeline Three Phase Flows

Similar ideas may be adopted for other multi-phase flows. For example pipeline three-phase flows may be approached in the same way. For the third phase (typically water), the conservation equations for mass and momentum are of the same form as those in equations (5)-(8) above. The six governing partial differential equations can be reduced to four equations supplemented with two algebraic equations by eliminating the two interfacial pressures. The same ideas as above may be adopted for the resulting set of equations and numerical results are described in Watson (1992).

Granular Flows

For the flow of granular materials (eg Savage, 1988), simple one-dimensional models lead to equations very similar in form to Dressler's equations for shallow water flow, but with more complicated frictional terms. Once again, the same ideas can be adopted and similar results could be expected for appropriate values of the physical parameters.

Dispersed Flows

The equations for dispersed two-phase flows are exactly of the same form as those above for separated two-phase flows, but with modified frictional terms and the hold-ups need to be interpreted as particle concentrations. Elimination of the interfacial pressure again leads to equations of the form (1) and (2). Once more, similar results are expected to hold. Indeed, the book by Wallis (1969) contains photographic plates of experiments by Thomas (1964) showing periodic phenomena of spherical particles in horizontal pipes. If the hold-up results shown in Figures for separated two-phase flows are re-interpreted for particle concentrations they immediately give a qualitative description of the periodic concentration profiles with the sharp concentration shocks of Thomas (1964).

CONCLUSIONS

This paper has concentrated on pipeline, gravitationally separated, two-phase flows which are of great importance and interest to the oil and gas industries. For constant inlet oil and gas flowrates the flow often develops instabilities on the interface leading to large amplitude waves and slugs which are often an unwelcome problem in production facilities, especially offshore.

Simple one-dimensional models for pipeline two-phase flows have been shown to lead to governing equations of the form (1) and (2) which may be analysed in exactly the same way as for Dressler's model for liquid flow down an open, inclined channel. It has been shown that the solutions are perfectly analogous to those discussed by Dressler. However, in order to demonstrate this, it is necessary to manipulate the usual governing equations into a suitable form enabling the analogy to be made easily. This manipulation of the equations is also in a form suitable for efficient computation using well established numerical techniques.

The governing equations for other multi-phase systems may be written in the vector form given by (1), such as pipeline three-phase flows, granular flows and dispersed two- (and three-)phase flows. Similar types of solutions may be expected to exist, indicating that the physical mechanism leading to the formation of periodic, large amplitude waves with sharp gradients, or shocks, is the same for a wide variety of multi-phase flows as for the Dressler problem of liquid flow down open, inclined channels.

ACKNOWLEDGEMENTS

The author wishes to thank The British Petroleum Company plc for permission to publish this paper.

REFERENCES

Cercignani, C. and Battara, V., 1983, A mathematical model for slug frequency, Proc Int Conf on the Physical modelling of Multi-Phase Flow (Coventry, England, 19-21 April, 1983), Cranfield, UK, BHRA, The Fluid Engineering Centre, Paper D3.

Dressler, R.F., 1949, Mathematical solution of the problem of roll-waves in inclined open channels, Commun Pure Applied Maths, 2:149.

Ferschneider, G., Lagiere, M., Bourgeois, T. and Fitremann, J.M., 1985, How to calculate two-phase flow of gas and oil in pipelines, Pipeline Industry, 63:33.

Miya, M., Woodmansee, D.E. and Hanratty, T.J., 1971, A model for roll waves in gas-liquid flow, Chem Eng Sci, 26:1915.

Savage, S.B., 1988, Flow of granular materials, Proc XVII Int Congress of Theoretical and Applied Mechanics (Grenoble, France, 21-27 August, 1988), North Holland.

Sod, G.A., 1985, "Numerical Methods in Fluid Dynamics," Cambridge University Press, Cambridge.

Taitel, Y. and Dukler, A.E., 1976, A model for predicting flow regime transitions in horizontal and near horizontal gas-liquid flow, AIChE J, 22:47.

Thomas, D.G., 1964, Periodic phenomena observed with spherical particles in horizontal pipes, Science, 144:534.
Wallis, G.B., 1969, "One-Dimensional Two-Phase Flow," McGraw-Hill, New York.
Watson, M., 1989, Wavy stratified flow and the transition to slug flow, Proc 4th Int Conf on Multi-Phase Flow (Nice, France, 19-21 June, 1989), Cranfield, UK, BHRA, The Fluid Engineering Centre, Paper G3.
Watson, M., 1990, Non-linear waves in pipeline two-phase flows, Proc 3rd Int Conf on Hyperbolic Problems (Uppsala, Sweden, 11-15 June, 1990), Studentlitteratur, Lund.
Watson, M., 1992, A model for pipeline three-phase flows, To appear in: Proc 4th Int Conf on Hyperbolic Problems (Taormina, Italy, 3-8 April, 1992).
Whitham, G.B., 1974, "Linear and Nonlinear waves," John Wiley & Sons, New York.
Wu, H.L., Pots, B.F.M., Hollenburg, J.F. and Meerhoff, R., 1987, Flow pattern transitions in two-phase gas/condensate flow at high pressure in an 8-inch horizontal pipe, Proc 3rd Int Conf on Multi-Phase Flow (The Hague, Netherlands, 18-20 May, 1987), Cranfield, UK, BHRA, The Fluid Engineering Centre, Paper A2.

INSTABILITIES AND THE FORMATION OF WAVE PATTERNS IN FLUIDIZED BEDS

Manfred F. Göz

Kernforschungszentrum Karlsruhe
Institut für Neutronenphysik und Reaktortechnik
Postfach 3640
D-7500 Karlsruhe 1
Germany

ABSTRACT

A two-phase flow model used for the description of fluidized beds is analyzed with the help of bifurcation theory. It is shown that the equations admit a variety of nonlinear concentration waves which may be assigned to experimentally observed flow patterns.

INTRODUCTION

The principal form of multi-phase flow equations is quite well-known now, whereas the detailed modeling, e.g. of interaction forces, is controverted. On the other hand, the analytical investigation of existing models is only at its beginning. In this contribution we will not bother about the reasoning of a standard model used for the description of the flow of two miscible materials – one fluid and one solid phase consisting of a huge number a small rigid particles – in a fluidized bed. Instead we shall use this exemplary model to show how such systems can be analyzed very effectively using bifurcation theoretical methods.

After introducing the model equations we first consider an infinitely extended one-dimensional bed. Using a special ansatz for travelling waves, the set of partial differential equations can be reduced to a two-dimensional dynamical system which can be studied extensively by phase plane methods. We find several branches of stationary and periodic solutions with different stability properties. The interaction of these branches leads to global bifurcations, namely homoclinic and heteroclinic saddle point connections. These solutions correspond to solitary waves and solitons, respectively.

Generalizing this ansatz to include two-dimensional effects, a similar analysis can be performed and several possibilities of bifurcating solutions are found. Here, double-periodic boundary conditions are employed. This setting is far-reaching enough to yield a thorough insight into the behavior of such complicated systems. It has also the advantage that a lot of details can be computed explicitly.

Thus, we obtain standing travelling waves and oblique travelling waves which are connected by the symmetry of the system. Further instabilities due to mode interactions between these waves will lead to quasi-periodic solutions. In addition, degenerate bifurcation points exist, from which multiple solutions emanate. The corresponding mode interactions may lead to higher order bifurcations and to more and more complex wave patterns until a point is reached where a transition to turbulence takes place.

MODEL EQUATIONS

The two-fluid model commonly used in fluidized bed studies can be written in the following form[1,2]:

$$\partial_t(1-\phi) + div\,[(1-\phi)\mathbf{v}] = 0, \tag{1.a}$$

$$\partial_t\phi + div(\phi\mathbf{u}) = 0, \tag{1.b}$$

$$\begin{aligned} Fr\,(1-\phi)(\dot{\mathbf{v}} + \mathbf{v}\cdot\nabla\mathbf{v}) = &-(1-\phi)\mathbf{k} + \hat{B}(\phi,\mathbf{u}-\mathbf{v}) - (1-\phi)\nabla p \\ &- \nabla H(\phi,\mathbf{u}-\mathbf{v}) + \frac{Fr}{Re}(\Delta + \kappa\nabla\nabla\cdot)\mathbf{v}, \end{aligned} \tag{1.c}$$

$$\begin{aligned} Fr\,\delta\phi(\dot{\mathbf{u}} + \mathbf{u}\cdot\nabla\mathbf{u}) = &-\delta\phi\mathbf{k} - \hat{B}(\phi,\mathbf{u}-\mathbf{v}) - \phi\nabla p \\ &+ \nu\frac{Fr}{Re}(\Delta + \bar{\kappa}\nabla\nabla\cdot)\mathbf{u}. \end{aligned} \tag{1.d}$$

Here, ϕ denotes the voidage, i.e., the volume fraction occupied by the fluid phase; correspondingly, $\mathbf{u}$ denotes the fluid and $\mathbf{v}$ the particle phase velocity. p is a voidage-weighted partial pressure, namely that of the fluid phase. The partial pressure of the particle phase has been eliminated by introducing heuristically the function of $H(\phi,\mathbf{u}-\mathbf{v})$. The equations have been made non-dimensional by scaling of the space variables using a characteristic length scale, namely, the radius r of the fluidized particles, and the velocities with the terminal free-fall velocity of a single particle in a stationary viscous fluid,

$$u_t = (1-\delta)\rho_s g \left[\frac{9}{2}\frac{\mu_f}{r^2}\right]^{-1}. \tag{2}$$

This means, that the Froude number and the Reynolds number are given by $Fr = u_t^2/gr$ and $Re = \rho_s u_t r/\mu_s$, respectively. The two phases are considered being incompressible continua, hence, $\delta = \rho_f/\rho_s$ denotes the ratio of the specific densities of the two phases. It is constant and usually small. The same is true for the ratio of the kinematic viscosity coefficients, $\nu = \mu_f/\mu_s$. The following relation among these parameters can be verified easily:

$$\frac{9}{2}\nu\frac{Fr}{Re} = 1-\delta\,. \tag{3}$$

Of course, g is the constant of gravity, which is directed against the unit vector $\mathbf{k}$. Finally, $\kappa = (\lambda_s + \mu_s/3)/\mu_s$ and $\bar{\kappa} = (\lambda_f + \mu_f/3)/\mu_f$ are viscosity related coefficients. For simplicity all viscosity coefficients are considered being constant; in fact, a non-constant behavior would not yield any significant contribution to and within a weakly nonlinear theory. The natural scaling of the pressure terms p and H is then carried out using $\rho_s gr$; the drag force has been scaled by $\rho_s g/u_t$. In our approach the drag force is assumed to be of the simple Stokes-like form of

$$\hat{B}(\phi,\mathbf{u}-\mathbf{v}) = B(\phi)\cdot(\mathbf{u}-\mathbf{v}), \qquad B(\phi) = \frac{1-\phi}{\phi^n}(1-\delta) \qquad (n\approx 3). \tag{4}$$

To be specific, let us consider a two-dimensional fluidized bed in the (y, x)-plane (x is the vertical direction) and impose periodic boundary conditions on $(\phi, \nabla p, \mathbf{u}, \mathbf{v})$ in each direction. Then the basic state of homogeneous fluidization is (up to Galilei transformations) given by

$$\phi = \phi_0, \ \mathbf{v} = 0, \ \mathbf{u} = \phi_0^{n+1}\mathbf{k} \equiv u_0\mathbf{k}, \ \nabla p_0 = -[1 - \phi_0(1-\delta)]\,\mathbf{k}. \tag{5}$$

ONE DIMENSIONAL TRAVELLING WAVE ANALYSIS

Equations

It is well known that the basic state is most unstable to longitudinal perturbations and that this instability is of an oscillatory type[3]. Thus, it is reasonable to look for periodic solutions in a moving coordinate system[4]. Using the travelling wave ansatz of

$$\phi(x, t) = \phi(x - \omega t) \equiv \phi(z), \ etc., \tag{6}$$

the system (1) is reduced to a system of ordinary differential equations which can be further reduced to a two-dimensional dynamical system for the voidage alone.

Incorporating the travelling wave ansatz (6), we first obtain

$$\frac{d}{dz}\left[(1-\phi)(v-\omega)\right] = 0, \tag{7.a}$$

$$\frac{d}{dz}\left[\phi(u-\omega)\right] = 0, \tag{7.b}$$

$$Fr\,(1-\phi)(v-\omega)v' = -(1-\phi) + \hat{B}(\phi, u-v) - (1-\phi)p' - \frac{d}{dz}H(\phi, u-v) + \frac{Fr}{Re}(1+\kappa)v'', \tag{7.c}$$

$$Fr\,\delta\phi(u-\omega)u' = -\delta\phi - \hat{B}(\phi, u-v) - \phi p' + \nu\frac{Fr}{Re}(1+\bar{\kappa})u''. \tag{7.d}$$

The first two equations can readily be integrated yielding

$$(1-\phi)(v-\omega) = c_1, \tag{8.a}$$

$$\phi(u-\omega) = c_2, \tag{8.b}$$

where c_1, c_2 are arbitrary constants which may be specified by demanding that the basic state is a solution of the equations. In this case

$$c_1 = -\omega(1-\phi_0), \quad c_2 = \phi_0(\phi_0^{n+1} - \omega). \tag{9}$$

Eliminating p', v and u leads to a single second order equation for ϕ, which is conveniently written as a first order system:

$$\begin{cases} \phi' = \psi, \\ \psi' = \left[f_2(\phi,\alpha)\psi^2 + f_3(\phi,\alpha)\psi + f_4(\phi,\alpha)\right]/f_1(\phi,\alpha), \end{cases} \tag{10}$$

with

$$f_1 = \frac{Fr}{Re}\phi(1-\phi)\left[-(1+\kappa)c_1\phi^3 - \nu(1+\bar{\kappa})c_2(1-\phi)^3\right], \tag{11.a}$$

$$f_2 = -2\frac{Fr}{Re}\left[-(1+\kappa)c_1\phi^4 + \nu(1+\bar{\kappa})c_2(1-\phi)^4\right], \tag{11.b}$$

$$f_3 = -\phi(1-\phi)\{Fr\left[c_1^2\phi^3 + \delta c_2^2(1-\phi)^3\right] + G(\phi; c_1, c_2)\phi^3(1-\phi)^2\}, \tag{11.c}$$

$$f_4 = \phi^3(1-\phi)^3\left[\hat{B}\left(\phi, (u-v)(\phi; c_1, c_2)\right) - \phi(1-\phi)\right], \tag{11.d}$$

$$u - v = \frac{c_2}{\phi} - \frac{c_1}{1-\phi}, \tag{11.e}$$

and

$$G(\phi;c_1,c_2)\phi' := \frac{d}{dz}H\left(\phi,(u-v)(\phi;c_1,c_2)\right). \qquad (11.f)$$

The complexity of the problem is expressed by the form of the functions f_i, $i = 1, \ldots 4$, and the appearance of the numerous parameters summarized in

$$\alpha = (c_1, c_2, \delta, Fr; Re, \nu, \kappa, \bar{\kappa}). \qquad (12)$$

This system can be analyzed extensively in the phase plane[5]. The steady states are all located on the ϕ-axis; their values ϕ_i, say, are given by the zeroes of f_4 and depend on the parameters c_1 and c_2. The stability of the steady states $(\phi_i, 0)$ is determined by the eigenvalues of the matrix

$$A(\phi_i,\alpha) = \begin{pmatrix} 0 & 1 \\ f_4'/f_1 & f_3/f_1 \end{pmatrix}(\phi_i(\alpha),\alpha). \qquad (13)$$

From this, stability changes corresponding to bifurcations of nontrivial solutions can be inferred easily. In particular, if a parameter value α^* exists such that

$$f_4'(\phi_i(\alpha^*);\alpha^*) = f_3(\phi_i(\alpha^*);\alpha^*) = 0 \qquad (14)$$

hold simultaneously, both eigenvalues equal zero which means that a bifurcation of codimension 2 takes place. From this observation the occurrence of a global bifurcation in the form of an infinite period bifurcation to a homoclinic saddle connection can be deduced.

Description of the Bifurcation Results

To be specific, the forms of (4) for the drag force and (9) for c_1 and c_2 will be used in the following sections. In addition, it will be assumed that $G(\phi) < 0$ applies at least in the vicinity of $\phi = \phi_0$. This usual and physically reasonable assumption is essential for obtaining the stability and bifurcation results described below (see, however, the recent results of Hernandez and Jimenez[6]). As c_1, c_2 depend on the travelling wave velocity ω and the basic voidage ϕ_0 only, ω and ϕ_0 may be considered the basic parameters of the bifurcation problem. It must be noted that a non-vanishing fluid (or particle) viscosity introduces a singularity into the equations (10) in the range of $\omega \in (0,1)$. Two additional singularities are located at the boundaries of the physically admissible domain of $\phi \in (0,1)$.

In addition to $(\phi_0, 0)$, up to two other steady states $(\phi_{1,2}, 0)$ are found to exist in the admissible range of $\phi_{1,2} \in (0,1)$; their location depends on ω (Figure 1). Whereas the steady state corresponding to the solution $\phi = \phi_0$ of $f_4(\phi;\omega,\phi_0) = 0$ exists for all values of ω, the other two exist in the intervals of $\omega \in (0, \omega_m)$ and $\omega \in (1, \omega_m)$, respectively. At ω_m, the two branches merge and disappear in a saddle-node bifurcation point. Let us denote the upper branch $\phi_1 = \phi_1(\omega)$, the lower branch $\phi_2 = \phi_2(\omega)$, and the voidage value common to both branches ϕ_m. Obviously, ϕ_m is determined by $f_4(\phi_m;\omega_m,\phi_0) = 0$. The line $\phi = \phi_0$ intersects either $\phi = \phi_1(\omega)$ or $\phi = \phi_2(\omega)$ at $\omega_* = (n+2)(1-\phi_0)$. If $\phi_0 = \phi_m$ holds, then a pitchfork bifurcation takes place at $\omega = \omega_m$ (Figure 1b).

The stability properties of $(\phi_{1,2}, 0)$ above all depend on the location of ϕ_0 relative to ϕ_m. In principle, the part of the branch of ϕ_1 or ϕ_2 that extends between ϕ_0 and ϕ_m corresponds to a stable node or spiral; in the other parts the steady states are saddle points. For $\phi_0 > \phi_m$, however, the stable part of the ϕ_1-branch may become unstable, such that a branch of periodic solutions emerges from the steady state $(\phi_1, 0)$. The nonlinear behavior of these periodic solutions and their relevance to applications has not yet been studied.

The Hopf bifurcations from $(\phi_0, 0)$ have been analyzed in more detail[5]. In case of $\omega > \omega_*$, ϕ_0 is a saddle point. Turning to smaller values of ω, ϕ becomes a stable node or spiral first and then loses stability at some value of $\omega_+ < \omega_*$.

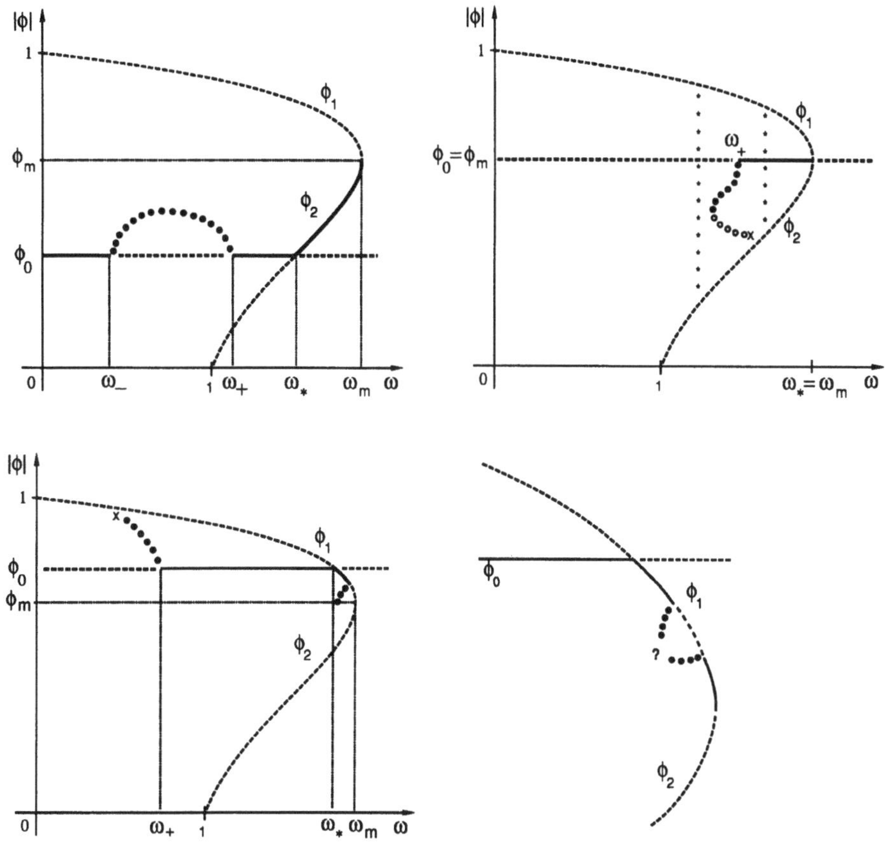

Figure 1. Stationary and periodic solutions of the travelling wave system (10) parametrized by their amplitude over ω: a) $\phi_0 < \phi_m$, b) $\phi_0 = \phi_m = \frac{n+1}{n+3}$, c) $\phi_0 > \phi_m$.
$|\phi|$ = amplitude, —— stable stationary solution, – – –– unstable stationary solution, •••• stable periodic solution, o o oo unstable periodic solution, × infinite period bifurcation, * * * heteroclinic connections.

At this point a branch of periodic solutions bifurcates from $(\phi_0, 0)$. It may either return to the basic or another steady state or terminate in a homoclinic orbit.

The first possibility results in a restabilization of the basic state at some value of $\omega_- \in (0, \omega_+)$ (Figure 1a). Such a scenario depends particularly on the value of δ; in fact, it cannot be realized within the limit of vanishing δ, which is often considered in fluidized bed investigations. It is not difficult to evaluate the ranges of existence of Hopf bifurcations at ω_+ and ω_- in dependence of the system parameters[5].

Whereas the second of the above mentioned possibilities has not yet been evaluated or confirmed, respectively, the third is complementary to the first and can be concluded from the unfolding of a codimension 2 bifurcation. The latter occurs, if the Froude number is varied such that ω_+ is shifted to ω_-. In this case the matrix $A(\phi_0, \alpha^*)$ has two zero eigenvalues. This Takens-Bogdanov bifurcation is well described in literature[7]. For this reason we shall omit the details and instead refer to the illustrations (Figs. 1b,c). The homoclinic orbit may develop further to a heteroclinic orbit or even a heteroclinic cycle in case of a certain symmetry of the equations.

Normal Form Analysis. A better understanding of such a behavior can be achieved by performing a normal form analysis[5]. In the present case, the essential features of the system (10) are represented by the normal form of

$$\begin{cases} p' = q, \\ q' = ap + bq + cp^2 + dpq + ep^3 + fp^2q. \end{cases} \tag{15}$$

The coefficients depend on the functions f_i and their derivatives with respect to ϕ, evaluated at $(\phi_0, 0)$, and, hence, on α.

Assuming $c^2 > 4ae$, the three steady states of (15) are given by

$$q_{0,1,2} = 0, \quad p_0 = 0, \quad p_{1,2} = \frac{1}{2e}\left(-c \pm \sqrt{c^2 - 4ae}\right). \tag{16}$$

Using an appropriate scaling, (15) can be viewed a perturbation of a Hamiltonian system containing a continuum of closed orbits surrounded by a homoclinic orbit. This Hamiltonian system contains two steady states only, the third has been transformed to $+\infty$ or $-\infty$. The two remaining steady states determine the possible homoclinic orbits. To obtain a heteroclinic saddle point connection, three steady states and, thus, a third order term must be present. Unfortunately, such a perturbation argument is not applicable in the general case. If, however, some of the parameters $(b,\ d,\ f)$ vanish at a certain value of α, then the system (15) is reduced to a Hamiltonian system. If $c = 0$ applies in addition, the two non-trivial steady states are located symmetric with respect to the origin and the system contains a heteroclinic cycle around the origin. Unfolding this degeneracy by claiming $c \neq 0$ but small, the heteroclinic cycle is destroyed and a homoclinic orbit is left. It depends on the sign of c, from which saddle point the homoclinic orbit emanates (Figure 2).

A bifurcation of codimension 3 occurs, if all terms except the p^3-term vanish in (15), i.e., if all steady states fall together. This is satisfied for the concrete system (10), if $\phi_m \equiv \phi_0 = \phi_1 = \phi_2$ holds at $\omega = \omega_m$. Hence, it is not surprising that while changing w several heteroclinic, homoclinic and closed orbits can be found numerically in this case (see Figure 1b).

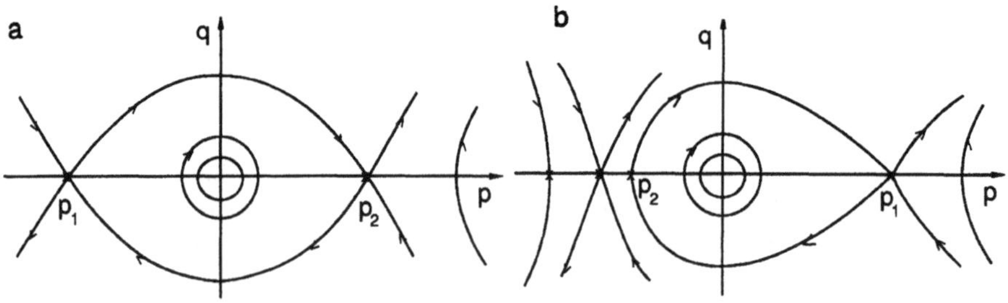

Figure 2. Phase flow of the system (15) with $b = d = f = 0$: a) Heteroclinic cycle ($c = 0$), b) homoclinic orbit ($c > 0$).

Discussion

To sum up our findings, it can be said that several branches of stationary and periodic solutions exist. Except for the basic state $(\phi_0, 0)$, all these solutions are confined in ϕ between 0 and 1, because these are the singular values of (10), and in ω between 0 and ω_m, because it can be shown that no periodic solutions exist for $\omega \leq 0$; on the other hand, $\phi = \phi_0$ is the only steady state for $\omega > \omega_m$ and represents a saddle point there. Hence, no periodic solutions are possible for $\omega > \omega_m$, too. Moreover, the amplitude of the periodic solutions cannot become larger than $max_\omega\{\phi_1(\omega), \phi_0\}$, as can be seen from Figure 1.

Therefore, the branches of periodic solutions must return to one of the steady states $(\phi_{0,1}, 0)$ or terminate in a homoclinic orbit. A third interesting possibility consists in $\psi = \phi'$ becoming very large while the amplitude, ϕ, remains finite. This could happen in the range of $\omega \in (1, \omega_m)$ only[5] and would yield approximately square waves oscillating between two steady states and around the intermediate steady state. Its evolution and approximative shape is illustrated in Figure 3.

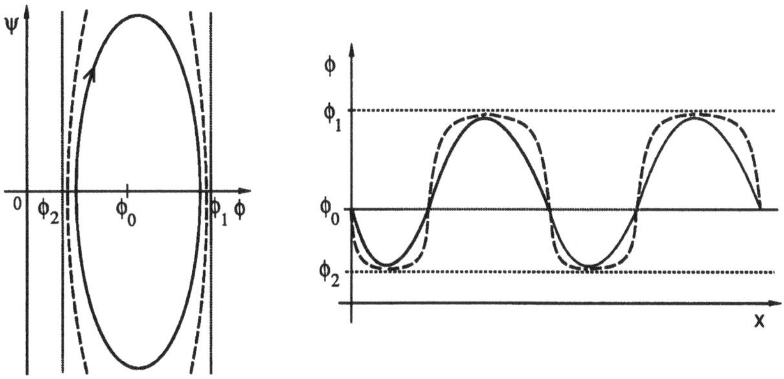

Figure 3. Closed orbits with steep gradients correspond to approximately square waves and are reminiscent of slug flows.

As such a solution is periodic and quasi-steady in the original (laboratory) coordinate system and contains steep gradients, it is well suited to explain the well-known slug flow behavior of two-phase systems.

In contrast to this, a homoclinic orbit represents a single distortion the form of a hump travelling steadily through the bed and leaving the bed undisturbed a long time after its passage. A heteroclinic orbit, however, monotonically connects two limiting voidage values of the bed, i.e., after the passage of such a soliton the bed is found to be in another stationary state. Moreover, a second heteroclinic saddle connection will close the cycle bringing the system back into its original state. Such orbits may be considered the limiting orbits which are reached when closed orbits around ϕ_0 hit one of the steady states ϕ_1 or ϕ_2, or both (cf. Figs. 2, 3).

TWO-DIMENSIONAL EFFECTS

The above type of analysis can be extended to incorporate oblique travelling waves of the form

$$\phi(x, y, t) = \phi(x - \omega t \pm ky). \tag{17}$$

Then, k enters the multi-parameter α as an additional parameter in the form of k^2 due to the reflectional symmetry of the system with respect to y. It turns out that periodic solutions bifurcate from the basic state only, if k^2 is not too large (and, eventually, not too small either). This means that the propagation direction of plane waves is restricted to a cone with axis directed downstream[5].

In view of the existence of oblique travelling waves it may be wondered whether 'standing travelling waves' exist which correspond to an interaction or – locally – a superposition of a pair of plane waves travelling through the bed diagonally to the left and right, respectively. Such waves are travelling downstream but are stationary in the spanwise direction; in the first approximation they are given by

$$\phi = \varphi_+ e^{i\lambda(x-\omega t)+iky} + \varphi_- e^{i\lambda(x\omega t)-iky} + c.c. + h.o.t. \tag{18}$$

Indeed, the ansatz of (17) can be enlarged again to include more general, truly two-dimensional travelling waves of the form

$$\phi(x,y,t) = \phi(x-\omega t, y) \equiv \phi(z,y). \tag{19}$$

A linear analysis of the resulting equations reveals the existence of up to three different modes with a given periodicity in the spanwise direction but various downstream (and time) periodicities and wave velocities. It is suggested that the nonlinear interaction of these modes – and/or of modes with different spanwise periodicity also – leads to bifurcations of higher order and to quasi-periodic solutions[2]. Since degenerate bifurcation points exist, from which multiple solutions may emanate, a rich variety of solutions seems to be admitted by the equations and first the occurrence of a solution with three independent frequencies and then that of a strange attractor might be conjectured. In this manner, more and more complex wave patterns would be obtained which might be assigned to observed flow regimes, until a point is reached where the transition to turbulence takes place.

It should be noted, however, that a primary bifurcating standing travelling wave of the form of (18), which we would like to associate with the frequently observed bubbles, shows a circular symmetric pattern[2] and is thus too regular compared to the experimental observations which show a particular distortion in the streamwise direction. The reason for not being realized in nature would then, of course, be its possible instability. Another reason might be due to the fact that the above mentioned bifurcation scenario is not possible, if $G(\phi_0) > 0$. It is quite common to assume that $H(\phi, \mathbf{u}-\mathbf{v})$ depends on ϕ only and that in this case $G(\phi) < 0$ holds generally[2] ($G(\phi)$ is defined by $G(\phi)\nabla\phi = \nabla H(\phi)$). The latter might not be valid, however, if H is assumed to depend on the relative velocity of the two phases, too, whence a different bifurcation behavior might be obtained[6].

A more promising approach to the bubble problem may consist in the instability of the plane streamwise travelling voidage wave train against transversal perturbations[8]. The stability of this wave can be calculated perturbatively and is determined by the sign of $\sigma = A + Bk^2 + O(k^4)$, where A and B are complicated real functions of the physical parameters of the system and k is the wave number of the transversal perturbation (it is assumed that the perturbation has the same streamwise periodicity as the perturbed wave). The solution of the linearized problem can be written in the form of

$$\phi = f(x-\omega t)\, e^{iky} + c.c. + h.o.t., \tag{20}$$

where f is a periodic function of $z = x - \omega t$. In comparison to (18), this yields a similar but more distorted shape of a solution periodic in all its variables, which is very reminiscent of real bubbles. Indeed, this bifurcation scenario has been confirmed both experimentally[9] and numerically[6]. In the latter work bubbles of quite a realistic shape have recently been obtained.

CONCLUSION

It has been shown that two-phase flow models can be treated very effectively using tools from bifurcation theory. The equations seem to admit a rich variety of solutions which have by no means been explored exhaustively. The sequences of instabilities and the bifurcating solutions obtained and conjectured, respectively, have to be compared carefully with the transitions and pattern formations observed in the experiments. This will yield more detailed information about the proper modeling of fluidized beds and multi-phase flows.

ACKNOWLEDGMENT

The presented results are based on the author's Ph. D. thesis written under the supervision of Prof. W. Jäger within the framework of the Sonderforschungsbereich 123 at the Institut für Angewandte Mathematik der Universität Heidelberg and supported by the Deutsche Forschungsgemeinschaft.

REFERENCES

1. S. K. Garg and J. W. Pritchett, Dynamics of gas-fluidized beds, *J. Appl. Phys.* 46:4493 (1975).
2. M. F. Göz, On the origin of wave patterns in fluidized beds, *J. Fluid Mech.* 240:379 (1992).
3. G. M. Homsy, A survey of some results in the mathematical theory of fluidization, *in*: "Theory of Dispersed Multiphase Flow", Academic Press, New York-London (1983).
4. D. J. Needham and J. H. Merkin, The existence and stability of quasi-steady periodic voidage waves in a fluidized bed, *J. Appl. Math. Phys. (ZAMP)* 37:322 (1986).
5. M. F. Göz, Bifurcation of plane voidage waves in fluidized beds, submitted to *Physica* **D**.
6. J. A. Hernandez and J. Jimenez, Bubble and channel formation in dense fluidized beds, *in*: "The global geometry of turbulence", J. Jimenez (ed.), NATO ASI ser. B, Vol. 268, Plenum Press, New York (1991).
7. J. Guckenheimer and P. Holmes, "Nonlinear Oscillations, Dynamical Systems, and Bifurcation of Vector Fields", Springer, Berlin-Heidelberg-New York (1986).
8. M. F. Göz, in preparation.
9. A. K. Didwania and G. M. Homsy, Flow regimes and flow transitions in liquid fluidized beds, *Int. J. Multiphase Flow* 7:563 (1981).

MONTE CARLO SIMULATION OF BUBBLING FLUIDIZED BEDS

Massamba Thioye, René Darrigo and Gérard Gouesbet

Laboratoire d'Energétique des Systèmes et Procédés
URA CNRS 230
INSA ROUEN
B.P. 08
76131 Mont Saint Aignan Cédex

INTRODUCTION

Gazeous fluidization is a unit operation of wide spread use in chemical engineering. The difficulty of mastering the behaviour of such multiphase flow may be increased by the development of instabilities like bubbling, channelling, spouting or slugging. With bubbling bed the knowledge of bubble properties such as size and velocity distributions is therefore required for a proper design of plants.

Like most systems which are constituted of two phases, here one continuous and one dispersed phases, fluidized bed may be modelled by deterministic approaches relying on mechanics or by stochastic methods which account for the apparently random nature of observed phenomena. Stochastic methods are mostly interesting in the present case of bubbling fluidized beds because they may more appropriately describe a big amount of elementary events requiring statistical descriptions. For instance, the population balance technique akin to the Boltzmann point of view in kinetic theory of gases is of very common use. A review and general formulation of this technique are available in Thioye (1988) chapter V, but cannot be reproduced here to avoid any undue lengthy development. In any case, population balances require to model each involved basic

Instabilities in Multiphase Flows, Edited by G. Gouesbet
and A. Berlemont, Plenum Press, New York, 1993

phenomenon and the produced complexity also depends on the number of state variable retained to describe the bubbles. In the present study, the complexity is reduced to a manageable although efficient enough size by only considering the following bubble state parameters : bubble volume v_{Bi} (or diameter d_{Bi}), location defined by altitude Z_i, shape characterized by horizontal and vertical dimensions d_{Hi} and d_{vi} respectively, and global number N of bubbles in the bed. All these quantities are assumed to be random variables. The aim is to predict the evolution of bubble volumes v_{Bi} in connection with their locations Z_i.

Correlations leading to the possibility of bubble mean size predictions with different kinds of particles are available from Johnson et al (1974), Hillgard and Werther (1987), Werther (1976), Miwa et al (1972) or Horio and Nonaka (1987). However population balances enable more refined descriptions, allowing us to reach mean values and fluctuations of bubble state variables with the further advantage of relying on a deeper understanding of the involved phenomena. In particular, the spatial evolution of the bubble size distribution function is obtained by solving the population balance equation after parameter modelling implementation or use of Janossy's probability density functions. An alternative is Monte Carlo simulation which is considered in this paper. The implementation of such a simulation also require the identification of elementary phenomena and features associated with probability law which are discussed in the next subsections.

Bubble shapes, coalescences, break-ups, are assumed to be random. The combination of elementary phenomena in a randomly behaving bed has been the focus of our attention. For a bubbling bed there is a superimposed randomness and so stochastic treatment is necessary.

INPUT FROM EXPERIMENTS

Bubble shape

Shape characterization is carried out by studying a few thousands photographs of bubbles by means of the 2D set described in Thioye (1988). We conclude that (i) bubble shapes essentially do not depend on the position in the bed but are mainly size dependant, (ii) small bubbles exhibit a circular projected area, (iii) big bubbles may be described by elliptic projected areas with vertical dimension d_v generally smaller than horizontal dimension d_H. On the bases of observations carried out for different fluidization velocities and at different altitudes, we have looked for a correlation between d_v and d_H. All representative points, plotted on a bilogarithmic diagram (d_H, d_v) are found to be located between two lines A and B. Each of these lines is defined by two coefficients depending on the kind of particle and on the state of fluidization. A bubble of given volume may be represented by a point located on a segment of slope (-1) (owing to the relation to $d_H \cdot d_v \quad V_B$) limited by two points (1) and (2) at the intersections with lines A and B. The shape determination of a bubble starts with a random sampling for vertical axis d_{vi} between the limits d_{v1} and d_{v2}, and then d_{Hi} is calculated.

Spatial distribution of bubbles

During their upward motion, bubbles migrate toward the central part of the bed. This phenomenon is connected with a smaller pressure drop due to the presence of large bubbles near to axis in the bed top level and with the observed downward net flow of particles in the vicinity of lateral walls.

To describe this migration phenomenon, the bed is subdivided into 14 sections of the same height (25 mm). For each section, camera observations enable us to determine a lateral segment of width y_B, symmetric with respect to bed axis in which the presence of bubbles is highly probable, i.e. the probability of the presence of bubbles outside the range y_B is essentially 0. The value of y_B is found to depend on altitude z according to the relation :

$$y_B = a - b \left[1 - \exp(-cz)\right] \tag{1}$$

in which constants a, b, c depend on the fludization conditions. For example with u_f = 6.7 cm/s, we found a = 26 cm, b = 10 cm and c = 0.1 cm^{-1}.

We make the assumption that the size of a bubble is irrespective of its lateral position along y_B.

Rise velocity of bubbles

The rising motion of bubbles is often discussed within two different extreme frames, none of which being fully satisfactory, namely a mechanical approach for an isolated bubble which does not incorporate collective interactions, or two phase flow approach in which the averaging process prevents a refined enough description. In the present work, in agreement with Clift and Grace (1985), we use a fairly accurate point of view which assumes (i) that the velocity of one bubble significantly depends on its position in the bed and (ii) that it is mainly influenced by other bubbles whose centers lie above that of the bubble being studied. The expression to evaluate the bubble rising velocity u_{Bi} then reads as a summation of two contributions representing (i) the velocity of a free bubble, (ii) the influence of all bubbles whose altitude is higher, according to :

$$u_{Bi} = \frac{1}{2} g^{0.5} \left(\frac{V_{Bi}}{e\pi}\right)^{0.25} + \frac{\sum_{i=1}^{N} \left[V_{Bi} \left(z_i > z_j\right)\right]}{\sum_{i=1}^{N} V_{Bi}} \cdot \frac{Q_B}{e.l} \qquad (2)$$

in which e and l are the bed thickness and width respectively and Q_B is the global gas bubble flow rate given by

$$Q_B = N_e \int_0^{\infty} v_B \phi_{vB} dv_B \qquad (3)$$

in which ϕ_{vB} is the volume distribution at the entrance section which is experimentally known, and N_e the probability for a bubble to pass through the bed section per unit of time. We must point out that the smallest bubble diameter is here taken to be ten times of a solid particle one.

Coalescence

When bubbles become closer before coalescence, they depart from their isolated shapes. After coalescence, the volume of the resulting combined bubble is larger than the sum of the volumes of the original bubbles. These features are not incorporated in our simulations. We however introduce the duration of coalescence which is the time delay between the entry of a bubble in the wake of the preceding one and the coalescence completeness. Assuming that bubble 2 is drawn into bubbles, the duration of coalescence is given by Farrokhalaee's correlation (Clift, Grace 1985) :

$$t_c = \frac{r_1^{\,2}}{0.8\, u_{B\infty 1}\, r_2}\left\{\left[1 + \left(\frac{r_2}{v_1}\right)\right]^2 - 0.67\right\}$$

in which r_1 and r_2 are the radius of bubbles 1 and 2 respectively and $u_{B\infty 1}$ is the isolated rise velocity of bubble 1.

To characterize the stochastic nature of the coalescence process we introduce the probability $q(B_1 \,;\, B_2 \,;\, t)$ of the coalescence of bubbles 1 and 2 during time t, according to

$$q\left[B_1 \,;\, B_2 \,;\, t\right] = 0 \qquad (4)$$

$$\text{if } \left(z_1 - z_2\right) > \frac{1}{2}\left(d_{v1} + d_{v2}\right) + \left(u_{B2} - u_{B1}\right) t \qquad (5)$$

$$q\left[B_1 \,;\, B_2 \,;\, t\right] = \left[1 - \exp\left(-4.69\,\frac{t}{tc}\right)\right] \frac{d_{H1} + d_{H2}}{\inf\left(y_1,\, y_2\right)} \qquad (6)$$

$$\text{if } \left[\begin{array}{c}\left(z_1 - z_2\right) < \frac{1}{2}\left(d_{v1} + d_{v2}\right) + \left(u_{B1} - u_{B2}\right) t \\ \text{and } \left(z_1 - z_2\right)\left(u_{B1} - u_{B2}\right) < 0\end{array}\right] \qquad (7)$$

This probability of coalescence is a modification of the Shah et al analysis (1977) by introduction of (i) the bubble velocity differences associated with a characteristic coalescence time, (ii) the stochastic nature of bubble shape, (iii) a bed central part in which the presence of bubble is highly probable.

The notation inf (y_{B1}, y_{B2}) designates the smallest lateral extension among y_1 and y_2 corresponding to altitudes z_1 and z_2 at which bubbles 1 and 2 are located respectively. This term accounts for the observation that coalescence takes place at the upper bubble location where y is minimum. Coefficient 4.69 in rel (6) is chosen in such a way that for t = tc, probability q is :

$$q = 0.99\, \frac{d_{H1} + d_{H2}}{\inf\left(y_{B1},\, y_{B2}\right)} \qquad (8)$$

In Monte Carlo simulations, coalescence is globally considered for the whole bed, i.e. at each time increment δt the bubble repartition in the whole bed is computed. Assuming that time increment δt it small with respect to the duration of coalescence t_c, the probability of coalescence per unit of time is found to be :

$$Q_t = \frac{1}{2} \sum_{\substack{i,j \\ i \neq j}} q\left[B_i \; ; \; B_j \; ; \; t\right] \tag{9}$$

Break up

When the size of a bubble increases, it eventually becomes unstable mainly because of particle motions in the front region, leading to break up. We define the splitting break-up occurence of a bubble when particles start moving from the nose to the wake indicating roof collapse. Although many authors observed that there is no simple correlation between particle and bubble sizes, according to Horio and Nonaka (1987), the break-up frequency is given by

$$f_s^* = \alpha \; u_{mf}^{-p} \tag{10}$$

in which p ranges from 1 to 1.2 and α is given by

$$\alpha = k \; u_B \; / \; \sqrt{g \; d_B} \tag{11}$$

in which constant k is 2.3 $10^{-2}ms^{-2}$ for p = 1 and 6.5 $10^{-3}m^{1,2}s^{-2,2}$ for p = 1.2 respectively.

For want of a precise and general correlation including the effect of particle and gaz properties, mainly the viscosity which has been found not insignificant by different investigators, in our simulation following Horio and Nonaka (1987) we have taken p = 1.1 and k = 6.5 10^{-3}.

If the bed contains N bubbles, then the break-up probability per unit of time is :

$$N_s = N \; f_s^* \tag{12}$$

Bubbles entering the bed

Lateral and upper boundaries of the bed significantly influence its behaviour. However, the distributor through which fluid is supplied is much more important. Since it is difficult to fabricate a completely uniform porous finite distributor, upstream boundary conditions for Monte Carlo simulations should not be taken at altitude 0. Another issue is to define what a bubble actually is. Since simulations will track bubble evolutions, we must own a criterion to define such an entity.

Therefore, we assume that a bubble is identified as such when its diameter is at least ten times the particle diameter and the upstream boundary conditions for simulations are taken at 5 cm above the distributor.

These conditions are experimentally defined by processing many thousand bubbles by means of a cine photography technique, leading to the determination of the entrance distribution law $\phi(v_B)$, in which $\phi(v_B)$ is the probability for a bubble to have its volume within v_B and $(v_B + dv_B)$. We then obtain initial cumulative distribution functions which will be used in simulations for the evaluation of initial bubble diameters by random sampling.

MONTE CARLO SIMULATIONS RESULTS AND DISCUSSION

Simulations are carried out for a given fluidization velocity, namely $u_f = 6.7$ cm/s. For this velocity, initial conditions of the whole bed at $t = 0$ are obtained from cinephotography and upstream boundary conditions are taken at $z = 5$ cm. Monte Carlo simulations then proceed by time steps. At each time step the fluidized bed state E_t, where N bubbles are present, is defined by : z_i, v_i with $i \in (1, N)$. To introduce the random character we consider : bubble entrance into the phase domain (z_B, v_B), coalescence and scission and the probability of appearance of one of them which is the ratio of its frequency to the sum of all event frequencies, i.e. Ne, Q, N_s respectively. If we then introduce the latence time τ, i.e. the time interval during which there is no modification of the state E_t the free bubble evolution is the solution of $\frac{dz}{dt} = u_B$ if $t < \tau$. A direct procedure gives the latence time distribution function

$$F\left(\tau|E_t\right) = 1 - \exp\left[-\left(Q + Ne + Ns\right)\tau\right] \tag{13}$$

A first random sampling gives F and time τ is deduced from (13). Therefore the time incrementation is random. A second random sampling is carried out to determine the phenomenon to be considered and a third random sampling gives the result associated with the phenomenon. After each event the global population N is accordingly modified. All random sampling are produced by using the function

$$Al = 93.al + 2581 - \text{trunc}\left[\left[al\ 93 + 2581\right] / 16\ 384\right] . 16384 \qquad (14)$$

which randomly and uniformly generates an integer in the range [0, 16 384], with al beeing the seed of the random sequence. Expression (14) allows random sampling to work with integer variables.

Results

After transcient killing, relevant parameters are recorded in 14 bed sections of the same height (25 mm). The bubble volume distribution results are shown in Fig. 1. The progressive shifting when z increases from small to big bubbles through a net balance between coalescence and break-up events is clearly evidenced.

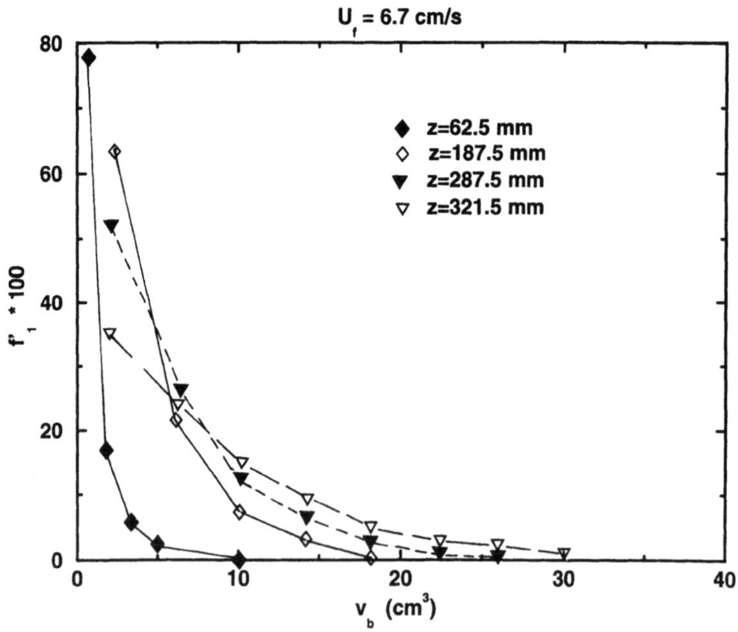

Figure 1. Volume distribution function : simulation results.

Fig. 2 indicates the drastic reduction of $\bar{n}$, bubble mean number per vertical length unit of bed, when z increases. Coalescence effects are essentially significant in the lowest part of the bed in agreement with Miwa et al observations (1972). However, the n-variation coefficient $\frac{\sigma_n}{\bar{n}}$

(standard deviation divided by mean value) is found to be approximatively constant.
Also, standard deviations of the random variable n(z) is often greater than mean values, i.e. fluctuations are dramatically significant and mathematical expectations of n(z) are therefore not sufficient for a good enough bubbling characterization. The n-variation coefficient increases with altitude in correlation with the decrease of bubble number-density.

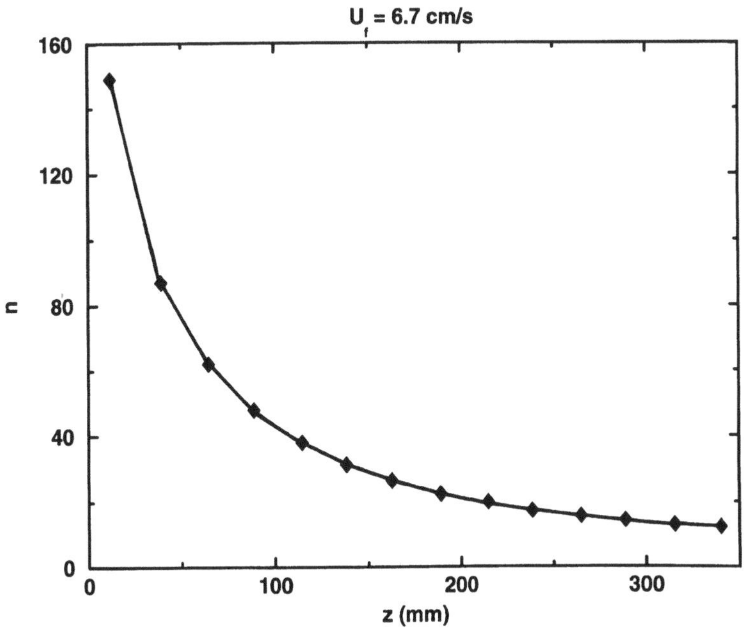

Figure 2. Bubble mean number, simulation results.

In the end we clearly observe that interactions between all bubbles present in the bed cannot be neglected. In other words, the classical assumption of independance between any two kinds of bubble states is not realistic.

Furthermore, the bubble mean volume $\bar{v}_B$ enables the determination of mean diameter $\bar{d}_B$ evolutions which are presented on Fig. 3 as a spatio-temporal mean value, i.e. averaging is carried out on time measurements and on bubbles present in each of the 14 sections.

In correlation with these results, bubble frequencies decrease when z increases (Fig. 4).

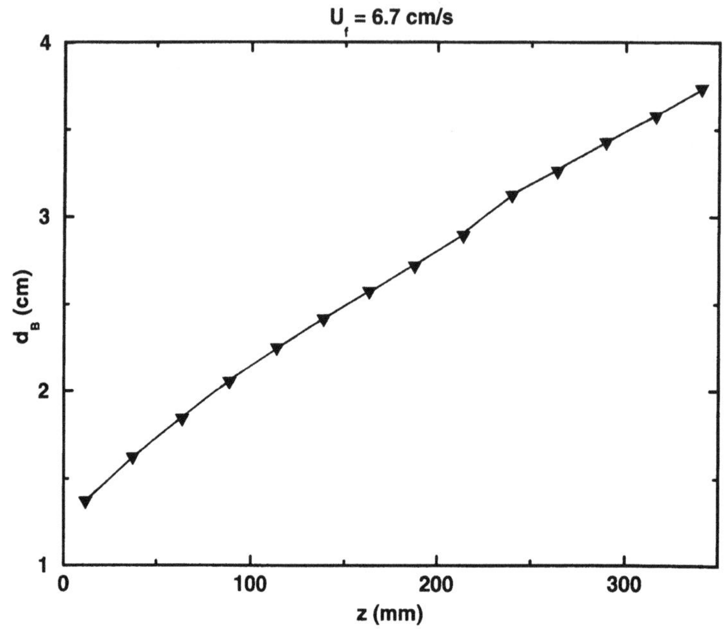

Figure 3. Bubble mean diameters versus z : simulation results.

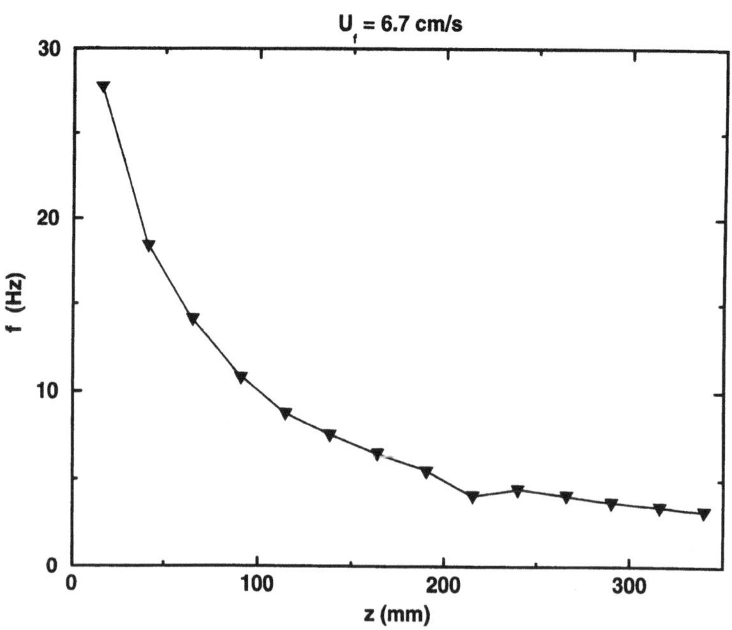

Figure 4. Simulation results for bubble frequencies

The bubble volume distribution function $f(z, v_B)$ is represented by the approximate solution f_1 obtained by using the first term when expanding $f(z, v_B)$ into a Laguerre serie, i.e. the distribution of bubble volume is represented by a Γ distribution according to :

$$f_1\left[z, v_B\right] = \bar{n} \frac{x^{y+1}}{\Gamma(y+1)} \exp\left[- a v_B\right] . v_B^{y} \tag{15}$$

in which $v_B = v'_B - v_{Bmini}$

in which v'_B is the real bubble volume and $v_{Bmin} = 1.34\ cm^3$

Furthermore :

$$x = \frac{\bar{v}_B}{\sigma^2_{v_B}} = \frac{\lambda}{a} \qquad y = \frac{\bar{v}_B}{\sigma^2_{v_B}} = \lambda - 1$$

here λ and a are coefficients introduced by Argyriou et al (1971).

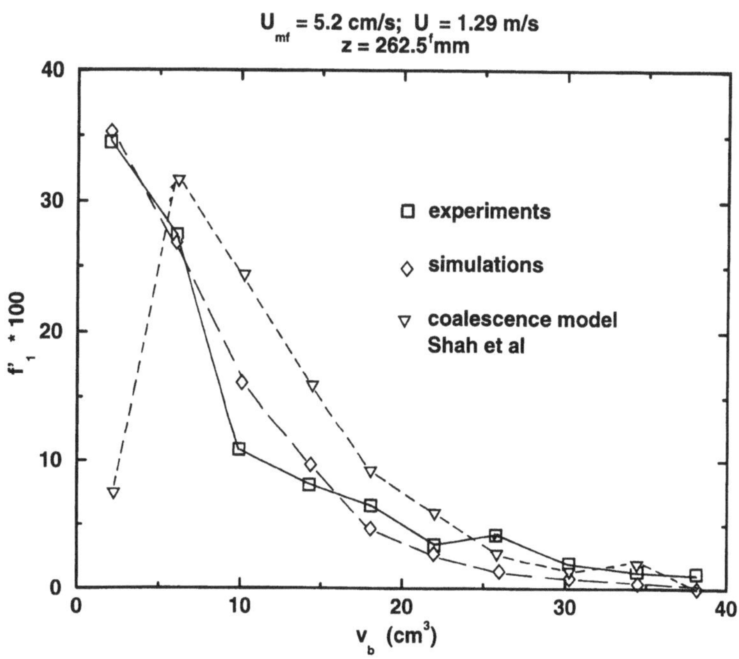

Figure 5. Normalized distribution function. Experimental and simulation results.

The normalized function :

$$f'_1\left[z, v_B\right] = \frac{f_1\left[z, v_B\right]}{\int_0^{\infty} f_1\left[z, v_B\right] dv_B}$$

is shown in Fig. 5 for which comparison with analytical results is satisfactory.

Comparison with other models

Our simulations are compared with some results from models presented by Werther, J. (1976), Argyriou et al (1971) and Agarwal (1985). We here point out restrictive conditions in these models that we have eliminated in Monte Carlo simulations by considering (i) relative bubble speeds for coalescence frequencies, (ii) the stochastique nature of bubble shape, (iii) the fact that some parts of the bed are free of bubbles, (iiii) the bubble break-up.

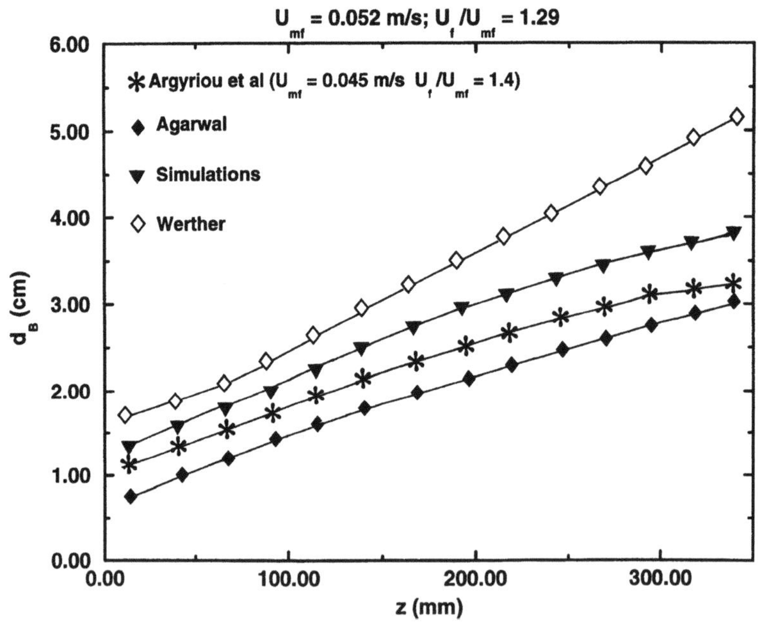

Figure 6. Mean bubble diameter comparisons (z_0 = 50 mm). z_0 is the height origine for computations.

Obtained double mean diameter Fig. 6 are close to the results of Argyriou et al model whereas Argarwal's model produces small sizes, and Werther's model, which assumes a 15 % increase in volume during coalescence, leads to bigger sizes, the difference enlarging with z. The difference of u_{mf}, minimum fluidization speed directly connected to particle size does not seen to give a satisfactory explanation for observed differences.

Bubble mean concentration (Fig. 7) are determined for two limit cases (i) with an experimental initial distribution obtained by a global cinephotography of the bed and (ii) by assuming that all bubbles at starting time t = 0 have a volume equal to the bubble minimum volume. From level z = 150 mm the two representative curves are similar and present only little difference with those of Argyriou et al and Argarwal models. At the lowest bed height, the distributor nature and the non zero double sizes at distributor entrance bed, seen to be determinating factor.

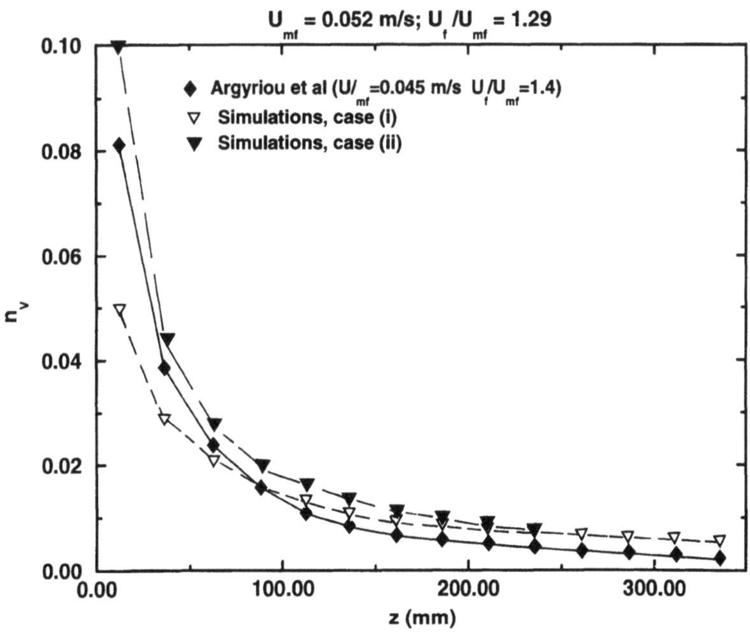

Figure 7. Bubble mean concentration : comparison with other models.

A discrepancy appears when we consider $\sigma^2 v_B / \bar{v}_B^{\,2}$ versus z as displayed in Fig. 8. Argyriou model values are from 2 to 3 whereas they are near 1 from our simulations. In order to test the relative significance of the different ingredients of our simulations, computations are also carried out by successively neglecting break-ups, assuming spherical bubbles, considering uniform size bubble distribution at entrance altitude and finally neglecting bubble relative velocities. From the comparisons between results obtained under these different specifications, we conclude that break-up is not essential and that bubbles can safely be assumed to be spherical. Entrance conditions greatly influence the first histogram size classes but observed differences decrease for bigger sizes. We point out that coalescence is essentially due to vertical velocity differences between the bubbles. The bubbling characterization must take into account velocity difference mainly in the lower part of the bed.

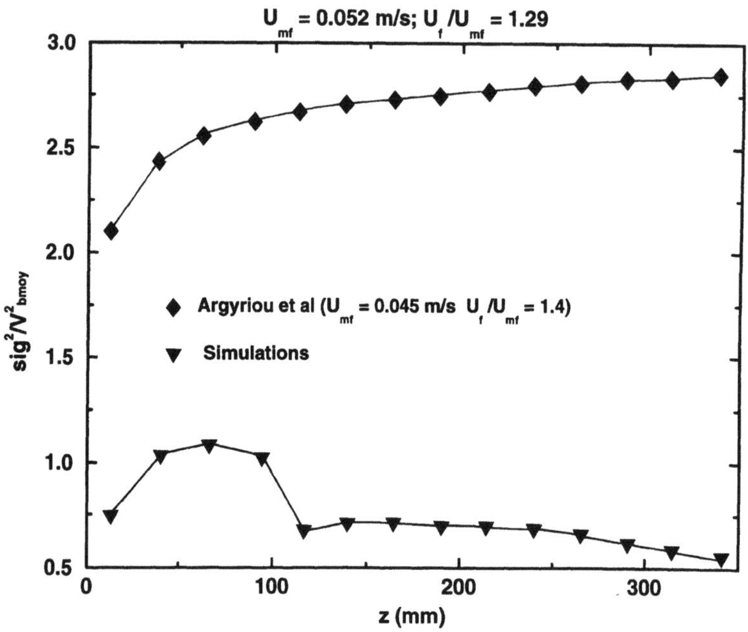

Figure 8. Bubble volume coefficient variation : our model and Argyriou model.

Comparison with experimental results

Comparison between our simulations and experimental data are carried out for a minimum fluidization velocity equal to 5.2 cm/s and fluidization velocity equal to 1.29 m/s. Large bubbles allow easy measurements, but bigger their sizes are the more particle calculations are significant at the boundaries. Therefore the fluidization conditions are not really favorable but, on the other hand, the field of action is relatively large for the validation of the measurement technique.

We may consider that agreement between simulations and experimental data is globally satisfactory for z = 162.5, 212.5 and 225.5 mm (see Fig. 5 relative to normalized distribution function f'_1). A significant discrepancy appears for z = 312.5 mm for which we notice that simulations systematically produce smaller bubble sizes. Direct observations can explain this discrepancy as follows. Bubble size distributions are quite different near the distributor where bubbles are sensibly uniformly distributed on the whole bed section, and for upper levels where large bubbles have a central location and lateral smaller bubbles cannot be detected by the sensor because of the too strong inclination of bubble path. Although possible in practice, experimental characterizations of such small bubble lateral distribution to refine comparisons with Monte Carlo simulations would lead to unaffordable time-consuming experimental runs.

CONCLUSION

These Monte Carlo simulations provide us with a realistic bubble coalescence and evolution model. These computations are satisfactorily validated by comparison with other model results and also with experimental data.

REFERENCES

Argawal, P.K., 1985, Bubble characteristics in gas fluidized beds, *Chem. Eng. Res. Des.*, 63, **9**, 323-335.

Argyriou, D.T., List, H.L., Shinnar, R., 1971, Bubble growth by coalescence in gas fluidized bed, *AICHE J.*, 17, **1**, 122-130.

Horio, M., Nonaka, A., 1987, A generalized bubble diameter correlation for gas-solid fluidized beds, *AICHE J.*, 33, 11, 1865-1871.

Johnsonn, J.E., Clift, R., Grace, J.R., 1974, Prediction of bubble distribution in freely bubbling two-dimensional fluidized bed, *Inst. Chem. Eng. Symp. Ser.* 38, Pap. B5.

Miwa, K., Mori, S., Kato, T., Muchi, I., 1972, Behaviour of bubbles in a gaseous fluidized bed, *Intern. Chem. Engin.* 12, 1, 187-194.

Shah, B.H., Ramkrishana, D., Bornwamker, J.D., 1977, Simulation of bubble population in a gas fluidized bed, *Chem. Eng. Sci.* 32, 1419-1425.

Thioye, M., 1988, Contribution à l'étude des lits fluidisés bouillonnants, Thèse Université de Rouen.

Werther, J., 1976, Bubble growth in large diameter fluidized beds, *Fluidization technology*, Vol. I, 215-235, Edited by Keairns DL. Hemisphere Publishing Corporation.

THE THERMOCONVECTIVE INSTABILITY IN HYDRODYNAMICS OF RELAXATIONAL LIQUIDS

I.Sh.Akhatov, and R.G.Chembarisova

Department Of Continuous Media Mechanics
Bashkir State University
450074 Ufa, Frunze 32, Russia

Abstract: The main purpose of this paper is to show the influence of relaxational characteristics on the beginning and development of thermal convection in non-Newtonian liquids. The thermoconvection in horizontal plane layer of viscoelastic liquid is considered. It shows that in this case the convective motion is described by Lorenz system with modified Prandtl number, which depends on relaxation times of liquid. Therefore the development of thermal convection in relaxational medium take place as well as in viscous Newtonian liquid. Only the critical values of Rayleigh number are changed quantitatively. The beginning and development of convective motion of relaxational liquid in horizontal porous layer is considered. The proper nonlinear dynamical system is deduced and it is showed that when relaxation about pressure gradient is absent this new system comes to the classical Lorenz system. The analytical and numerical research of this new system solutions depending on the Rayleigh number and relaxation time values is realized. In particular is noted that stability disturbance for Rayleigh number less than classical Darcy-modified Rayleigh critical value is possible in certain cases. The toroidal-shaped porous medium saturated with relaxational liquid is considered. It shows that the solution of thermoconvective problem in this case in approximated by above-stated nonlinear dynamical system exactly, so that the higher garmonics decreases exponentialy in time and their influence on lowest garmonics is equal to zero.

1. INTRODUCTION

It is known (Rayleigh, 1916) that the stable mechanical equilibrium of liquid heated from below is possible only for small degree of temperature difference (Rayleigh number). If the temperature difference is high, then the equilibrium is getting unstable and the convective motion is generated (Benard, 1900). For the approximate mathematical simulation of nonlinear evolutional stage of thermal convection in classical hydrodynamics the Galerkin's expansions method is used usually. As a result an initial partial equations system reduces to nonlinear dynamical system. In the original work of Lorenz (1963) was shown that for high temperature gradients the stable convection becomes unstable and the convective motion is detting chaotic. Later such phenomena were named "deterministic chaos". The Lorenz's work was based on the original numerical results of Saltzman (1962), that in some cases all high garmonics of Galerkin's expansion decreases in time and their influence on lowest garmonics is negligible. But it is not always correct. Subsequently the Lorenz's result was specified by Curry, Herring, Loncaric and Orszag (1984). In their work the using of greater set of garmonics (more than 19) led to disappearance of chaotic behaviour. But the problem of "deterministic chaos" was not closed for this reason because by that time several very important examples of hydrodynamical systems with chaotic behaviour had been discovered. In particular by Creveling, de Paz and Schvenhals (1975) it was noted that thermoconvection of Newtonian liquid in thin toroidal-shaped region disposed on the vertical plane and heated from below is exactly described by the classical Lorenz's dynamical system, so that the increasing of garmonic number should not lead to disappearance of chaotic motion.

The thermoconvective motion in plane horizontal layer of porous media saturated by viscous Newtonian liquid was investigated by Horton, Rogers (1945) and Lapwood (1948). In these works the Darcy-modified Rayleigh number was obtained and its critical values were predicted. These theoretical results were corroborated by experiments of Katto, Masuoka (1967).

In connection with problems of intensification of some technological processes of oil industry and chemistry it is important to investigate the peculiarities of thermoconvective motion in liquids which possesses relaxational properties.

2. THERMOCONVECTION IN HORIZONTAL PLANE LAYER OF VISCOELASTIC LIQUID

The set of differential equations of two-dimensional motion of liquid is as follows (y-axis is directed upwards)

$$\rho\left(\frac{\partial u}{\partial t} + u\frac{\partial u}{\partial x} + v\frac{\partial u}{\partial y}\right) = -\frac{\partial p}{\partial x} + \frac{\partial \tau_{xx}}{\partial x} + \frac{\partial \tau_{xy}}{\partial y} \tag{2.1}$$

$$\rho\left(\frac{\partial v}{\partial t} + u\frac{\partial v}{\partial x} + v\frac{\partial v}{\partial y}\right) = -\frac{\partial p}{\partial y} + \frac{\partial \tau_{yx}}{\partial x} + \frac{\partial \tau_{yy}}{\partial y} - \rho g \tag{2.2}$$

$$\frac{\partial \rho}{\partial t} + u\frac{\partial \rho}{\partial x} + v\frac{\partial \rho}{\partial y} + \rho\left(\frac{\partial u}{\partial x} + \frac{\partial v}{\partial y}\right) = 0 \tag{2.3}$$

Here $\vec{v} = (u,v)$ - velocity, p - pressure, ρ - density, g - acceleration of gravity, τ- components of a stress deviator tensor. For analysis of thermoconvective motion Boussinesq

(1903) approximation is usually used. According to this approximation the compressibility of liquid is considerable only for calculation of gravity term in motion equation (2.2). Then mass balance equation (2.3) can be written as

$$\frac{\partial u}{\partial x} + \frac{\partial v}{\partial y} = 0 \tag{2.4}$$

In streamfunction formulation

$$u = -\frac{\partial \psi}{\partial y}, \qquad v = \frac{\partial \psi}{\partial x} \tag{2.5}$$

mass conservation equation (2.4) is satisfied exactly, motion equations (2.1),(2.2) are transformed to one equation as follows:

$$\rho_0 \left(\frac{\partial \Delta\psi}{\partial t} + \frac{\partial \psi}{\partial x}\frac{\partial \Delta\psi}{\partial y} - \frac{\partial \psi}{\partial y}\frac{\partial \Delta\psi}{\partial x} \right) = -g\frac{\partial \rho}{\partial x} + \left(\frac{\partial^2}{\partial x^2} - \frac{\partial^2}{\partial y^2} \right) \tau_{xy} + \frac{\partial^2}{\partial x \partial y}\left(\tau_{yy} - \tau_{xx} \right) \tag{2.6}$$

Here $\Delta = \partial^2/\partial x^2 + \partial^2/\partial y^2$, $\rho_0 = \rho(T_0)$. It is necessary to define concretely the rheological law, i.e. dependence of stress deviator tensor τ_{ij} on the rate of strain tensor $\dot{\varepsilon}_{ij}$, and the heat expansion law, i.e. dependence of ρ on temperature T.

For the description of rheological behaviour of liquid let us use the linear model (Oldroyd, 1964)

$$\left(1 + \lambda_1 \frac{\partial}{\partial t} \right) \tau_{ij} = 2\mu \left(1 + \lambda_2 \frac{\partial}{\partial t} \right) \dot{\varepsilon}_{ij} \tag{2.7}$$

Here λ_1, λ_2 - relaxation times which satisfy to inequality $\lambda_1 > \lambda_2$ (Gorotsov, Leonov, 1968).

The heat expansion law can be written as

$$\rho = \rho_0 [1 - \alpha(T - T_0)] \tag{2.8}$$

For mathematical simulation of convectional heat transfer let us use the ordinary equation

$$\frac{\partial T}{\partial t} + u\frac{\partial T}{\partial x} + v\frac{\partial T}{\partial y} = \kappa \, \Delta T \tag{2.9}$$

where α and κ - heat expansion and conductivity coefficients.

Let us consider the thermoconvection in plane horizontal layer with thickness equal to h under the following boundary conditions:

$$T\Big|_{y=0} = T_o + \delta T \ , \quad T\Big|_{y=h} = T_o , \quad v\Big|_{y=0,h} = 0 \tag{2.10}$$

The temperature distribution is

$$T(x,y,t) = T_o + \delta T(1 - \frac{y}{h}) + \theta(x,y,t)$$

$$\theta\Big|_{y=0,h} = 0 \tag{2.11}$$

Then the equations (2.5)-(2.9) lead to the system

$$\frac{\partial\Delta\psi}{\partial t} + \frac{\partial\psi}{\partial x}\frac{\partial\Delta\psi}{\partial y} - \frac{\partial\psi}{\partial y}\frac{\partial\Delta\psi}{\partial x} = g\alpha\frac{\partial\theta}{\partial x} + \nu\Delta\Delta(1+\lambda\frac{\partial}{\partial t})\psi \tag{2.12}$$

$$\frac{\partial\theta}{\partial t} - \frac{\partial\psi}{\partial y}\frac{\partial\theta}{\partial x} + \frac{\partial\psi}{\partial x}\frac{\partial\theta}{\partial y} - \frac{\delta T}{h}\frac{\partial\psi}{\partial x} = \kappa\ \Delta\theta \tag{2.13}$$

$$\nu = \mu/\rho \ , \ \lambda = \lambda_2 - \lambda_1$$

Taking after Lorenz (1963) let us expand functions ψ and θ on series of eigen functions of linearized problem (2.10)-(2.13)

$$\psi = \frac{\sqrt{2}\,(1+a^2)\kappa}{a} X(t)\sin(\frac{\pi a}{h}x)\sin(\frac{\pi}{h}y) \tag{2.14}$$

$$\theta = \frac{R_c\delta T}{\pi R}\left[\sqrt{2}\ Y(t)\cos(\frac{\pi a}{h}x)\sin(\frac{\pi}{h}y) - Z(t)\sin(\frac{2\pi}{h}y)\right] \tag{2.15}$$

$$R = \frac{\alpha g h^3\delta T}{\nu\kappa} \ , \quad R_c = \frac{\pi^4(1+a^2)^3}{a^2} .$$

Here R and R_c - Rayleigh number and its critical value for thermoconvection problem in horizontal plane layer of viscous Newtonian liquid. The second term in (2.15) represents higher garmonic which appears as a result of nonlinear infer action of first garmonics. If the expressions (2.14),(2.15) are use in (2.12),(2.13) then thermoconvection problem reduces to following nonlinear dynamical system

$$\dot{X} = -\frac{\sigma}{1+\sigma L}(X-Y)$$

$$\dot{Y} = -XZ + rX - Y \tag{2.16}$$

$$\dot{Z} = XY - bZ$$

Here

$$r = \frac{R}{R_c}, \quad \sigma = \frac{\nu}{\kappa}, \quad b = \frac{4}{1+a^2}, \quad L = \left(\frac{\pi}{h}\right)^2 (1+a^2)\kappa\lambda ,$$

and the point means the differentiation on dedimensionalized time

$$\tilde{t} = \left(\frac{\pi}{h}\right)^2 (1+a^2)\kappa t.$$

It is easy to see that system (2.16) represents Lorenz's system with modified Prandtl number

$$\sigma^* = \frac{\sigma}{1+\sigma L} > \sigma \qquad (L < 0)$$

Therefore the development of thermoconvection in relaxational medium is like the one in viscous Newtonian liquid. Only critical values of Rayleigh parameter are changed quantitatively. Stability disturbance of stationary temperature distribution in immovable medium happens to be like the one in Newtonian liquid when $r=1$ $(R=R_c)$, but stability disturbance of stationary convective motion takes place when

$$r > \varphi(\sigma^*) = \frac{\sigma^*(\sigma^* + b + 3)}{\sigma^* - b - 1}$$

Hence, we have that if

$$\sigma < \frac{b+1}{1-(b+1)L}$$

then stationary convection is stable for any values of r.

Analysis of function $\varphi(\sigma)$ behaviour shows that

$$\varphi'(\sigma) > 0 \quad \text{for} \quad \sigma > \sigma_+ = b+1+\sqrt{2(b+1)(b+2)} \qquad (2.17)$$

$$\varphi'(\sigma) < 0 \quad \text{for} \quad 0 < \sigma < \sigma_+$$

Since $\sigma^* > \sigma$ then from (2.17) follows that the relaxational properties of liquid lead either to stabilisation $(\sigma > \sigma_+)$ or to destabilization $(\sigma < \sigma_+)$ of convective motion.

3. HEAT CONVECTION OF RELAXATIONAL LIQUIDS IN POROUS MEDIA

When studying the motion of liquids in porous media it is usually supposed that the equilibrium between the pressure gradient and filtration velocity is settled instantly (Darcy's law). But for many rheologically anomalous liquids (clay solutions, polymer solutions, paraffin containing oils, gas-liquid mixtures) this equilibrium may be settled after some delay. The nonequilibrium effects of these systems lead to the necessity of using modified Darcy's law (Alishaev, Mirzadjanzade, 1975) as

$$(\tau_1 \frac{\partial}{\partial t} + 1)\ \vec{v} = - \frac{k}{\mu}\ (\tau_2 \frac{\partial}{\partial t} + 1)\ \vec{\nabla} p$$

where τ_1, τ_2 - relaxation times, k - permeability of porous medium, μ - viscosity of liquid, $\vec{v}$, p - filtration velocity and pressure.

Two-dimentional thermoconvective motion of such relaxational liquid in porous horizontal layer with thickness equal h in Boussinesq approximation is written by the following system (y - axis is derected upwards)

$$\operatorname{div} \vec{v} = 0$$

$$(\tau_1 \frac{\partial}{\partial t} + 1)\ \vec{v} = - \frac{k}{\mu}\ (\tau_2 \frac{\partial}{\partial t} + 1)(\ \vec{\nabla} p + \rho_1 g\ \vec{e}_y) \qquad (3.1)$$

$$\frac{\partial T}{\partial t} + c(\ \vec{v}\cdot\vec{\nabla}\ )T = \kappa\ \Delta T\ , \quad \rho_1 = \rho_{10}[1-\alpha(T-T_0)]$$

$$c = \frac{\rho_1 c_1}{m\rho_1 c_1 + (1-m)\rho_2 c_2}\ , \quad \kappa = \frac{\lambda_1 m + (1-m)\lambda_2}{\rho_1 c_1 m + (1-m)\rho_2 c_2}$$

Here the temperatures of liquid and porous medium are supposed to be equal, m - porosity, ρ_i, λ_i, $c_i (i=1,2)$ - density, heat conductivity and thermal heat capacity of liquid and porous medium respectively. The system of equations (3.1) is considered under the boundary conditions (2.10).

In streamfunction formulation (2.5) after some transformations the system (3.1) comes to equations on ψ and θ (see (2.11))

$$(\tau_1 \frac{\partial}{\partial t} + 1)\Delta\psi = \frac{k\alpha g\rho_{10}}{\mu}\ (\tau_2 \frac{\partial}{\partial t} + 1)\frac{\partial\theta}{\partial x}$$

$$\frac{\partial\theta}{\partial t} + c\left[- \frac{\partial\psi}{\partial y}\frac{\partial\theta}{\partial x} + \frac{\partial\psi}{\partial x}\frac{\partial\theta}{\partial y} - \frac{\delta T}{h}\frac{\partial\psi}{\partial x}\right] = \kappa\ \Delta\theta \qquad (3.2)$$

Let us expand (see the second section) functions ψ and θ on series of eigen functions of linearized problem (3.2), (2.10),(2.11)

$$\psi = \frac{\sqrt{2}\ (1+a^2)\kappa}{ac}\ X(t)\ \sin(\frac{\pi a}{h}\ x)\ \sin\ (\frac{\pi}{h}\ y) \qquad (3.3)$$

$$\theta = \frac{R_c\,\delta T}{\pi R}\ \left[\sqrt{2}\quad Y(t)\ \cos(\frac{\pi a}{h}\ x)\ \sin\ (\frac{\pi}{h}\ y) - \right.$$

$$-Z(t)\sin(\frac{2\pi}{h}y)\Big]$$

Here $R_c = \pi^2(1+a^2)^2/a^2$ – critical value of Rayleigh number for thermoconvection problem of viscous Newtonian liquid in porous medium.

Then the system (3.2) reduces to nonlinear dynamical system

$$\begin{aligned} \dot{X} &= -\frac{D_2}{D_1}XZ - \frac{1-rD_2}{D_1}X + \frac{1-D_2}{D_1}Y \\ \dot{Y} &= -XZ + rX - Y \\ \dot{Z} &= XY - bZ \end{aligned} \tag{3.4}$$

where $D_i = \tau_i\kappa(\pi/h)^2(1+a^2)$, $i=1,2$; parameters r,b and dedimensionalized time are calculated samely as for system (2.16) in the second section.

It is easy to see that for $D_2 = 0$ the system (3.4) is transformated to the Lorenz-like system (instead of σ we have D_1^{-1} in this case).

The system (3.4) have the stationary solution $X=Y=Z=0$ which corresponds to immovable liquid. Characteristic equation for this stationary solution is written by

$$(\lambda+b)\left[\lambda^2 + \left(1 + \frac{1-rD_2}{D_1}\right)\lambda + \frac{1-r}{D_1}\right] = 0 \tag{3.5}$$

Hence we have stability criterion

$$r < \min(1, r_*),\quad r_* = \frac{1+D_1}{D_2} \tag{3.6}$$

It is easy to see that for $r_* < 1$ it is possible the stability disturbance for Rayleigh number less than its classical critical value ($R < R_c$). But for $r_* < r < 1$ there are no new stationary solutions of system (3.4). For $r>1$ it appears to be two new stationary solutions

$$X = Y = \pm\sqrt{b(r-1)}\quad,\quad Z = r - 1 \tag{3.7}$$

which coincide with the respective solutions of Lorenz's system. For each of these solutions the characteristic equation is as follows:

$$\lambda^3 + \left(\frac{1}{D_1} + b + 1 - \frac{D_2}{D_1}\right)\lambda^2 + b\left[\frac{1}{D_1} + r + \frac{D_2}{D_1}(r-2)\right]\lambda + \\ + \frac{2b(r-1)}{D_1} = 0 \tag{3.8}$$

From Gurwits criterion follows that new stationary solutions (3.7) are stable if the inequalities are satisfied

$$B > 0, \quad \beta > 0 \tag{3.9}$$

$$B = 1+(b+1)D_1 - D_2 \ , \ \beta = \frac{(1-2D_2)B}{D_1} - \left[2 - B\left(1 + \frac{D_2}{D_1}\right)\right]r + 2$$

The analysis of these inequalities gives the following possible situations:

1) $r_* < 1$, $B > 0$, $\beta >< 0$;

2) $r_* > 1$, $B > 0$, $\beta >< 0$;

3) $r_* < 1$, $B < 0$, $\beta >< 0$

(the situation when $r_* > 1$, $B < 0$, is impossible).

Let us consider some calculation variants when $b=8/3$, $D_1=0.1$ (it conforms with case $b=8/3$, $\sigma=10$ which was considered by Lorenz (1963)). When $D_2=0$ the system (3.4) coincides with Lorenz's system. And so, just as expected, for small values of D_2 the thermoconvection development takes place according to the same scenario like for Lorenz's system.

For example, when $D_2=0.01$ we have $r_* > 1$, $B > 0$, and (for $r < 30.128$) $\beta>0$. If $r>30.128$ then the stable convective motion becomes the chaotic. The strange attractor and its onedimentional reflection in this case do not differ from the Lorenz results.

The subsequent increase of D_2 render some stabilizing influence on convective motion. When $D_2=0.5$ we have $B > 0$, $\beta > 0$. And so convective motion which arises in system when $r>1$ is stable for any values of r .

When $D_2=1.2$ then $r_*<1$ ($r_*=0.92$), $B>0$, and β becomes positive only for $r>2$. Therefore the stability disturbance takes place for Rayleigh number less than its critical value. After stability disturbance for $r_*<r < 1$ the attracting manifold is limit cycle placed on environment of point $X=Y=Z=0$. When $1<r<2$ we have two unstable equilibrium positions in phase space. And phase trajectory is limit cycle too. But structure of cycle in this case differs from cycle for $r_*<r < 1$. The equilibrium positions become stable when $r>2$, but region of attraction for points (3.7) is very small.

When $D_2=2.2$ convection begins for Rayleigh numbers less than its critical values too ($r_*=0.5$). For $r=r_*$ the limit cycle appears. When parameter r increases this cycle undergoes the quaint befurcations (the period of oscillations increases two, three, etc., times). We should note that this process is accompanied with strong dependence of solution on the initial data. Insignificant change of initial point coordinates leads to one or another limit cycle. The number of possible limit cycles increase with value of r.

The question about accuracy of solution of equations (3.2) by this modification of Galerkin's approximation may be answered correctly only using higher garmonics, as it was done by Curry, Herring, Loncaric and Orszag (1984) for classical Benard-Rayleigh convection.

4. HEAT CONVECTION OF RELAXATIONAL LIQUIDS IN TOROIDAL-SHAPED POROUS MEDIUM

The main purpose of this section is to show that the thermoconvection of relaxational liquid in toroidal-shaped porous medium is approximated by dynamical system (3.4) exactly, so that the higher garmonics decrease exponentialy in time and their influence on lowest garmonics is equal to zero.

Let us consider toroidal-shaped porous medium disposed on the vertical plane and heated from below. Let us assume that the thickness of tor is very small in comparison with its dimension. Then thermoconvective motion after averaging on tor's section will be characte- rized only by temperature distribution over the angle $T=T(\varphi,t)$ and by speed of liquid rotation $\omega(t)=v(t)/R$. The system of equations for discription of such motion is the following:

$$\left(\tau_1 \frac{\partial}{\partial t} + 1\right)\omega = -\frac{k}{\mu}\left(\tau_2 \frac{\partial}{\partial t} + 1\right)\left(\frac{1}{R^2}\frac{\partial p}{\partial \varphi} + \frac{\rho_1 g \sin\varphi}{R}\right)$$

$$\frac{\partial T}{\partial t} + c\omega \frac{\partial T}{\partial \varphi} = \kappa_1 \frac{\partial^2 T}{\partial \varphi^2} - \lambda[T - T_1(\varphi)] \qquad (4.1)$$

$$\rho_1 = \rho_{10}[1-\alpha(T-T_0)], \quad T_1(\varphi) = T_0 + \frac{\delta T}{2}(\cos\varphi + 1)$$

$$\lambda = \frac{2\lambda_3}{a[m\rho_{10}c_1 + (1-m)\rho_{20}c_2]}, \qquad \kappa_1 = \frac{\kappa}{R^2}.$$

Here c,κ - calculated by (3.1); λ_3- heat transfer coefficient from the surroundings to porous medium; $T_1(\varphi)$ - temperature of surrounding medium; R - tor radius, a - radius of tor thickness.

Fourie series of temperature distribution is written by

$$T(\varphi,t) = u_0(t) + \sum_{n=1}^{\infty} [a_n(t)\sin(n\varphi) + b_n(t)\cos(n\varphi)] \qquad (4.2)$$

From (4.1),(4.2) we have set of equations which after averaging

$$X = \frac{\omega}{\omega_*}, \quad Y = \frac{a_1}{a_*}, \quad Z = \frac{\tilde{b}_1}{b_*},$$

$$Y_n = \frac{a_n}{a_*}, \quad Z_n = \frac{b_n}{b_*} \quad (n \geq 2), \quad \tilde{b}_1 = -b_1 + \frac{\lambda\delta T}{2(\lambda+\kappa)}$$

$$a_* = b_* = \frac{2h(\lambda+\kappa)R\mu m}{k\alpha\rho_{10}gc}, \quad \omega_* = \frac{\lambda+\kappa}{c}$$

is transformed into the following form

$$\dot{u}_0 = -\lambda u_0 \tag{4.3}$$

$$\dot{X} = -\frac{D_2}{D_1} XZ - \frac{1-rD_2}{D_1} X + \frac{1-D_2}{D_1} Y$$

$$\dot{Y} = -XZ + rX - Y \tag{4.4}$$

$$\dot{Z} = XY - Z$$

$$\dot{Y}_n = nXZ_n - \frac{\lambda+\kappa n^2}{\lambda+\kappa} Y_n$$

$$\dot{Z}_n = -nXY_n - \frac{\lambda+\kappa n^2}{\lambda+\kappa} Z_n \tag{4.5}$$

$$D_1 = \tau_1(\lambda+\kappa), \; D_2 = \tau_2(\lambda+\kappa), \; r = \frac{c\lambda k\alpha\rho_{10}g\delta T}{4R\mu m(\lambda+\kappa)^2}$$

The point means the differentiation on dedimensionalized time

$$\tilde{t} = (\lambda+\kappa)t$$

The behaviour of system (4.3)-(4.5) solutions depends on only the solutions of system (4.4). It really leads from (4.3) that $u_0(t)$ decreases exponentially in time. It is easy to show by system (4.5) that the "energy" of higher garmonics $E_n = Y_n^2+Z_n^2$ decreases exponentially too

$$E_n = E_n^0 \exp\left(-2\frac{\lambda+\kappa n^2}{\lambda+\kappa}\tilde{t}\right).$$

System (4.4) offers system (3.4) in which b=1. Therefore all results received in third section for system (3.4) may be applied to system (4.4) without any alternations.

REFERENCES

Alishaev, M.G., and Mirzadjanzade, A.Kh., 1975, For the calculation of delay phenomenon in filtration theory, *Izvestya Vuzov, Neft i Gaz*. 6:71.

Benard, H., 1900, Less tourbillons cellulaires dans une nappe liquide,*Revue Generale des Sciences,Pures et Appliquees*.12:1261, 1309.

Boussinesq,J., 1903,"Theorie analytique de la chaleur," Vol.2, Paris.

Creveling,H.F., de Paz,J.F., Baladi,J.Y.,and Schvenhals,R.J., 1975, Stability characteristics of a single-phase free convection loop, *Journal of Fluid Mechanics*.67:65.

Curry,J.H., Herring,J.R., Loncaric,J., and Orszag,S.A., 1984, Order and disorder in two- and three- dimensional Benard convection,*Journal of Fluid Mechanics*.147:1.

Gorodtsov,V.A., and Leonov,A.I.,1968, About kinematics, nonequilibrium thermodynamics and rheological correlations in linear theory of viscoelasticitiy, *Prikladnaya Matematika i Mehanika 1*.32:70.

Horton,C.W., and Rogers,F.T.,1945, Convections currents in a porous medium, *Journal of Applied Physics*.16: 367.

Katto,Y., and Masuoka,T., 1967, Criterion for the onset of convective flow in a fluid in a porous medium, *Int. J. Heat and Mass Transfer 3*.10: 297.

Lapwood,E.R., 1948, Convection of a fluid in porous medium, *Proc. Camb. Phil. Soc.4*.44: 508.

Lorenz,E.N., 1963, Deterministic Nonperiodic Flow, *Journal of the Atmospheric Sciences*.20: 130.

Oldroyd,J.G., 1964, Non-linear stress, rate of strain relations at finite rates of shear in socalled "linear" elastico-viscous liquids, *in* : "Second order effects in elasticity, plasticity and fluid dynamics, Proc. Int. Symp. 1962," Pergamon Press.

Rayleigh.,1916, On convection currents in a horizontal layer of fluid, when the higher temperature is on the under side, *Phil.Mag*.32:529.

Saltzman,B.,1962, Finite amplitude free convection as aninitial value problem, *Journal of the Atmospheric Sciences*.19:29.

Sparrow,C., 1982, The Lorenz Equations: Bifurcations, Chaos and Strange Attractors, *in*: "Applied Mathematica Sciences.41," Springer-Verlag, New-York -Heidelberg -Berlin.

Yorke,J.A., Yorke,E.D., and Mallet-Paret,J., 1987, Lorenz-like chaos in a partial defferential equations for a heated fluid loop, *Phisica D*.24D: 279.

DYNAMIC PROPERTIES OF THERMAL CONVECTION IN POROUS MEDIUM

Dmitry V. Lyubimov

Theoretical Physics Department
Perm State University
15, Bukirev Street
614600, Perm, Russia

1. INTRODUCTION

Thermal convection in porous media is considered in most of the works on the base of Darcy law and Boussinesq approximation. The main distinction of the Darcy-Boussinesq equations from that of the uniform fluid convection is the lower order of the velocity field spatial derivatives. This leads to the change of the boundary conditions: at the impermeable boundary one can impose only the condition of impermeability (and not the no-slip condition). The second distinction is of less importance: the inertial terms are absent in the equation of motion. This should prevent the appearance of the oscillatory states. The similar situation takes place in the uniform fluid when the value of the Prandtl number is high enough.

From the first glance the above distinctions should not cause considerable changes in the general scenario for the onset and development of convection. In fact in many publications it is shown that in the case of heating strictly from below the mechanical equilibrium is possible and the stability of this equilibrium is defined by the dimensionless parameter R_p, which is the analog of the Rayleigh number. If the value of the Rayleigh number is small so that $R_p < R_p^*$, the equilibrium is stable. At $R_p > R_p^*$ it is unstable with respect to the monotonously increasing perturbations. Besides, at the bifurcation point $R_p = R_p^*$ two stationary solutions arise corresponding to two different directions of the fluid circulation. With the growth of R_p the number of possible solutions increases, this increase being faster for the cavities with longer horizontal size. In the limit case of infinite horizontal layer an infinite set of the stationary solutions exists but in finite regions the number of the solutions remains finite. With the growth of R_p the stationary solutions may lose stability, thus leading to the onset of oscillations and eventually to chaos. In most of earlier works the authors considered either the infinite horizontal layer or the region with the adiabatic conditions at the sidewall, which is the model of the layer cell. We found out [1] that the case of high thermal conductivity of the boundaries, which is not less realistic, demonstrates the considerably more complicated properties than the above cases. It has been shown that for supercritical values $R_p > R_p^*$ the infinite number of stationary solutions may exist even in the case of non-symmetrical finite region. In present work the influence of small violations of the boundary conditions on the dynamics of the system is studied and the complication of the behavior with the growth of R_p is investigated.

2. BASIC EQUATIONS: BOUNDARY CONDITIONS

Let us consider two-dimensional convective motion ($v_y = 0, \frac{\partial}{\partial y} = 0$, $\vec{v}$ is the filtration velocity, the axes x,y of the Cartesian coordinate system are horizontal and z is the vertical axis) in the simply connected region Ω with the boundary Γ. In the Darcy-Boussinesq approach the equations of the convection have the form:

$$-\frac{1}{\rho}\nabla p - \frac{\nu}{K}\vec{v} + g\beta T\vec{\gamma} = 0 \tag{1}$$

$$(\rho C_p)_m \frac{\partial T}{\partial t} + (\rho C_p)_f \vec{v}\nabla T = \kappa \Delta T \tag{2}$$

$$\operatorname{div} \vec{v} = 0 \tag{3}$$

where ρ is the fluid density at the temperature T_o , p is the convective additive to the pressure, $\vec{v}$ is the filtration velocity, T is the deviation of the temperature from the mean value T_o, β is the coefficient of the fluid thermal expansion, ν is the fluid viscosity, K is the permeability, g is the acceleration of gravity, $\vec{\gamma}$ is the unit vector along the vertical direction, $(\rho C_p)_f$ is the fluid heat capacity, $(\rho C_p)_m$ is the heat capacity of the saturated porous medium.

We choose the values of L, L^2/χ_m , χ/L, θ, $\rho\nu\chi/K$ ($\chi_m = \kappa/(\rho C_p)_m$, $\chi = \kappa/(\rho C_p)_f$) as the units of the length, time, velocity, temperature and pressure correspondingly. The equations (1)-(3) take the form:

$$-\nabla p - \vec{v} + R_p T\vec{\gamma} = 0 \tag{4}$$

$$\frac{\partial T}{\partial t} + \vec{v}\,\nabla T = \Delta T \tag{5}$$

$$\operatorname{div} \vec{v} = 0 \tag{6}$$

where $R_p = g\beta\theta LK/\nu\chi$.
The boundary conditions are:

$$v_n|_\Gamma = 0, \quad T|_\Gamma = f(\vec{r})|_\Gamma \tag{7}$$

where $f(\vec{r})$ is the known function.

Our assumption of two-dimensional character of the flow means from the physical point of view that we consider the convection in the horizontal cylinder, which is uniform along the y axis and confined in this direction by thermoinsulating walls. These walls can be removed into the infinity but in the experiments the 2-dimensional character of convection could be more expected if the size of the region along the y axis is of the same order or even less than the other sizes. We will be most interested in the case of heating from below. In this case function $f(\vec{r})$ in (7) is the linear dependence on the vertical coordinate: $f(\vec{r}) = -z$. In the experiment such a situation can be created if the porous medium is surrounded by the array of high thermal conductivity.

3. AMPLITUDE EQUATIONS AT SMALL SUPERCRITICALITIES; CO-SYMMETRY PROPERTY

The problem (4)-(7) admits the solution $\vec{v} = 0$, $T = -z$, which corresponds to the mechanical equilibrium. Let us formulate the problem for the deviations from this solution. We present the temperature in the form $T = -z + \theta$. The velocity $\vec{v}$ can be expressed through the stream function ψ: $v_x = -\frac{\partial \psi}{\partial z}$, $v_z = \frac{\partial \psi}{\partial x}$. In this case the equation of continuity is satisfied automatically. We eliminate the pressure and arrive at the problem:

$$\Delta\psi - R_p \frac{\partial \theta}{\partial x} = 0 \tag{8}$$

$$\frac{\partial \theta}{\partial t} + J(\psi,\theta) = \Delta\theta + \frac{\partial \psi}{\partial x} \tag{9}$$

$$\theta\,|_\Gamma = 0\,, \quad \psi\,|_\Gamma = 0 \tag{10}$$

Here $J(\psi,\theta)$ is the Jakobian of the fields ψ,θ over the variables x, z. We have taken into account, when formulating the boundary conditions for ψ, that the region Ω is simply connected.

We consider the problem of small perturbations of the mechanical equilibrium. As in the case of uniform fluid it is easy to show that for the heating from above the equilibrium is stable and for the heating from below the oscillatory perturbations are not presented in the spectrum. The problem for the critical monotonous perturbations for the heating from below can be written in the form:

$$\Delta\psi = C \frac{\partial \theta}{\partial x} \tag{11}$$

$$\Delta\theta = -C \frac{\partial \psi}{\partial x} \tag{12}$$

$$\theta\,|_\Gamma = 0\,, \quad \psi\,|_\Gamma = 0 \tag{13}$$

where the notation $C = \sqrt{R_p}$ was introduced and the scale of v was changed. The problem (11)-(13) has the remarkable property - it is invariant with respect to the change of the variables ($\psi \to \theta$, $\theta \to -\psi$). One easily sees that the two solutions obtained in this way are linear-independent, i.e. all the critical eigenvalues are at least double-degenerated for any shape of the region [1]. By special choice of the geometry it is possible to reach even higher degeneracy for some eigenvalues, but, as it was shown in [2], the degeneracy of the principal eigenvalue always equals 2.

Let us consider now the behavior of the non-linear system (8)-(10) at small supercriticalities. In the case of convection in the uniform fluid the double degeneracy of the principal eigenvalue (as for example in the case of rectangular region at some special selections of the aspect ratio) leads to the bifurcation of two branches of steady states from the equilibrium, with only one of these branches being stable. In the porous medium the situation is qualitatively different. It can be shown by the standard Lyapunov-Schmidt method that in all orders of the expansion only free parameter exists [1]; this means that the non-trivial steady solutions of (8)-(10) form the one-parameter family. The normal form for the reduction of (8)-(10) on the central manifold can be easily derived. We rewrite the solution for small supercriticalities in the form:

$$\psi = \alpha\tilde{\psi} + \beta\tilde{\theta}$$

$$\theta = \alpha\tilde{\theta} - \beta\tilde{\psi}$$

where $\tilde{\psi}$,$\tilde{\theta}$ is solution of the problem (11)-(13) normalized by right way, use the method of multi time scales and assume α,β to be the functions of average time τ. Then we obtain:

$$\frac{d\alpha}{d\tau} = \alpha\,(1 - \alpha^2 - \beta^2) \tag{14}$$

$$b\frac{d\beta}{d\tau} = \beta\,(1 - \alpha^2 - \beta^2) \tag{15}$$

where the positive coefficient b is defined by the geometry of the region. The phase portrait for the system (14)-(15) is plotted in Figure 1. The stationary solutions are located on the ring $\alpha^2 + \beta^2 = 1$ and all of them are stable.

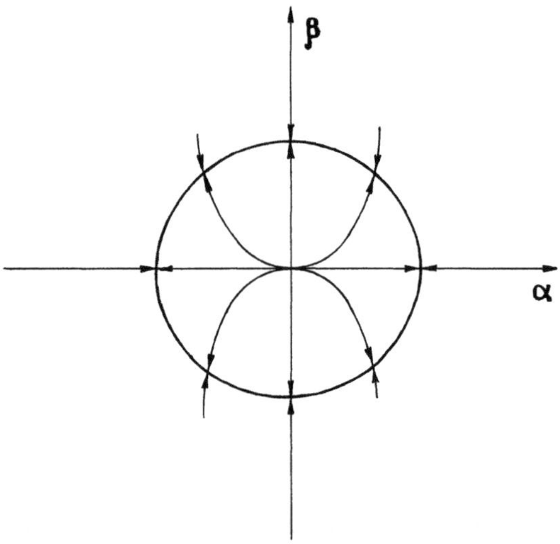

Figure 1. Phase portrait for the system (14)-(15): the slow motions space in the case of heating from below at small supercriticalities.

It must be noted that the problem (8)-(10) does not have any symmetry. In this situation the existence of the stationary states family looks strange. The origin of this phenomenon was explained recently in [2] with the help of the co-symmetry theory. It is known that two operators $\hat{L}$ and $\hat{N}$ acting in the Hilbert space H form a co-symmetrical pair if $\langle \hat{L}\theta, \hat{N}\theta \rangle = 0 \ \forall\ \theta \in H$, where < , > denotes scalar product. It is shown in [2] that the existence of non-trivial co-symmetry leads to the following result: if the additional degeneracies are absent and $\hat{N}\theta$ is non-zero, then the solutions of the equation $\hat{L}\theta = 0$ are not isolated and form the one-parameter family. Note that the existence of symmetry implies the existence of co-symmetry but the inverse is true only in case of holonomous co-symmetry, which for the above result is not necessary.

The existence of the family of solutions in the presence of co-symmetry can be easily understood in the finite-dimensional case. Here the co-symmetry property means the existence of the vector field L orthogonal to the considered one N. If L is not equal to zero then the condition of orthogonality puts on the only bound for the system of the equations defined the pieces of the field N, which, generically, leads to the appearance of the family of the solutions.

The existence of co-symmetry for the problem (8)-(10) can be easily shown. Multiplying (9) by ψ and integrating over the volume we obtain:

$$<\psi \frac{\partial\theta}{\partial t}> = <\psi\, J(\theta,\psi)> + <\psi\, \Delta\theta> + <\psi \frac{\partial\psi}{\partial x}> \tag{16}$$

The first and the last integrals in the righthand-side of (16) vanish due to the boundary conditions and the second integral can be found zero by using the equation (8).

4. SYMMETRY BREAKDOWN

At least two group of the questions arise in the discussed situation:
1. What happens at small perturbations of the problem destroying the co-symmetry;
2. What happens to the one-parameter family and the solutions in its vicinity with the change of the parameters (in our case with the change of R_p).

The first group of the questions was partially answered in [1,3]. From general reasons it is evident that one-parameter family should be destroyed and only the finite number of the stationary solutions and slowly evolving non-stationary ones would persist. Exactly this result was found in [1,3] but the behavior of the system at different violations of the co-symmetry proved to be different. In the case of heating non-strictly from below the normal form describing the behavior near the critical value of the Rayleigh number R_p* has a form:

$$\dot{\alpha} = \alpha\,(r - \alpha^2 - \beta^2) - \delta \tag{17}$$

$$b\,\dot{\beta} = \beta\,(r - \alpha^2 - \beta^2) \tag{18}$$

Here r - is the supercriticality, δ is the parameter characterizing the sidewall-heating. The typical phase portrait is shown in Figure 2.

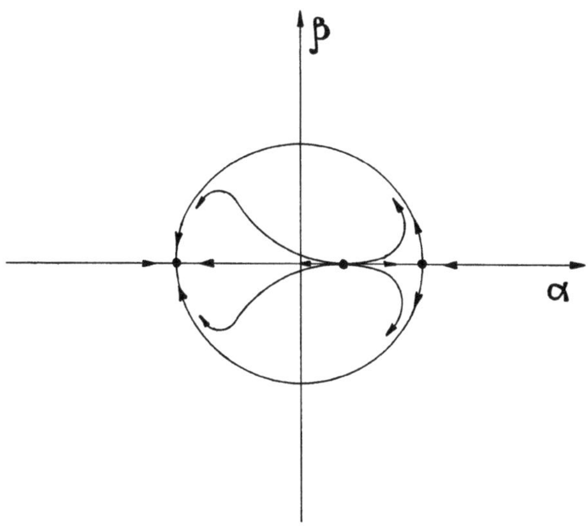

Figure 2. Phase portrait for the system (17)-(18): the case of the weak heating from the sidewall.

Thus in the above case only one stable stationary state remains.

Another situation emerges in the case of heating strictly from below but at some deviation from the infinite thermal conductivity of the boundaries. In this case the normal form is

$$\dot{\alpha} = \alpha\,(r - \alpha^2 - \beta^2) \tag{19}$$

$$b\,\dot{\beta} = \beta\,(r - \delta - \alpha^2 - \beta^2) \tag{20}$$

where δ is proportional to the ratio of thermal conductivities of the porous medium and the walls. The phase portrait (in the case $r > \delta > 0$) is given in the Figure 3.

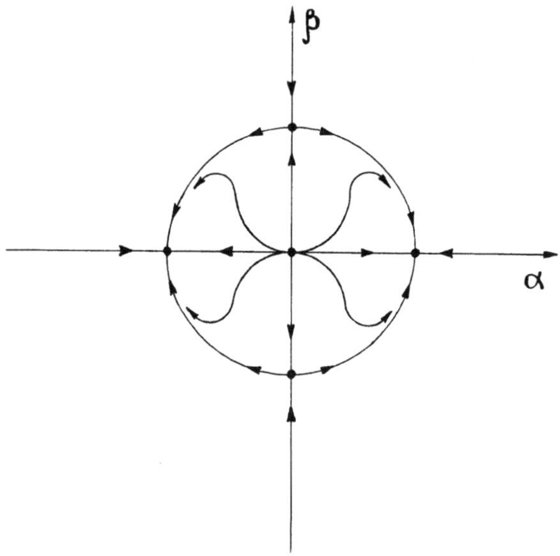

Figure 3. Phase portrait for the system (19)-(20): the case of finite thermal conductivity boundaries.

Let us consider now the case when the region is heated from below and the walls are of the infinite thermal conductivity but the hydrodynamical boundary conditions are different: namely, the slow seeping of the fluid through the sidewalls is performed in the horizontal direction. In this case the normal form is

$$\dot{\alpha} = \alpha\,(r - \alpha^2 - \beta^2) - \delta\beta \tag{21}$$

$$b\,\dot{\beta} = \beta\,(r - \alpha^2 - \beta^2) + \delta\alpha \tag{22}$$

where δ is proportional to the velocity of the seeping. The phase portrait for the system (21)-(22) is plotted in Figure 4. As one sees, in this case no stationary states from the one-parameter family survive; instead the limit cycle appears. The absence of non-trivial stationary solutions in this situation can easily be shown directly also from the full system (8)-(10).

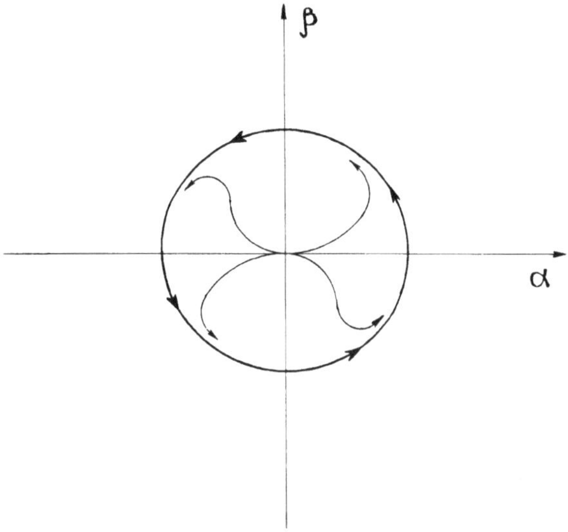

Figure 4. Phase portrait for the system (21)-(22): the case of the seeping through the sidewalls.

For the investigation of bifurcations of the stationary solutions bifurcations with the change of R_p, the problem for the concrete shape of the region - the square one - was considered with the help of the Galerkin method. The eigenfunctions for the problem of square membrane fixed on the contour were used as the basis functions for ψ and θ. The obtained finite-dimensional dynamical system was investigated numerically. It has been found that with the growth of R_p up to the certain critical value R_1 all the solutions from the family remain stable. At $R_p > R_1$ part of the solutions loses stability and with the further growth of R_p the portion of the unstable solutions increases. It is interesting to note that the loss of the stability is not accompanied by the appearance of the other fixed points or limit cycles in the phase space: the evolution of the small perturbations of the unstable solutions ends up in those stationary solutions from the family which have not yet lost their stability. At R_p large enough the chaotic set emerges in the phase space. The mechanism of the generation of this set is analogous to the case of the Lorenz model, but since the unstable manifold of the zero solution is two-dimensional, this chaotic set is not an attractor. It becomes attractive only at still higher values of R_p.

ACKNOWLEDGMENT

The author is grateful to Prof. G.Z.Gershuni and Prof. V.I.Yudovich for the useful discussions.

REFERENCES

1. D.V. Lyubimov, On the convective flows in the porous medium heated from below, *Journal of Applied Mechanics and Technical Physics*, 2:131 (1975) (in Russian).
2. V.I. Yudovich, Co-symmetry, degeneracy of the operator equations solutions, onset of the filtration convection, *Mathematical notes*, 49:142 (1991) (in Russian).
3. A.F. Glukhov, D.V. Lyubimov and G.F. Putin, Convective motions in a porous medium near the equilibrium instability threshold, *Sov. Phys. Dokl.*, 23:22 (1978)

NON LINEAR THREE DIMENSIONAL MISCIBLE VISCOUS FINGERING IN POROUS MEDIA

Jean-Claude Bacri, Nicole Rakotomalala,
Dominique Salin, and Robert Wouméni

Laboratoire d'Acoustique et Optique de la Matière Condensèe (*)
Univerité Pierre et Marie Curie, Tour 13, Boîte 78
4 Place Jussieu, 75252 Paris, Cedex 05, France

I. - INTRODUCTION

Viscous fingering resulting from unstable fluid displacements in porous media has been studied extensively over the last forty years since the pioneering experiments of Hill[1]. Recent reviews on this subject by Homsy[2] and Yortsos[3] are available as well as on the special issue of the Saffman-Taylor[4] finger in a Hele-Shaw cell by Bensimon et al.[5]. Most of the papers on viscous fingering deal with immiscible fluids, but indeed the problem involving miscible fluids deserves at least as much attention as the immiscible case : as in the immiscible case, the unfavorable viscosity ratio (displacing fluid less viscous than the displaced one) generates the instability but here the stabilizing effect is due to the hydrodynamic dispersion which tends to spread out growing fingers. Dispersion is more subtle than interfacial tension. Further, it is anisotropic and flow dependent which leads to new predictions[6-12] such as a cross-over between diffusive and linear growth regimes[7,12] and an enhancement of the instability due to the interplay between the large viscosity ratio and the velocity dependent hydrodynamic dispersion[9]. Experiments are scarce[1,13-18] and deal generally with a pseudo 2D geometry involving qualitative visualization. In this paper, we use a newly developed acoustic technique[19-21] to carry out the first study of the profiles of viscous fingers in 3D porous media. Our experiments have been performed out on three different porous media with a wide range of viscosity ratios and flow rates. Both the diffusive and the linear growth are observed including the cross-over from one to the other. Taken together, our data are best understood in terms of a new instability parameter, that characterizes the main features of viscous fingering. Our determination of the dependence of this parameter on the viscosity ratio, the flow rate and on the porous medium when placed in the context of existing theory leads to new physical insights on this rich and varied problem.

(*) Associated with the Centre National de la Recherche Scientifique

Instabilities in Multiphase Flows, Edited by G. Gouesbet
and A. Berlemont, Plenum Press, New York, 1993

II. - THEORETICAL FORMULATION OF THE PROBLEM

As we are mainly interested in the early stages of the experimental development of viscous fingering, we will summarize, in this section, the theoretical results of linear stability analysis, compare the results due to different authors[6-10] and try to get instability growth criteria suitable for comparison with experimental data. We will follow the approach of Tan and Homsy[7] but including gravity (Wooding[6]) and the non linear dispersion effects of Yortsos and Zeybek[9].

A. - Governing equations

We analyse the flow stability in a rectilinear, homogenous and isotropic porous medium, initially saturated with a viscous fluid (viscosity μ_+) with concentration 1 ; at time t = 0 another incompressible miscible less viscous fluid (viscosity μ_-) is injected at the inlet at a uniform velocity U in the z-direction. At a macroscopic scale, large compared to any details of the porous medium, the PDE describing mixing in miscible fluid flow is a convection-dispersion equation (CDE) :

$$\frac{\partial C}{\partial t} + \vec{U} \bullet \vec{\nabla} C = \vec{\nabla} (\bar{\bar{D}} \bullet \vec{\nabla} C) \tag{1}$$

where C (x,y,z,t) is the concentration and $\bar{\bar{D}}$ is the velocity dependent dispersion coefficient tensor ($D_{/\!/}$ in the flow direction, $D_\perp$ in the transverse direction). Fluid incompressibility holds. Darcy's law in a porous medium of permeability k with gravity $\vec{g}$ taken into account is :

$$\vec{\nabla} p = - \frac{\mu}{k} \vec{v} + \rho \vec{g} \tag{2}$$

p is the pressure, $\mu(C)$ and $\rho(C)$ are respectively the concentration dependent viscosity and density. We note that this set of equations (1-2) involves neither a characteristic time nor a length ; so the proper length and time scales are $D_{/\!/}/U$ and $D_{/\!/}/U^2$ in which case the only parameters involved in (1-2) are the viscosity-concentration and the density-concentration relations. These equations admit a base state solution ($\vec{v} = \vec{U}$) :

$$C_0 (x,y,z,t) \equiv C_0 (z,t) = \frac{1}{2}\left[1 + \operatorname{erf} \frac{z - Ut}{2\sqrt{D_{/\!/}t}}\right] \tag{3}$$

where erf is the error function[20]. The time dependence of this base state is due to longitudinal dispersion. Experimental determinations of concentration profiles $C_0(z,t)$ have demonstrated[19] the relevance of (3) (and then of the Gaussian dispersion statement involved in the CDE (1)) and allowed the measurement of $D_{/\!/}(U)$ (for stable flow).

B. - Stability analysis

In order to analyse the linear stability of the base state, we assume small disturbances in the transverse direction. ψ and Φ are respectively the small disturbances of the concentration and the velocity field U in the z direction. ψ and Φ are time t and z dependent, their growth or decay will determine the instability or the stability of the flow. Following references[6,7,9] but including all terms we get :

$$\left(\frac{\partial}{\partial t} + q^2 D_\perp - D_{/\!/} \frac{\partial^2}{\partial z^2}\right)\psi = \left(- \frac{\partial C_0}{\partial z} + \frac{dD_{/\!/}}{dU} \frac{\partial}{\partial z} \frac{\partial C_0}{\partial z}\right)\Phi \tag{4}$$

$$\left(\frac{\partial^2}{\partial z^2} - q^2 + \frac{1}{\mu_0}\frac{\partial \mu_0}{\partial C}\frac{\partial C_0}{\partial z}\frac{\partial}{\partial z}\right)\Phi = \left(\frac{U}{\mu_0}\frac{d\mu_0}{dC} - \frac{g\,k}{\mu_0}\frac{d\rho}{dC}\right)q^2\psi \qquad (5)$$

Such a set of equations (4-5) had been previously written down but never with all terms : Wooding[6] and Tan and Homsy[7] neglected the derivative of the dispersion. Chouke's work[22] is similar to the recent one of Tan and Homsy[7] but with only a linear viscosity concentration relationship and small viscosity ratio. Yortsos and Zeybeck[9] used the same approach as Tan and Homsy but emphasized the role of a velocity dependent dispersion coefficient. One key issue in the miscible fluid case is that the base state is time (and space) dependent : a quasi steady state approximation (QSSA) has to be assumed[7,9] : the base state profile is frozen at time t_0 in which case partial derivatives relative to z transform into total ones and a growth rate $\sigma(t_0)$ can be defined ($\sigma(t_0) = \partial/\partial t$) . Numerical solutions can be found[7,9] but analytical solutions of the initial growth rate (time t = 0) that are very useful for physical discussions can be determined under slight restrictions : Tan and Homsy[7] assume an exponential viscosity-concentration relationship, $\mu(C) = \mu_- e^{RC}$, $M = \mu_+/\mu_- = e^R$; this convenient relationship is not too far away from the experimental law. In addition Yortsos and Zeybeck[9] included the dispersion flow velocity dependence by mean of a parameter L :

$$L = \frac{U}{D_{/\!/}}\frac{dD_{/\!/}}{dU} \qquad (6)$$

We note that the term in brackets on the RHS of eq. (5) can be simply written[7,23] : $URGq^2\psi$, with $G = 1 - (V_c/U)$ and the critical velocity $V_c = gk\,(d\rho/d\mu_0)$. An exact analytical calculation of the initial growth rate and critical velocity for arbitrary concentration dependences of μ and ρ is available[24].

Let us first give the initial growth rate in the simple case L = 0 , i.e. as in Tan and Homsy[7] but including gravity terms :

$$\sigma = \frac{1}{2}R\,G\,U\,q - D_\perp q^2 - \frac{1}{2}D_{/\!/}q^2\left[\left(1 + \frac{2RGU}{D_{/\!/}q}\right)^{1/2} - 1\right] \qquad (7)$$

The occurance of instability ($\sigma > 0$) is determined by the lowest order q term : injection of a low viscous fluid into a more viscous one leads to instability provided gravity does not stabilize the flow (G >0). At large q, the instability is stabilized by both transverse and longitudinal dispersion ; there is then a maximum initial growth rate σ_m at a wavevector q_m and a cut-off wavevector ($\sigma(q_c) = 0$) above which the instability is impossible (fig. 1).

The main features of this analysis at time t = 0 remain still valid as time proceeds[7] (QSSA still assumed) ; the spreading of the base state yields a decrease of σ_m and a shift of q_m and q_c to lower values ; analytical expressions are not available but numerical calculations[7] show that the t = 0 picture is still valid but with smaller σ_m and q_m ; for the sake of the simplification, we will use as a guide the initial growth analysis. Indeed this is this more efficient growth on which the instability should be linked.

For $L \neq 0$, the discussion is less simple but gives rise to a new effect discovered by Yortsos and Zeybeck[9] : for a large enough combination of L and R, there is no cut-off wavevector q_c (fig. 1) ; all wavevectors are unstable because transverse stabilization is overcome by the destabilizing interplay of viscosity and the longitudinal dispersion velocity dependence. The cross-over between the two regimes is given by the parameter :

$$\eta = L\,\frac{R}{2}\,\mathrm{th}\,\frac{R}{2} - 1 - \sqrt{\varepsilon} \qquad (8)$$

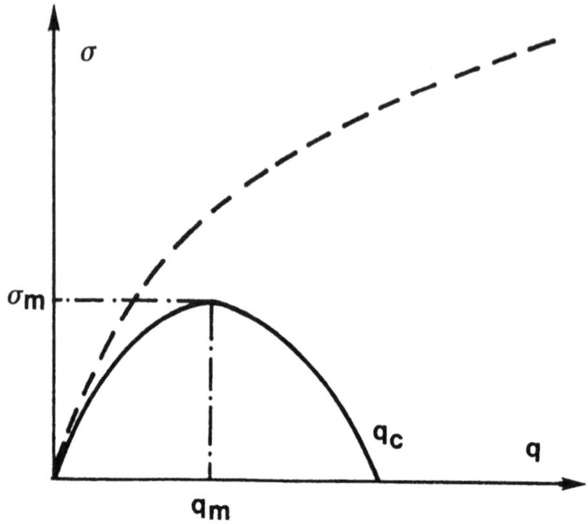

FIGURE 1.
Dispersion relation, i.e. growth rate σ versus wave vector q, for classical instability ($\eta < 0$, -) and enhanced instabilty ($\eta > 0$,...) due to coupling between viscosity and velocity dependent hydrodynamic dispersion.

where $\varepsilon = D_{\perp}/D_{/\!/}$. For $\eta < 0$ the instability is identical to the L = 0 case ; for $\eta > 0$ the instability is harder ($q_c = \infty$). Above the threshold ($\eta = 0$) the growth σ is roughly proportional to q ; indeed q should be bounded by geometric limitation, wavelengths smaller than the pore size, or l_D , are not allowed in this continuum medium approach : then the overall effect of this transition ($\eta = 0$) would be a large enhancement of the growth rate characterized by σ (q_M) where q_M is the maximum allowed wavevector. Anyway, in a mechanical dispersion regime, in which $D_{/\!/}$ is proportional to U [19,25] ($D_{/\!/} = U\, l_D$) , L = 1 and the thresholds corresponding respectively to $\varepsilon \leq 0.1$ and $\varepsilon = 1$ (isotropic dispersion) are $M \simeq 20$ ($R \sim 3$) and $M = 55$ ($R \simeq 4$) . Experiments are needed in this region to test the predictions.

As discussed by the authors[9] this drastic effect is only present at very short time (i.e. very straight profile) and decreases (the transition is "rounded") as time goes on. To end up with the linear stability analysis, we note that in an experiment q is in between q_{min} and q_{max}, which are set by the sample width and the pore size or the dispersion length.

C. - Further development of viscous fingers

The linear stability analysis predicts the main features of the very beginning of the instability. The initial exponential growth of small disturbances at q_m and σ_m require one to take into account non linearities which allow shielding splitting and fading of fingers[2,7] ; numerical calculations[7,11,12] of these basis phenomena have been performed in 2D and will be soon available in 3D [26]. One of the results of these calculations deals with the behaviour of the overall front width Δz as fingers develop. Starting from a step front at time t = 0, it grows first as in a diffusive process ($\Delta z \sim \sqrt{t}$) , and then in a convective way ($\Delta z \sim t$) : fingers spread linearly with time as for the Saffman-Taylor[4] instability in a Hele-Shaw cell ; this behaviour has also been observed in effluent technique experiments[13,17] in porous media leading to phenomenological models of viscous fingering processes[27,28].

III. - EXPERIMENT

An acoustic technique described elsewhere[21] is used to determine the space and time dependence of the concentration C(z,t) in our samples of typical size $4 \times 4 \times 30\ cm^3$. Concentration measurements are derived from the velocity variations of a sound wave in ten cross-sections of the sample (10 z values). The acoustic wave is generated and analysed using ten transmitter-receiver pairs of transducers laid every 2.1 cm along the larger side of the samples. The spatial resolution in this direction is 3 mm. The ten positions are scanned in less than 1 minute through an automatic procedure. Taking care of temperature stabilization (to better than 0.5°C), our calibration curve of the velocity variation with the mixture concentration and the accuracy of relative velocity measurements (better than 10^{-4}) give an overall precision in C of better than 0.1 %.

We measured [19,25] previously the longitudinal dispersion $D_{/\!/}$ versus the flow velocity U in a limestone (permeability $k = 15 \times 10^{-15}\ m^2$), a millstone ($k = 11 \times 10^{-14}\ m^2$) an a fireproof brick ($k = 75 \times 10^{-12}\ m^2$) : as U increases, $D_{/\!/}$ increases from a molecular diffusive regime (L = 0) to a mechanical dispersion regime ($D_{/\!/} \sim U$,L = 1) with in between, for the fireproof brick an enhancement of L (L ~ 1.5). $D_{\perp}$ estimates[29] show an evolution from an isotropic dispersion ($\varepsilon = 1$) at low flow rate in the molecular diffusion regime (with also L = 0) to an anisotropic regime at larger rate ($\varepsilon \leq 0.05$.)

The displacing fluid is pure water ($\mu_- = 8.0 \times 10^{-4}$ Pa.s. at 30°C) ; the displaced fluids are water-glycerol mixtures of different concentrations C_0, and viscosity from $\mu_+ = 8.0 \times 10^{-4}$ to 0.6 Pa.s. at 30°C ; density varies as well from $1.0 \times 10^3\ kg m^{-3}$ for pure water to $1.26 \times 10^3\ kg\ m^3$ for pure glycerol. From the viscosity concentration relationship we can compute for each experiment the rescaled R value[24] : $R = (\mu'_+ + \mu'_-) / (\mu_+ + \mu_-)$ where the values of μ and the derivative μ' are taken at concentration C = 0 for - and $C = C_0$ for + . Using the density concentration dependence we get the rescaled critical velocity[24] : $V_c = gk\ (\rho'_+ + \rho'_-) / (\mu'_+ + \mu'_-)$. Due to the permeability difference between the three porous media, the critical velocity in both the limestone ($V_c \sim 10^{-8}\ m\ s^{-1}$) and the sandstone ($V_c \sim 10^{-7}\ m\ s^{-1}$) is too tiny compared to the range of flow velocities (from 10^{-5} to $10^{-3}\ m\ s^{-1}$) to lead to gravity effects (then G ~ 1), whereas there is a measurable effect in the fireproof brick. The low permeability of the two natural samples limits the range of available viscosity ratios and flow rates due to the large pressure drop required to flow across the sample. We note that the flows are driven vertically ,with gravity as a stabilizing effect, the lighter and less viscous fluid (water) is injected from the top. Neglecting the gravity term, the rescaled η value is[24] : $\eta = -1 - \sqrt{\varepsilon + L}\,(\mu'_+ - \mu'_-) / 2\,(\mu_+ + \mu_-)$.

A. - Limestone

From the stable flow experiments on the limestone[29,30], we know that l_D and ε are nearly independent of the flow rate U [9,25] and that the critical velocity V_c is always too small to be within our experimental range. Thus the limestone is expected to be suitable to test the viscosity ratio effect (M or R) ; measurements have been performed for M = 1.18 ; 1.52 ; 1.75 ; 1.88 ; 2.18 ; 3.12 ; 3.4 ; 6.25 ; 12.5 ; 25.0 at flow rates ranging from 0.1 to 25 cm/h. Because of the small permeability, larger viscosity ratios are not possible. Figure 2 shows the resulting concentration profiles for M = 1.75 (R = 0.40) and U = 6.5 cm/h for ten z positions (z = 2.5 ; 2.5 + 2.1 ; 2.5 + 4.2 ; ... cm from inlet) : compared to stable flows[19] , we clearly see the instability growth as we go away from the inlet, starting with a more or less diffuse front to small growing fingers. For a larger viscosity ratio (M = 12.5 , R = 3.6 in fig. 3), the observed profiles (five out of ten) exhibit shoulders which is the signature of fingering ;

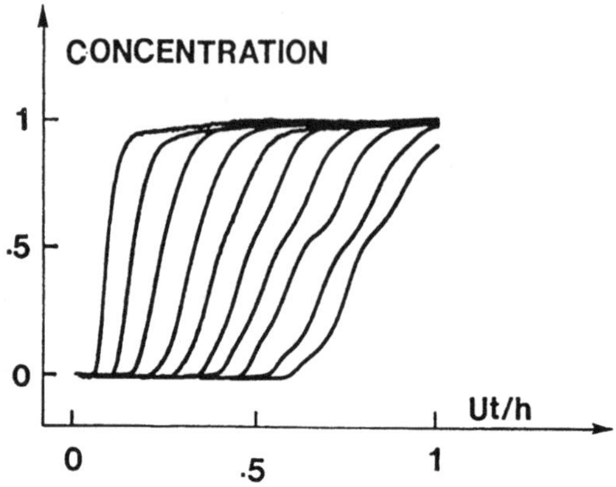

FIGURE 2.
Concentration profiles : concentration versus reduced time (Ut/h , h : sample length) at flow rate U = 6.5 cm/h and viscosity ratio M = 1.75 in a limestone ; from left to right each profile corresponds to increasing distance z from the inlet (2.5 , 2.5 + 2.1 , 2.5 + 4.2 cm ,...).

figures 3 1 top and 3 bottom correspond to the same viscosity ratio but two different flow velocities U = 3.2 cm/h and U = 9.6 cm/h . The plot against the pore volume (i.e. Ut/h , h sample length) shows their likeness. In the limestone, the instability growth is nearly independent of the flow rate. It is obvious that experimental profiles are more unstable in

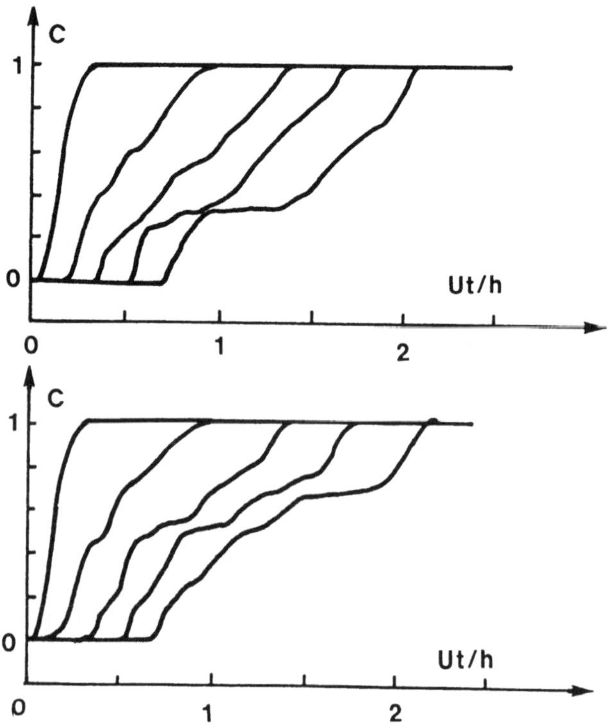

FIGURE 3.
Concentration profiles for M = 12.5 but for different flow rates (top : U = 3.2 cm/h , bottom : 9.6 cm/h). The use of reduced time (Ut/h) emphasizes the likeness of the two series of profiles.

figure 3 (M = 12.5) than in fig. 2 (M = 1.75). However we need a more quantitative characterization of the degree of instability. As suggested by Tan and Homsy[7], the profile width is characteristic of the overall instability ; since we have ten profiles at ten z values, we measure for each profile the time width between C = 0.1 and C = 0.9 (arbitrary criterion which is not very critical) ; from the stable flow measurements we also know the stable time width due to dispersion which is the base state width ; so we characterize each instability profile by the excess time width δt which is the difference between the unstable and stable widths. Moreover, to take into account the flow velocity independence (fig. 3), we will plot $U\delta t$ versus z as the instability characteristic. The resulting analysis from the data of figures 2 and 3 is given in figure 4 for the two viscosity ratios M = 1.75 and M = 12.5 : for the low viscosity ratio, as can be guessed from the direct profile (fig. 2), we do observe a cross-over between a diffusive regime ($U\delta t \sim \sqrt{z}$, dashed-dotted line) at low z values to the fully developed fingers regime ($U\delta t \sim z$, dashed line) , whereas the large viscosity experiment only exhibits the latter one. The low viscosity ratio experiment is more interesting because it clearly demonstrates the cross-over between the diffusive and the developed fingers regimes in agreement with the numerical simulations[7,12] ; the initial small disturbances grow, shield, spread leading to first an enlarged front which still looks like a diffusive front ; when the disturbances become large enough, non linearities stabilize the fingers which just spread further linearly with time.

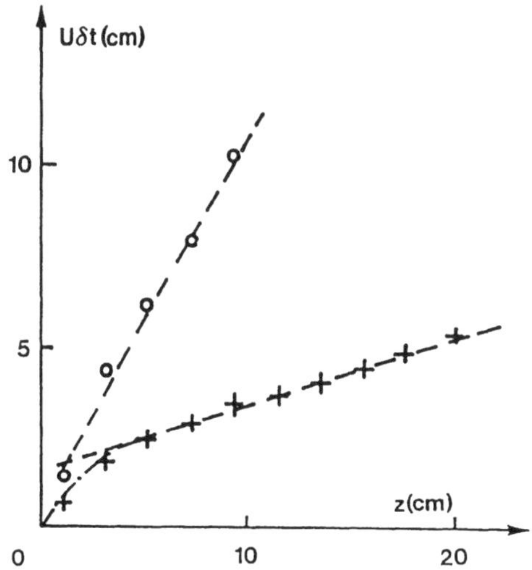

FIGURE 4.

Excess width ($U\delta t$) of the profile versus distance z from the inlet. + and **o** correspond to fig. 2 and 3. The dashed dotted line is a eye guide for the diffusive regime. The dashed line is the linear temporal spreading regime ; the slope of the dashed line is our experimental parameter p.

The linear temporal spreading of the fingers when the instability is fully settled is only characterized by the slope of $U\delta t$ versus z ; we have noticed that the larger M the larger $U\delta t/z$. As the relevant effect of M in the stability analysis is through R, in figure 5 we give a plot of $p = U\delta t/z$ against R : the data are well aligned, meaning that our experimental parameter p used to describe the fingering is proportional to R ; this indeed gives some support to the continuum approach in which the viscosity of the mixed fluid does only enter as R especially in the linear stability analysis. We have to test in the two other porous media this type of experimental analysis.

B. - Millstone

We have studied[29,30] the viscosity ratios M = 3.12 ; 12.5 ; 135 and 760 . The higher viscosity ratios have been obtained at the cost of a limited range of flow rates (1.2 cm/h for M = 135 and only 0.3 cm/h for M = 760) which has the disadvantage of going out of the mechanical hydrodynamic dispersion regime (L = 1 , ε small) to the molecular diffusive regime in which case L decreases and ε increases up to 1. At low viscosity ratios, a transition between a diffusive and a convective regime is also observed. For the two lowest viscosity ratios (3.12 and 12.5) similarly to the limestone, in the convective regime (U ranging from 5 to 25 cm/h) the instability features are flow independent and the fingers develop with a linear time dependence ($U\delta t \propto z$) . The corresponding slopes p are given in figure 5 versus R ; the two measurements at M = 135 and 760 plotted in the same figure 5 align nicely with the others. This gives more confidence in the proportionality to R of our experimental parameter p, but on the other hand we do not observe any special enhancement of the fingering as M (or R) goes in the range of the Yortsos and Zeybeck[9] L effects . At M = 135 (R = 11.2) the flow rate is small (1 cm/h) and thus[19] we are out of the convective regime with L decreasing and ε increasing : both effects contribute to increase the viscosity ratio threshold to the vicinity of 100 ; at M = 760 the situation is worse essentially because L is vanishing and $\eta < 0$ still . Then in the millstone, our experiment allows measurements with $\eta < 0$, and no special enhancement of the instability has to be expected. The larger slope of p versus R obtained for the limestone than for the millstone deserves a comment ; the millstone is much more homogenous than the limestone. In the millstone fluctuations of permeability are of the order of 25 % whereas for the heterogenous limestone we measure up to 200 % of fluctuations[31]. The viscosity ratio reveals the heterogeneity of the velocity field, smoothening them for M < 1 and enlarging them for M > 1 .

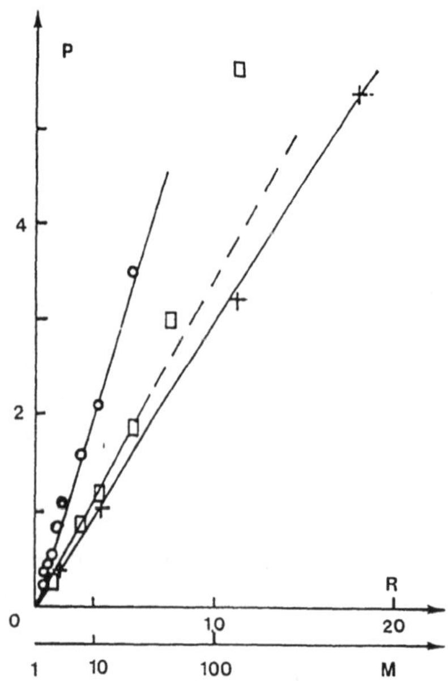

FIGURE 5.

Experimental parameter p versus the viscosity variable R for the limestone (o), the millstone (+) and the fireproof brick (□). M is the viscosity ratio.

C. - Fire proof brick

In the range of available velocities (U from 0.1 to 500 cm/h) both ℓ_D and ε vary[19]. For a given viscosity ratio M, we get R and the critical velocity V_c required for the instability threshold. For a given M, a series of experiments at different flow-rates involve different gravity contributions G (expression (7)) as well as ε and ℓ_D. For small M, V_c is large, thus we have not been able to probe the close vicinity of M ~ 1. We have used M = 1.88 ; 3.1 ; 6.25 ; 12.5 ; 25 ; 50 ; 135, with some limitations in the flow rate for large M due to the large pressure drop required. The critical velocity V_c is determined and compared to the theoretical one[30].

The procedure to analyse the data is the same as for the limestone ; for each M value and each flow rate U, we determine Uδt versus z. As opposed to the limestone and the millstone, in the fireproof brick the experimental parameter p = Uδt/z is U (and M) dependent. Figure 6 gives a log-log plot of this parameter p versus the flow velocity U. Let us first focus on the low viscosity ratio curves (up to M = 25) ; as U increases from V_c onwards, p (i.e. instability) increases monotonously giving a plateau at high flow-rate. These plateaus are reminiscent of the flow independent regime we get both in limestone and millstone. The millstone and the brick straight lines are close to one another in figure 5, indicating that these two porous media have small permeability heterogeneities variations compared to the limestone, assuming that the slope p versus R is linked to the permeability fluctuations. We note also that this set of low M values data are monotonously increasing and levelling off at a plateau. We do not see any significant change as ℓ_D passes through its minimum or increases or ε decreases : p is insensitive to L, ℓ_D and ε . Its only relevant dependence is R .

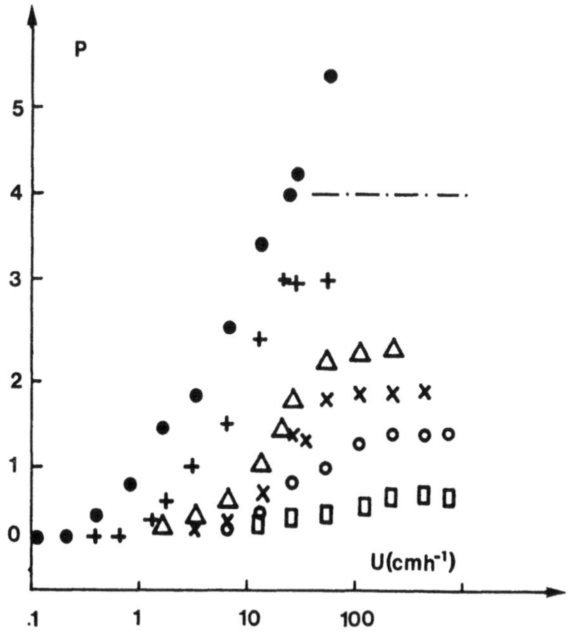

FIGURE 6.

Semilogarithmic plot of the experimental parameter p versus flow-rate U in the fireproof brick at different viscosity ratios M = 3.1 (□) , 6.25 (O) , 12.5 (x) , 25 (Δ) , 50 (+) , 135 (•). The dashed dotted line corresponds to the hypothetical plateau for M = 135 if p is proportional to R .

For the two larger M values (M = 50 and 135), one would expect in continuity with the lower values a higher plateau at lower U values. If such a tendency is seen for M = 50, this is absolutely not the case for M = 135, for which p keeps increasing over two decades ; furthermore if we plot on figure 5 the highest values of p for M = 50 and 135 (a kind of lower bound of an hypothetical plateau, dashed-dotted line in fig. 6) these data are well above the linear R relationship (especially for M = 135). Guided by the linear stability analysis one would think that this behaviour corresponds to the Yortsos and Zybeck[9] prediction of a harder instability for large M values. To support this we note that the conditions necessary for such a larger instability growth ($\eta > 0$) are fulfilled : U = 25 cm/h corresponds to the maximum slope of ℓ_D versus U (L = 1.5) and a low ε value ($\varepsilon \sim 0.1$) leading to $\eta \sim 7$ for M = 135 .

D. - Summary of the data analysis and discussion

We have studied the growth of profiles of viscous fingering in three different porous media. The experiments[29,30] yield the following features :

i) There is a cross-over between a diffusive regime ($\sim \sqrt{t}$ growth) to a convective regime ($\sim$ t growth) for the width of the profile.

ii) In the convective regime, in which fingers grow linearly with time, we have been able to define an instability parameter p which characterizes the overall instability of growth.

iii) In most of the cases analysed (excepted *iv)*) the only p viscosity dependence is linear with R (p $\sim$ R). The larger the heterogeneity of the medium, the larger the slope of p versus R.

iv) We have observed, in one instance, an instability growth faster than the p $\sim$ R law. This corresponds to a high viscosity ratio (M = 135) and the brick porous medium at a flow-rate U for which the slope of the velocity dependent longitudinal dispersion is maximum (and ε small).

We have made detailed comparisons of our data with theoretical treatments of linear stability analysis[6-10] and numerical simulations[7,11,12] both based on a continuum approach. Our analysis shows that :

i) The cross-over regime is observed experimentally in agreement with numerical simulation[7,12] of finger growth in a porous medium.

ii) The linear temporal growth regime has been predicted[7,12] but no parameter analogous to p has been used for the analysis and there are no viscosity tests available in this regime.

iii) The only parameter entering the theory are the viscosity through R and the gravity through G (in the linear stability analysis). It is not therefore surprising that this is what we get in the experiment (except for fingers with the gravity overriding).

iv) Yortsos and Zeybeck[9] have predicted an enhancement of the initial growth rate under the subtle interplay of large viscosity ratio and large velocity dependence of the hydrodynamic dispersion ; this is a way to account for the M = 135 data set in the fireproof brick which is well above the p $\sim$ R relationship and for which the theoretical condition is totally fullfilled. As we do not get this effect in millstone even with a larger viscosity ratio (M = 760) but a much lower L, these two porous media demonstrate that both large M and L are required to get this effect. As a last comment, its seems at first sight most surprising that features of the linear stability analysis apply so well to the fully developed regime of viscous fingering where non linear effects are expected to dominate. Our results indicate that the initial growth rate changes rapidly with the flow parameters (R, G and L mainly), and the memory of this early stage remains as the fingers develop .

IV. - CONCLUSION

Using a powerful acoustic technique we have analysed the growth of profile occuring in viscous fingering in three porous media. In the experimental analysis we have been able to define an instability parameter p which captures the essential features of the instability. Compared to continuum theory, our data demonstrates a cross-over regime between diffusive and convective growth, the dependence of the instability on the viscosity ratio and on the effect of gravity. Moreover some of our data are consistent with the theoretical predictions of an enhancement of the growth rate due to the coupling of large viscosity contrast to velocity dependent hydrodynamic dispersion.

ACKNOWLEDGEMENT

This paper benefits of stimulating discussions with J. BANAVAR, G. M. HOMSY, A. LIBCHABER, F. M. ORR, H. TCHELEPI, Y. C. YORTSOS and W. ZIMMERMAN. This work has been partly supported by NATO grant No 0209.1988.

REFERENCES

1 HILL S., *Chem. Eng. Sci.* **1** : 247 (1952).
2 HOMSY G. M., *Ann. Rev. Fluid Mech.* **19** : 271 (1987) and references therein.
3 YORTSOS Y. C., *J. Phys. : Condens. Matter* **2** : 443 (1990), and references therein.
4 SAFFMAN P. G. and TAYLOR G. I., *Proc. Roy. Soc. London* **A24** : 312 (1958).
5 BENSIMON D., KADANOFF L. P., LIANG S., SHRAIMAN B. I. and TANG C., *Rev. Mod. Phys.* **58** :977 (1986).
6 WOODING R. A., *J. Fluid Mech.* **39** :477 (1969).
7 TAN C.-T. and HOMSY G. M., *Phys. Fluids* **29**: 3549 (1986) ; *ibid* **31**:1330 (1988).
8 HICKERNELL F. J. and YORTSOS Y. C., *Stud. Appl. Math.* **74** : 93 (1986).
9 YORTSOS Y. C. and ZEYBEK M., *Phys. Fluids* **31** : 3511 (1988).
10 YORTSOS Y. C. and HICKERNELL F. J., *SIAM J. Appl. Math.* **49** : 730 (1989).
11 CHRISTIE M. A. and BOND D. J., *SPE* **14896** (1986) ;
CHRISTIE M. A. (to be published).
12 ZIMMERMAN W. and HOMSY G. M., *Phys. Fluids A* **3** : 1859 (1991).
13 ARAKTINGI U. and ORR F. M., *SPE* **18095** (1988) and references therein.
14 BLACKWELL R. J., RAYNE J. R. and TERRY W. M., *Trans. Sco. AIME* **216** :1 (1959).
15 HABERMANN B., *Trans. AIME* **219** : 264 (1960).
16 SLOBOD R. L. and THOMAS R. A., *Soc. Pet. Eng. J.* **3** : 9 (1963).
17 PERKINS R. L., JOHNSTON O. C. and HOFFMAN R. N., *Soc. Pet. Eng. J.* **5** : 301 (1965).
18 BACRI J.-C., RAKOTOMALALA N. and SALIN D., in *Physics and Chemistry of Porous Media* (A.I.P. Publ., New-York, 1987)
19 BACRI J.-C., RAKOTOMALALA N. and SALIN D., *Phys. Rev. Lett.* **58** : 2035 (1987).
20 BACRI J.-C., RAKOTOMALALA N. and SALIN D., *Phys. Fluids* **A2** : 674 (1990).
21 BACRI J.-C., HOYOS M., RAKOTOMALALA N., SALIN D., BOURLION M., LENORMAND R. and SOUCEMARIANADIN A., *J. Physique III, Paris* **1** : 1455 (1991).
22 GARDNER J. W. and YPMA J. G. J., *SPE* **10686** : 163 (1982).

23 TRYGGVASON G. and AREF H., *J. Fluid Mech.* **154** : 287 (1985).
24 BACRI J.-C., SALIN D. and YORTSOS Y, *C.R.Acad.Sc. Paris* (to appear in 1992).
25 HULIN J.-P. and SALIN D., in *Disorder and Mixing* , ed. by E. Guyon, J.-P. Nadal and Y. Pomeau, *NATO ASI* **E 152** : 89 (1988).
26 ZIMMERMAN W. and HOMSY G. M., submitted to Phys. of Fluids.
27 KOVAL E. J., *Soc. Pet. Eng. J.* **3** (1963) 145.
28 TODD M. R. and LONGSTAFF Q. J., *Soc. Pet. Eng. J.* **12** (1972) 874.
29 BACRI J.-C., SALIN D. and WOUMÉNI R., *Phys. Rev. Lett.* **67** : 2005 (1991).
30 BACRI J.-C., RAKOTOMALALA N., SALIN D. and WOUMÉNI R., *Phys. of Fluids A.* (to appear in 1992).
31 TCHELEPI H., Private communication.

VISCOUS FINGERING AND SURFACE STRUCTURE FORMATION IN POROUS MEDIA

Vladimir V.Kuznetsov, Oleg B.Bocharov, and Oleg V.Vitovsky

Institute of Thermophysics, Siberian Branch of the Rus. Ac. Sci.
630090, Novosibirsk, Russia

ABSTRACT

Finger-like instability with immiscible displacement of one viscous fluid with the other from porous material has been studied experimentally for two main cases: transparent two-dimensional model of porous medium with chaotic distribution of volumetric pores and thin 1x20x66 sm sample of arenaceous quartz placed horizontally. Data on selection of surface structures with instable displacement front in two-dimensional and three-dimensional pore systems has been obtained. In porous materials occur various phenomena related with the stability of interphase boundary in random medium when the capillary forces govern interphase boundary motion in small scale and viscous forces in a large one. As the result, capillary instability in small scale generating the noise with large amplitude and viscous instability in large scale are simultaneously developed. Surface structures development with viscous fingergrowing when there is large noise has been in detail studied and the experimental data are discussed from the standpoint of well-known theoretical models of viscous fingergrowing.

INTRODUCTION

Plain boundary instability and spontaneous structures formation in the originally unstable region is a common phenomenon occured in different fields of science. As a physically realized example of front propagation in unstable form, one can consider the displacement of more viscous fluid by the less viscous one in a simple hydrodynamic system designed as a quite narrow gap between two parallel plates (Hele-Shaw cell) or made of porous material (Saffman and Taylor, 1958). This type of instability is known as the

Instabilities in Multiphase Flows, Edited by G. Gouesbet
and A. Berlemont, Plenum Press, New York, 1993

Taylor-Saffman one, and it is governed by two major forces which are viscosity and gravitation. The fingers formation and growth in the Hele-Shaw cell have been examined in detail in a great number of original works, the detailed review of which is given in works (Bensimon et al., 1986; Homsy, 1987).

The problems of dynamic selection of surface structures originating therewith and their transition into fast growing fingers continue to play major part in theory of structures formations in non-linear dissipative systems and have been actively discussed in literature (Bensimon et al., 1986; Homsy, 1987). Fingers instability is of great importance in dynamic structure selection. Arising instability is supercritical bifurcation and emerges with finite amplitude noise (Bensimon, 1986).

Interphase boundary separating two immiscible fluids or fluid and gas is formed in porous medium by capillary menisci which separate the phases, and their motion in porous channels is of random nature. Combined in this case are two phenomena related with the fact that capillary forces govern the boundary movement in a small scale and generate the noise while the viscous ones act in a large scale. The goal of this work is to determine the influence of small and large scales on dynamic structure selection with unstable displacement in porous medium, to discuss the influence of capillary and viscous forces on fingers growth and show the detailed experimental observation of unstable finger growth in porous material, and to check experimentally theoretical models.

EXPERIMENTAL EQUIPMENT

The experiments were performed in a transparent two-dimensional model of porous medium with a pore chaotic space and in a thin sample of arenaceous quartz placed horizontally, Fig. 1a, b. Two-dimensional model of porous medium (the modified Hele-Shaw cell) was made of two rough glass plates of 0.2 m length and 0.4 m width with the projections up to 80 μm fitted tightly to each other. The fluids were pumped in pore space between the plates through the inlet cells providing uniform feeding of a pumped fluid across the model transverse section. Before the experiments, the working region was evacuated, filled with immersion fluid (aqueous solution of NH_4CNS and $CoCl_2$) and photographed. The analysis of negatives has shown that the pore space of model represents the system of volumetrical pores randomly distributed by size and connected with pores necks. The average number z of pores necks fitted to one pore varied from 3 to 4. The model permeability is of ca $20 \cdot 10^{-12}$ m^2 and its porosity, $m = 0.39$.

Analyzed in the experiments were immiscible displacements of oils with different viscosities and water - glycerine solutions with water containing dye - stuffs and additives which change the value of wetting angle and with purified kerosine as well. Fluids pairs used in the experiments formed wetting angles from $52\text{-}80^0$ (wetting close to neutral) to $0-30^0$ (high wetting).

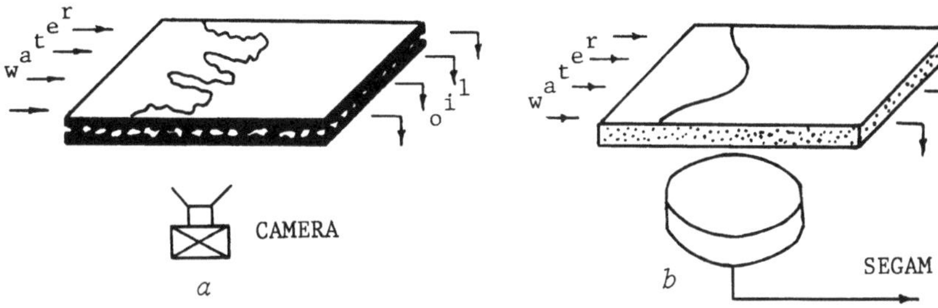

Figure 1. Schematic of the experimental apparatus

Immiscible displacement in pore three-dinemsional lattice was investigated in thin arenaceous quartz sample of 0.2 m width and 0.6 m length, Fig. 1b, with granule size of ca 160 μm. Because of sample thickness being 0.01 m, a large number of pores was placed in its cross-section and the flow governed with capillary forces in this small scale was three-dimensional while in a large scale viscous flow was two-dimensional.

Surface structures were registered by successive photographing of transparent model. With arenaceous quartz flowing out from the thin sample into displacing fluid, radioactive isotope of mercury was added and distribution of displacing fluid in the sample was measured with γ-camera which resolution was 16 mm^2 with detector diameter of 0.3 m.

EXPERIMENTAL RESULTS

Fig. 2 shows evolution of surface structures in two-dimensional transparent model of porous medium with unstable displacement and wetting of pores walls with fluids close to neutral. Direction of displacement is oriented along the normal to gravitational vector and it does not significantly influence on the evolution of surface structures. Location of a displacing fluid in porous medium before the beginning of the displacement is shown with dark region in the lower part of the photograph in Fig. 2 a, c while the black disks correspond to external system providing the uniformity of the gap between the rough glass plates.

Experiments performed on transparent model have shown that the boundary movement is always unstable and instabilities occur in two scales. There prevail the capillary forces in pore scale and chaotic microstructure of pore space brings to random jumping movement of interphase menisci forming the interphase boundary similar to classical results of percolation invasion theory (Chandler et al., 1982). Capillary instability develops just after the displacement begins, Fig. 2 c, and takes place at all evolution steps of surface structures. The peculiarity of percolation-type capillary instability appearing at large wetting angles θ for displacing fluid lies in that the region of two-phase flow in displacement front (two-phase zone) would not exceed 5-10 pore sizes for two-dimensional lattice of volumetric pores

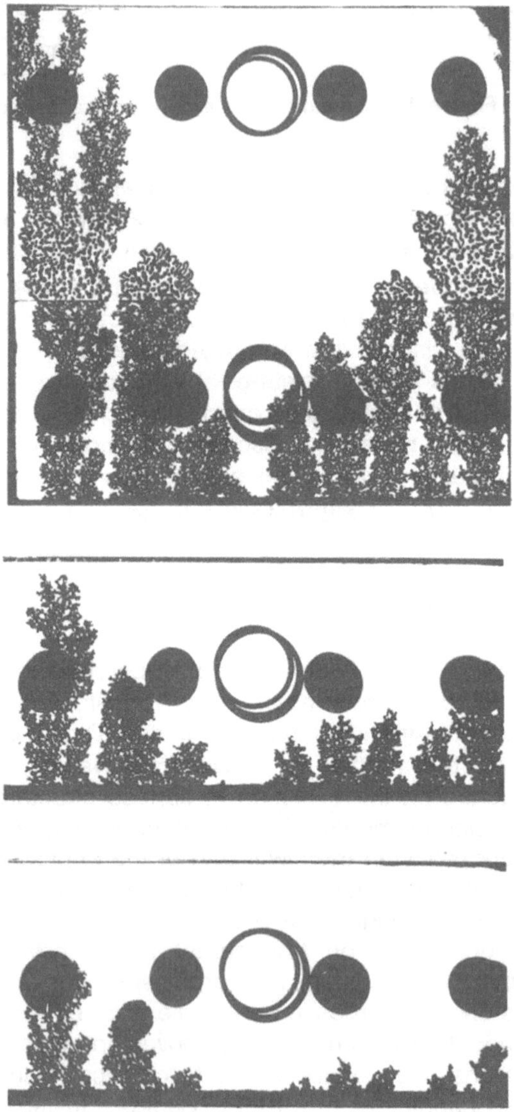

a b c

Figure 2. Evolution of surface structures.
$v_0 = 11.2 \cdot 10^{-5}$ m/s; $M_c = 4.8$; (a) $t = 365$ s, (b) $t = 240$ s, (c) $t = 190$ s.

and slightly depends on displacement velocity. This is stipulated by the sequence of filling up pore constrictions and expansions with the displacing fluid at large θ. Internal regulating movement results in fractal dimensionality of front D, Fig. 2 a, equal 1.47, that is substantionally less than the front fractal dimensionality with directed invasion percolation for nodes problem in two-dimensional lattice equal 1.82 (Wilkinson and Willemsen, 1983).

In this case displacement process is piston one with a perfectly distinguished displacement front and a large noise close to it. The permeabilities of displacing k_2 and displaced k_1 fluids differ because of the capillary pinching displacing fluid behind the displacement front but, if the fluids viscosities, however, differ quite enough, the mobility relation for the front $M_c = \mu_2 k_1 / \mu_1 k_2$ may significantly be more than the unit. Then the displacement process would be instable in a large scale and "viscous" fingering appears here, Fig. 2 a-c.

At the earlier stage of developing of large-scaled instability, "capillary" instability forms the front perturbations with different wavelength, Fig. 2 c, and perturbations with a wavelength of maximum growth λ_m have the greatest growth velocity (Chuoke et al., 1959) and form the fingers.

When amplitude of growing instabilities is comparable to their wavelengths, the displacement velocity is not too high, Fig. 3 a. The existance of cell structures in viscous fingers growing is one of more interesting phenomena observed with finite relation of fluids mobilities M_c.

At large displacement velocity or large mobilities relation the stationary cell structures are not observed, and fingers length grows with time, Fig. 3 b, c. In this case arises strong fingers interaction, suppression of slowly growing fingers, and further they grow extremely irregulary. Finger interaction in the Hele-Shaw cell is followed by broadening of a leading finger, and isolation of a finger with width close to half of displacement

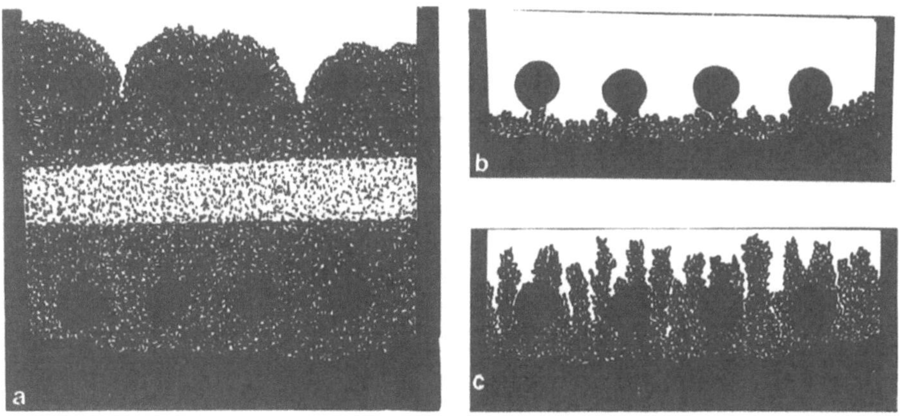

Figure 3. Cell structure (a) and finger-like structure (b) in developing boundary. $k = 18 \cdot 10^{-12}$ m^2; (a) $M_c = 6.4$, $v_0 = 1.3 \cdot 10^{-6}$ m/s, $t = 8.6 \cdot 10^4$ s; (b), (c) $M_c = 14.4$, $v_0 = 1.1 \cdot 10^{-4}$ m/s, (b) $t = 14$ s, (c) $t = 64$ s.

region width. When there is large noise near the front, this process interrupts. In porous medium with finger broadening there appears its instability, head unstable movement, and branching that stabilize the width of growing fingers, Fig. 4 b, c. Arising instability means super-critical bifurcations which take place at finite-amplitude perturbations of filtration velocity near finger head (Bensimon et al., 1986; Homsy, 1987; Bensimon, 1986). Dimensionless critical noise amplitude at which this instability arises is defined by the relationship of cell width in which the finger grows to wavelength of maximum growth of small perturbations (Bensimon, 1986).

$$v_c = 26 \frac{\lambda_m}{\lambda\sqrt{3}} e^{-2.3\lambda/\lambda_m} \tag{1}$$

Since there is the direct relationship between perturbation velocity near the finger boundary and perturbance amplitude of its boundary A^*, from (4) one can get limit cell width in which the finger grows without branching

$$\lambda^* = -\frac{\lambda_m}{2.3} \ln \frac{A^*\sqrt{3}}{26\lambda_m} \tag{2}$$

Critical cell width $\lambda' = \lambda^*/\lambda_m$ as a function of perturbation amplitude at finger boundary $A' = A^*/\lambda_m$ is shown in Fig. 4. Here are given experimental data for fingers without branching, points 1, and branching fingers, points 2.

After the beginning of fingers branching, surface structure development is defined by the relationship of transverse size of displacement region H and limit cell width in which the finger grows without branching. Typical modes of fingers evolution with different relations H/λ^* are shown in Fig. 5. With $H/\lambda^* > 20$ the active process of chaotic branching brings to statistical equalization of pressure field in vicinity of growing fingers because of unstable movement of finger head and hinders from fingers broadening which is larger than λ^*. In this case the advance process of some fingers by the others is weakly expressed, Fig. 5 d. With size decreasing of displace-

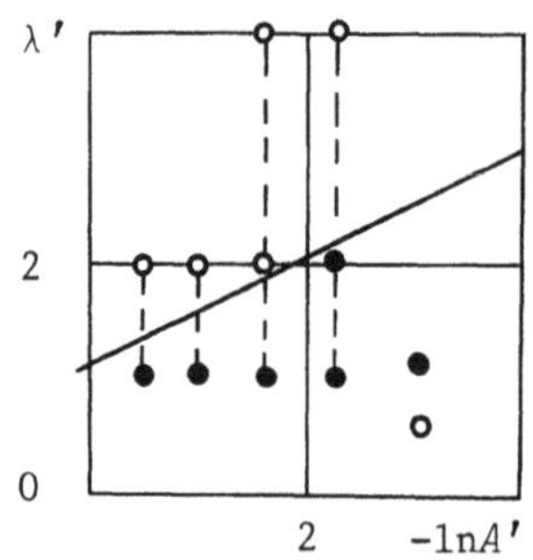

Figure 4. Critical cell width as function of noise amplitude at finger boundary

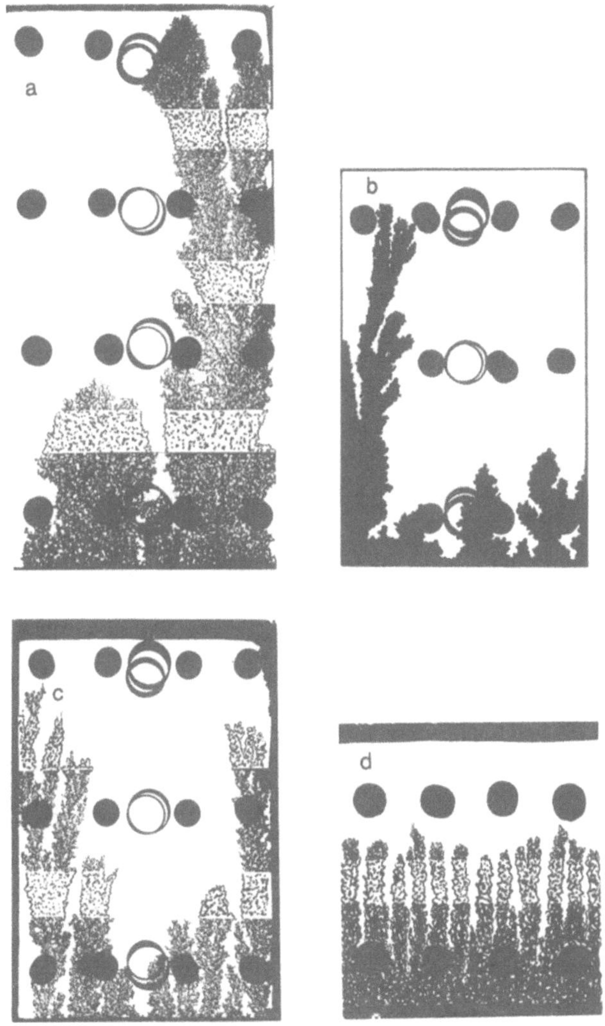

Figure 5. Typical modes of fingers evolution. (a) $H/\lambda^* = 2$; (b) $H/\lambda^* = 3 \div 4$; (c) $H/\lambda^* = 5 \div 6$; (d) $H/\lambda^* = 20$.

ment region up to $H/\lambda^* < 10$, no through-flow conditions on side walls of displacement region prevent from statistical equalization of pressure field, and the advancing fingers growing on side walls are observed, Fig. 5 a-c. At $H/\lambda^* = 2-4$ the branching finger grows on one of the walls supressing the other fingers, Fig. 5 a-b.

To describe branching modes of fingers, viscous finger formation was simulated on the basis of random markers walking in displacement region (Liang, 1986). Comparison of experimental results, Fig. 5 b, with numerical simulation data, Fig. 9 e (Liang, 1986), shows that with preset values $H/\lambda^* < 10$ there is good agreement between computational and experimental results, the predominant finger growth on side walls has been found at $H/\lambda^* < 10$. At the same time, fingers structures similar to that obtained with numerical simulation ($H/\lambda^* \sim 14$), Fig. 9 f (Liang, 1986), have not been obtained experimentally. At $H/\lambda^* > 10$ fingers growing in experiments is more uniform than in calculations, Fig. 5 c, d, and no branching finger has been observed. In real porous materials, for example, sandstones, oil fields, small-scaled capillary instability develops in three-dimensional pore structure. If the displacing fluid wets well pores walls, the flow thresholds for displaced and displacing fluids would be different (Kuznetsov and Dimov, 1991) and here is possible simultaneous flow of two fluids in the same pore system. In this case when the displacement begins, capillary instability develops not only in small scale but also in large one. As a result, in large scale arises the region of two-phase flow with microstructure which is defined with small scale; saturation S_1 of displacing fluid changes behind the displacing front and conception of fluids interphase boundary loses its sense. Saturation S_1 of displacing fluid means its volume per unit of pore space.

Distribution of displacing fluid saturation S_1 under immiscible displacement of viscous water-glycerine solution with kerosine is shown in Fig. 6. Fig. 6 a shows isoline for constant saturation of displacing fluid S_1

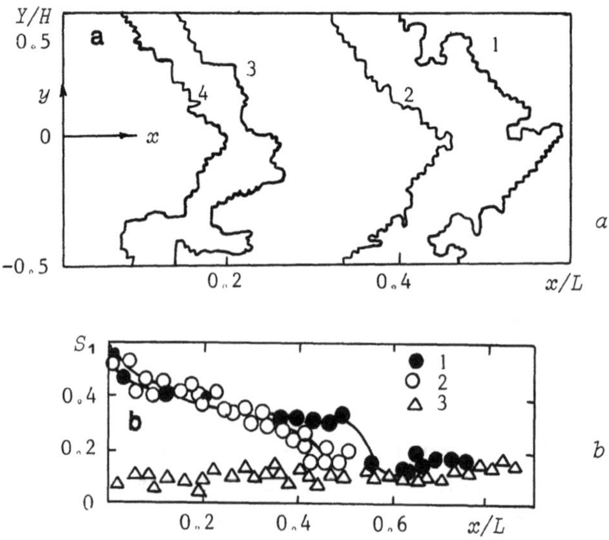

Figure 6. Saturation distribution of displacing fluid. $k = 12 \cdot 10^{-12}$ m^2; $v_0 = 3.3 \cdot 10^{-5}$ m/s; $m = 0.41$; $\mu_1/\mu_2 = 0.06$; $\tau = tv_0/mH = 0.5$, (here t means time); (a) isoline of constant saturation: 1. $S_1 = 0.2$, 2. $S_1 = 0.3$, 3. $S_1 = 0.35$, 4. $S_1 = 0.4$; (b) longitudinal saturation profiles: 1. $y/H = 0$, 2. $y/H = 0.4$, 3. $\tau = 0$.

while Fig. 6 b shows the longitudinal profiles of saturation S_1 in different cross-sections. With unstable displacement as a result of surface structures selection, has been isolated one finger with variable saturation of displacing fluid behind the displacement front and with large gradients S_1 in vicinity of the front. It is evident that flow fluids in large scale is defined not only by capillary invasion but also viscosity, and the continuous description of simultaneous flow of both fluids with their saturation gradients should include a set of averaging equations. For two-phase flows in porous media such equations are given (Collins, 1961), they include the generalized Darcy's laws

$$u_i = -k \frac{k_i(s)}{\mu_i} \nabla p_i \tag{3}$$

equation of mass conservation

$$\frac{\partial}{\partial t}(mS_i) + \mathrm{div}\, u_i = 0 \tag{4}$$

where $i = 1.2$ corresponds to displaced and displacing fluids (3) and (4) supplement relationship $S_1 + S_2 = 1$ and also the relationship which combines pressure jump in fluids with interphase menisci curvature in pore system

$$p_2 - p_1 = \frac{\sigma}{\sqrt{k/m}} J(S_1) \tag{5}$$

Relative phase permeabilities $k_i(S_1)$ and the Leverett function $J(S_1)$ are empirical dimensionless functional parameters defined by two-phase flow microstructure at developed capillary instability and they should be experimentally determined in special tests.

To analyze fingers evolution in porous medium, the equations (3)-(5) have been solved numerically with such variables as normalized saturation, and average effective pressure. The calculations were performed in two-dimensional rectangular region under no through-flow conditions on side boundaries. The problem of numerical approximation of boundary values in considered zone was performed on block-centred grid with uniform step. With the help of IMPES-method the solution has been found with elliptical equation for pressure solved with overrelaxation method. To generate the perturbances, the input flow rate depended on time in the form of

$$\begin{aligned} &v_1(0,y,t) = v_0[1-\alpha\cos(2\pi y/H)] && 0 \le t \le t_0 \\ &v_1 = v_0 && t > t_0 \\ &v_2(0,y,t) = 0 && t > 0. \end{aligned}$$

where α governs the perturbations amplitude at time t_0 after which flow rate at the inlet was uniform, H is width of displacement region. α has been selected so that at $x/L = 0.09$ where L is length of displacement region, that is at a small distance from the inlet, finger length coincides with the observed in the experiment.

Calculations were performed with parameters corresponding to experimental conditions. The relative phase permeabilities were defined experimentally by steady displacement while the Leverett function was determined by porous-diaphragm method.

$$k_1(S) = 0.305S^{2.3} + 0.31S^{1.9}$$

$$k_2(S) = (1+20S)^{0.125}[0.12(1-S)^{1.6} + 0.35(1-S)^{3.6}]$$

$$J(S) = -0.226(S+0.01)^{0.25}[1+(s+0.119)^3]$$

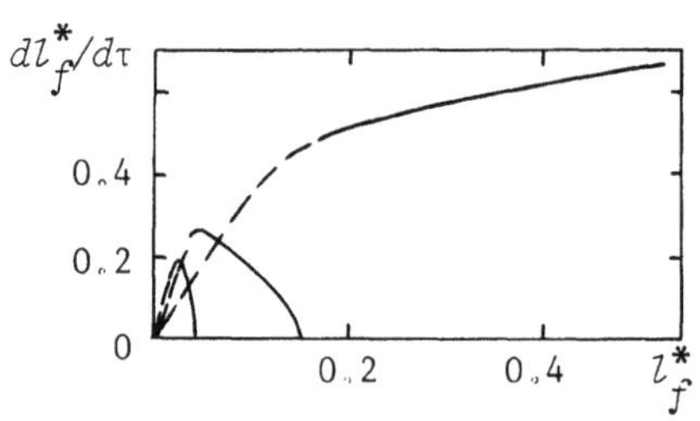

Figure 7. Velocity of amplitude growth as a function of finger amplitude.
1. $\lambda/H = 0.2$; 2. $\lambda/H = 0.3$; 3. $\lambda/H = 1$.

Fig. 7 shows velocity change of finger length as a function of dimensional lengths $l_f^* = l_f/H$ which were obtained for various wavelengths of perturbations λ. The data show that with the increase of fingers length their growth velocity is decreased and at λ/H and 0.3 is observed finite fingers length which increases with the increase of wavelength. The cause of decrease of finger growth velocity lies in capillary overflows from the head to tail parts of a finger in the field of large gradients of displacement front saturation which enlarges with the increase of front curvature. These overflows suppress viscous instability at non-linear stage of fingers evolution similar to surface tension in the Hele-Shaw cell and provide the stable cell front structure. Finite amplitude noise in large scale that is always present in the injects with fractal structure would, however, generate perturbances with various wavelengths including those which are close to displacement region width. Such perturbances for experiments conditions, Fig. 6, can not longer be stabilized with capillary overflows and provide a finger being developed, Fig. 7, that has been observed in experiments.

SUMMARY

The main conclusion deduced from the work lies in that the large amplitude noise in large and small scales defines surface structure selection at viscous fingergrowth in porous medium. Due to this phenomenon, viscous fingergrowing in porous medium differs substantially from that in the Hele-Shaw cell studied in detail earlier. Two major mechanisms which bring to noise formation with immiscible displacement are instability of pore space and capillary. The capillary instability in small scale generates simultaneous flow fluids in the vicinity of displacement front and behind it, and fluids capillary overflows appear from the head to tail finger parts. Capillary overflows instead of surface tension are the major mechanisms suppressing viscous fingergrowth. These overflows together with the noise in large scale define the selection of surface structures with unstable displacement.

REFERENCES

Bensimon, D., 1986, On the stability of viscous fingering, Phys. Rev. A 33: 1302.

Bensimon, D., Kadanoff, L.P., Liang, S., et al., 1986, Viscous flow in two dimensions, Rev. Modern Phys. 58: 977.

Chandler, R., Koplik, T., et al., 1982, Capillary displacement and percolation in porous media, J. Fluid Mech. 119: 249.

Chuoke, R.L., Van Meurs, P., and Van der Poel, C., 1958, The instability of slow, immiscible, viscous liquid - liquid displacements in permeable media, Trans. AIME 216: 188.

Collins, R.E., 1961, "Flow of Fluids through Porous Materials", Reinhold Publishing Corporation, New York.

Homsy, G.M., 1987, Viscous fingering in porous media, Annu. Rev. Fluid Mech. 19: 271.

Kuznetsov, V.V., and Dimov, S.V., 1991, Microstructure of two-phase flow in porous media, Proceedings of "International Seminar on Heat and Mass Transfer in Porous Media", Dubrovnic, Yugoslavia.

Liang, S., 1986, Random walk simulations of flow in Hele-Shaw cells, Phys. Rev. A 33: 2663.

Saffman, P.G., and Taylor, G.I., 1958, The penetration of fluid into a porous medium or Hele-Shaw cell containing a more viscous liquid, Proc. Royal Soc. Ser. A 245: 312.

Wilkinson, P., and Willemsen, J.F., 1983, Invasion Percolation: New Form of Percolation Theory, J. Phys. A, 16: 3365.

TRANSITION PROCESS OF LAMINAR CONFINED JET IN SOLID-LIQUID TWO-PHASE FLOW

Kazuma Fukuda, Koichi Hishida and Masanobu Maeda

Department of Mechanical Engineering
Keio University
3-14-1 Hiyoshi, Kohoku-ku, Yokohama
223 Japan

INTRODUCTION

Two-phase flows containing solid or liquid particles are frequently observed in many industrial processes. The analysis of the flow characteristic is required for an effective operation in the two-phase flow systems. Precise fluid mechanics in such two-phase flows appears to be unexplored due to the complex interactions between dispersed and continuous phases. In addition, the stability of flow is of significance for an stable operation of the two-phase flow system. The main purpose of the present study is to investigate the effects of particles on the transition process from laminar to turbulence.

In the past two decades, several investigations based on the spectrum analysis have been presented on the structure of the transition process of various single-phase flow fields (H.H.Minh, 1986, F.O.Thomas and H.C.Chu, 1989, et al.). R.W.Miksad (1972,1973) has investigated the transition process in a free shear layer with the spectrum analysis and pointed out that the interaction by disturbances which were excited artificially, leaded to the generation of subharmonics, harmonics and the numerous combinations of mode in the frequency of the velocity fluctuation.

However, there are few reports on multiphase flow such as gas-solid or liquid-solid two-phase flow. C.T.Crowe et al. (1991) have examined the effects of particles on the instability of a two-phase wake flow by numerical simulation. They concluded that the instability characteristics of the flow switches from the absolute instability to the convective instability due to the presence of particles.

The authors' group (1987,1989) has investigated the turbulence characteristics of the two-phase flow and reported that the turbulent kinetic energy generated in the shear layer decreased and the transport of turbulence energy was suppressed by the presence of solid particles in a highly turbulent flow.

The transition process should be occurred by the energy transport from mean flow to disturbance and the energy redistribution among disturbances, therefore, the interaction between fluid and particles must have been considered to affect such process in a two-phase flow. In the present study, an experimental investigation has been carried out concerning

Instabilities in Multiphase Flows, Edited by G. Gouesbet
and A. Berlemont, Plenum Press, New York, 1993

the transition of an initially laminar particle-laden circular confined jet flow, focusing on the effect of particle presence on the transition process.

EXPERIMENTAL APPARATUS AND PROCEDURE

Figure 1 shows the flow configuration in the present apparatus which is set in gravitational direction. X denotes the streamwise spatial coordinate and Y denotes the radius spatial coordinate. The origin of streamwise direction is located at the end of the inner pipe. The jet was confined with the inner pipe of D_1=14mm and the outer pipe of D_2=100mm in diameter. Liquid phase was water with the inner flow velocity of U_1=0.18m/s and the annular flow velocity of U_2=0.05m/s. Craya-Curtet number was 2.5 and the recirculating flow region did not exist. Reynolds number, based on U_1 and D_1, was Re=1255 so that the flow might be initially laminar.

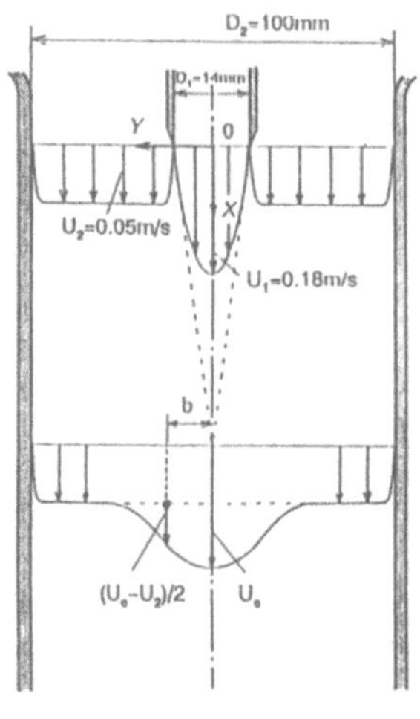

Figure 1. Flow configuration.

Figure 2 shows the size distribution of the employed spherical grass particles and the arithmetic mean diameter is 180.3μm, its standard deviation is 18.5μm and the density is 2590kg/m^3. The particles were mixed into the inner pipe at 150×D_1 upstream from the origin with the constant volumetric concentration of 0.10%. Tulk powder of about 1μm in diameter were added as a tracer for measuring the liquid phase velocity.

A laser Doppler velocimeter (LDV) was applied for simultaneous measurements of particles and liquid velocities, discriminating both phases by light scattered intensity difference. The signal processing system based on a Digital Signal Processor (DSP) is illustrated by the block diagram in Fig.3. The Doppler burst signal filtered by Analog Signal Conditioner was leaded into DSP board, digitized by an analog-to-digital (A/D) converter and dealt with the 128point FFT analyzing in the DSP56001 to obtain the frequency of Doppler burst signal. The host computer (80386) acquired some discrete frequency of Doppler burst signal and calculated the velocity information of the peak frequency by the extended Gaussian fitting (K.Kobashi et al., 1990). This fitting operation might cause within 0.1% errors for each velocity datum. Since a signal intensity and time interval of each datum were obtained simultaneously in addition to velocity information, the particle size discrimination was performed and the time series data of velocity of both phases could be obtained simultaneously. Because this DSP system could do high speed processing for each input signal, it was possible to obtain about 300 data per second.

In order to estimate the power spectra of velocity fluctuation from the time series data of velocity (random sampling data), Blackman-Tukey method was adopted (J.S.bendat and A.G.Piersol, 1971). This method first calculated Eulerian correlation function from time series data of velocity fluctuation, then, estimated the power spectrum by the cosine Fourie transformation of acquired correlation function.

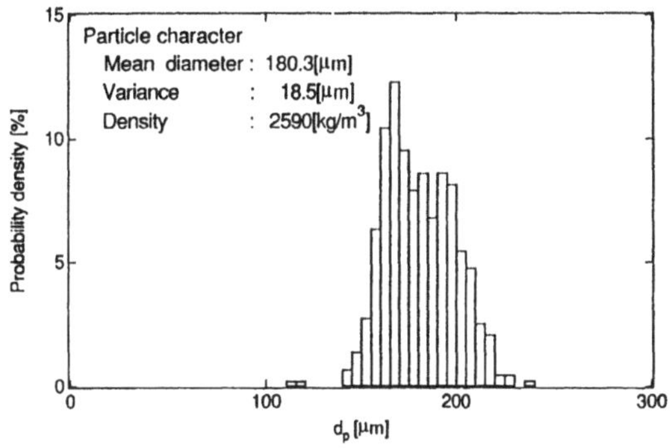

Figure 2. Particle size distribution of loading spherical glass particle.

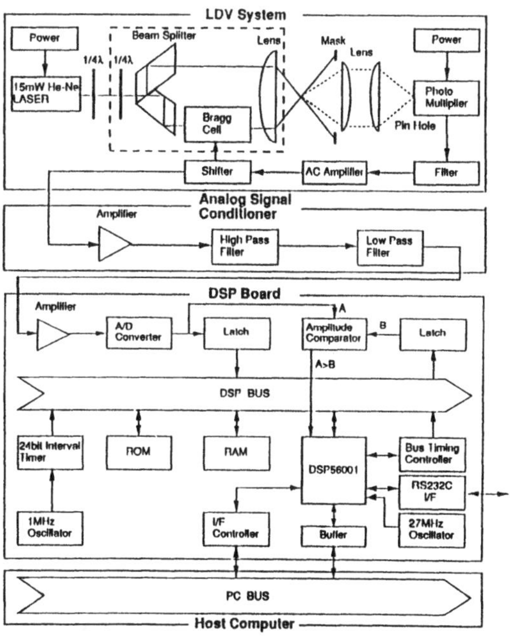

Figure 3. Optical arrangement of LDV and Block diagram of signal processing system.

RESULTS AND DISCUSSION

Figure 4 illustrates the time series data of streamwise centerline velocity at each position denoted by X/D_1. In the single-phase flow, partial small disturbance abruptly appears around at $X/D_1=8$, that is, the disturbance generated in the shear region at $X/D_1=0$ begins to diffuse toward the core of jet flow. After $X/D_1=8$, the disturbance occupies most of the data and laminar flow commences to break down to turbulence. In the two-phase flow, the same phenomena observed in the single-phase flow seems to be appeared. It is well observed that particles lead liquid velocity, because the velocity of particles tends to be larger than that of liquid phase due to the gravitation and these data showed the particle size discrimination must have been done successfully.

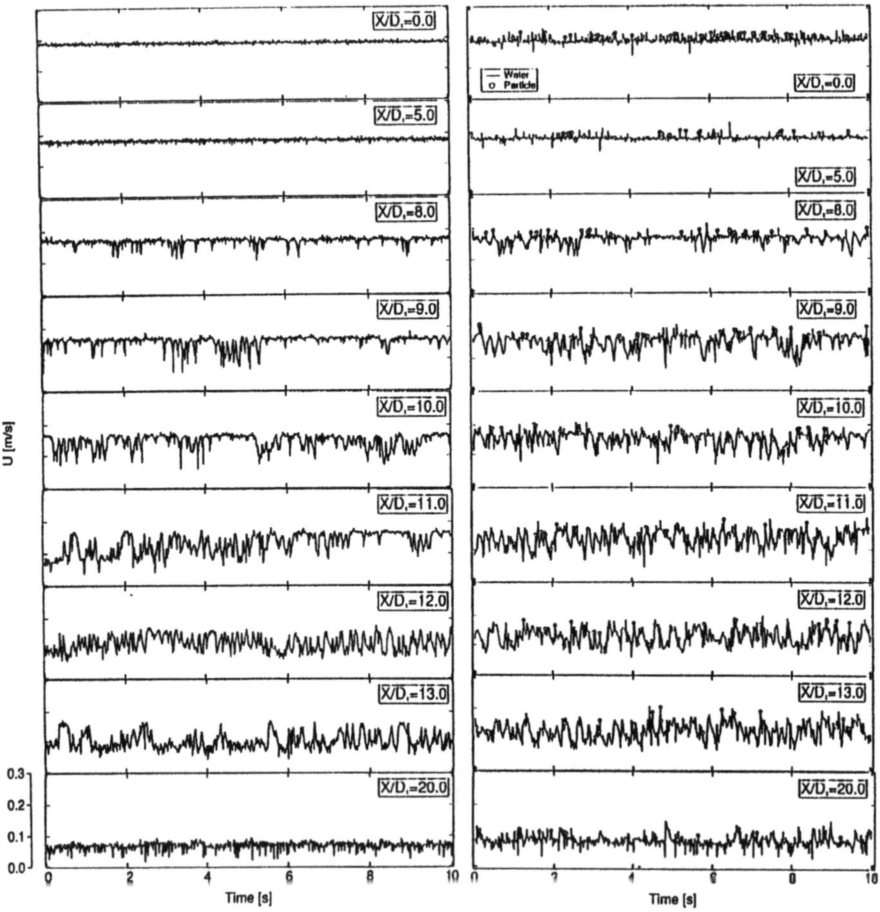

(a) single phase (b) two phase

Figure 4. Time series data of velocity for downstream locations at $Y/D_1=0$.

Cross sectional distributions of intermittency factor evaluated by the time series data of velocity are shown in Fig.5. These figures suggest that the flow in the shear layer has already been accompanied with some fluctuations at the inlet of the test section while the core of jet is initially laminar. It is noted that the intermittency factors around the core of the jet in the single-phase flow become larger gradually from the region $X/D_1=5$ to 10. In contrast, those in two-phase flow remains constant at $5<X/D_1<10$. The penetration of turbulence to the core of jet rapidly occurs in the location of $X/D_1=15$. The results described that the transport of turbulent energy generated in the shear layer to the core region is suppressed due to the presence of particles.

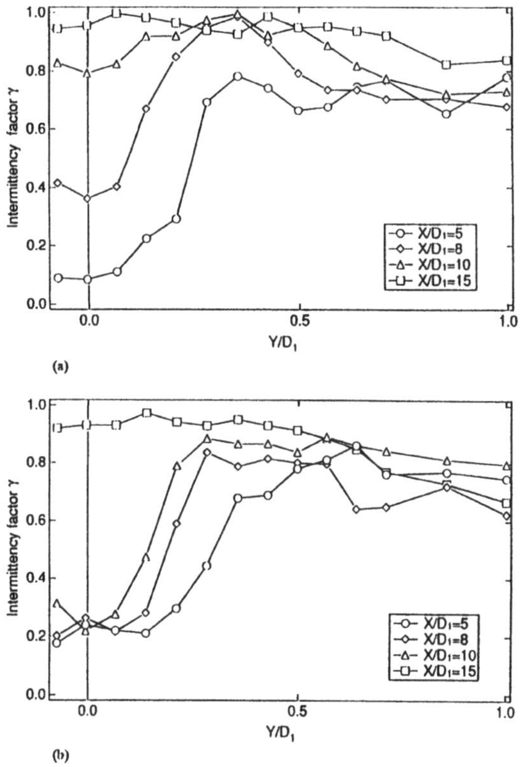

Figure 5. Cross sectional distributions of intermittency factor γ.
(a) single phase (b) two phase

The streamwise variation of mean velocity and velocity fluctuation along centerline of the jet flow are shown in Fig.6(a-c). In Fig.6(a-b), velocities and velocity fluctuations are normalized by U_1-U_2. In the single-phase flow, the mean velocity keeps almost constant in the range from the origin to X/D_1=8 where the potential core exists and decreases remarkably in the downstream after X/D_1=8. The velocity fluctuation in the liquid flow is also constant up to X/D_1=7, and increases steeply from X/D_1=7 to 12 and reaches the maximum value at X/D_1=12, and decreases in the downstream after X/D_1=12. In the two-phase flow, the mean velocity of liquid phase keeps constant from X/D_1=0 to 9 and decreases gradually after X/D_1=9. The velocity of particles is larger than that of liquid due to the gravitational direction flow. Consequently, the liquid phase velocity in the two-phase flow increases by adding solid particles and the liquid phase are accelerated in the downstream direction. The velocity fluctuation of liquid velocity remains constant at $0<X/D_1<6$, increases gradually from X/D_1=6 to 12 and reaches the maximum value at X/D_1=12, decrease in value follows gradually in the downstream direction. It is noted that the maximum absolute value of velocity fluctuation is decreased by the presence of solid particles.

In order to express the ratio of velocity fluctuation to mean flow velocity, the velocity fluctuation is normalized by U_c-U_2 in Fig.6(c), where U_c is centerline velocity of each phases. This figure also shows the transition begins at X/D_1=7 and develops into turbulence near X/D_1=13 in the single-phase flow. Furthermore, it is clear that the transition to turbulence is delayed and the maximum intensity of turbulent is decreased in the two-phase flow.

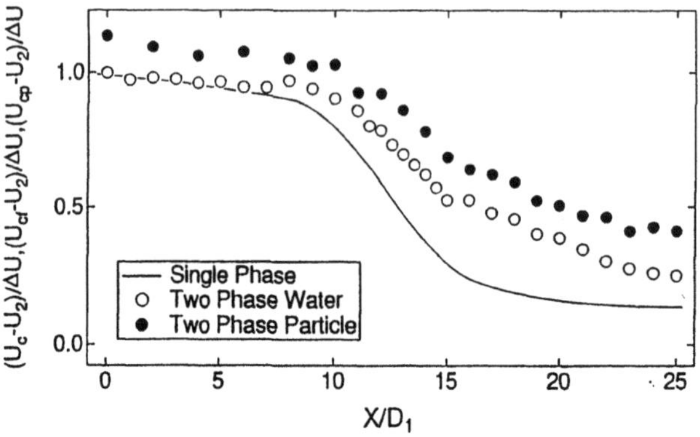

Figure 6 (a). Streamwise distribution of mean velocity along centerline of jet. (normalized by U_1-U_2)

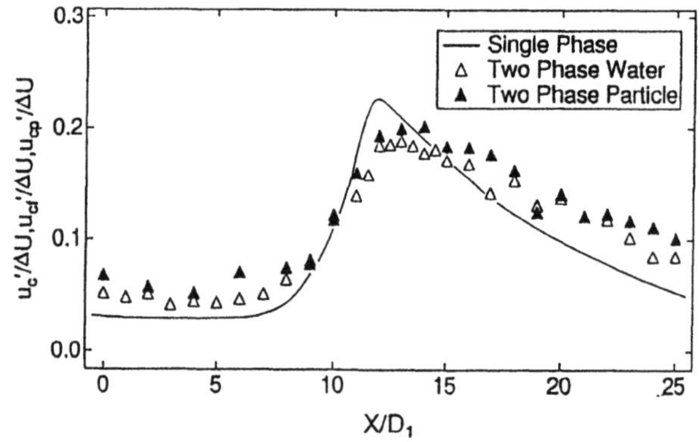

Figure 6 (b). Streamwise distribution of velocity fluctuation along centerline of jet. (normalized by U_1-U_2)

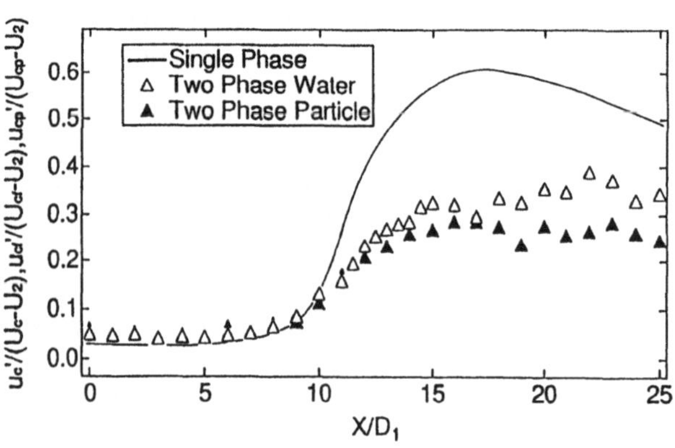

Figure 6 (c). Streamwise distribution of velocity fluctuation along centerline of jet. (normalized by U_c-U_2)

Figure 7 shows cross sectional distributions of local particle volumetric fraction. The volumetric fraction ϕ_p can be estimated by the measurement of number density count N, which is obtained by the counting the number of particle in a specific time interval per location. Due to the adjustment for sensitivity of the trigger level, the absolute volume can not be obtained, but a proportional number density distribution at each X/D_1 is calculated. The unknown proportional factor is calculated by integrating this distribution at each X/D_1. With single particle volume v_p, the ϕ_p is given by,

$$\phi_p = v_p \frac{N}{U_p} \tag{1}$$

This describes that the volumetric fraction remarkably increases near the wall at the outlet of the pipe due to the Saffman lift force caused by the velocity gradient and highly concentrated region exists in the shear layer of inner and outer flow up to $X/D_1=5$. After $X/D_1=5$, particles disperse uniformly with flowing downstream.

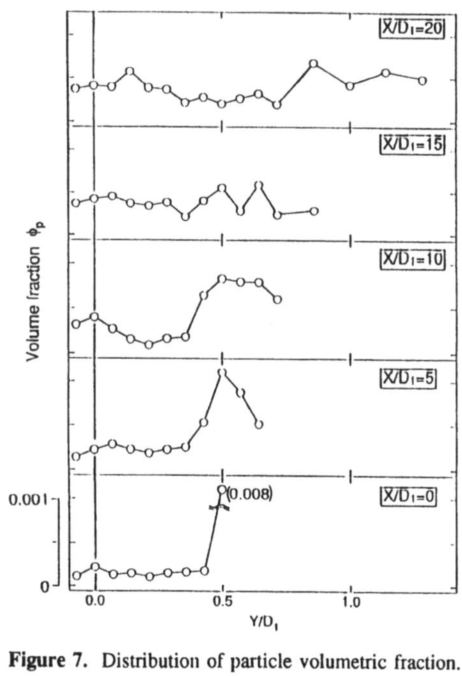

Figure 7. Distribution of particle volumetric fraction.

The cross sectional distributions of the streamwise mean velocity in the single and two-phase flows are shown in Fig.8, where the velocity is normalized by U_1-U_2. At the outlet of the inner pipe, the velocity profile of both phases are parabolic (Hagen- Poiseuille flow), i.e. the flow is initially laminar. Compared with velocity of the single phase, liquid phase velocity at $X/D_1=0$ and 5 increases near the wall region due to high particle concentration as mentioned in Fig.7. After $X/D_1=10$, velocity of liquid phase increases in the downstream direction because the entrainment to inner flow from outer flow occurs due to the presence of particles which leads the liquid-phase flow as mentioned in Fig.6(a).

Figure 9 shows the cross sectional distributions of the streamwise velocity fluctuation normalized by U_c-U_2. In the single-phase flow, the velocity fluctuations are almost constant in the radial direction at $X/D_1=0$. At $X/D_1=5$ and 8, the velocity fluctuation

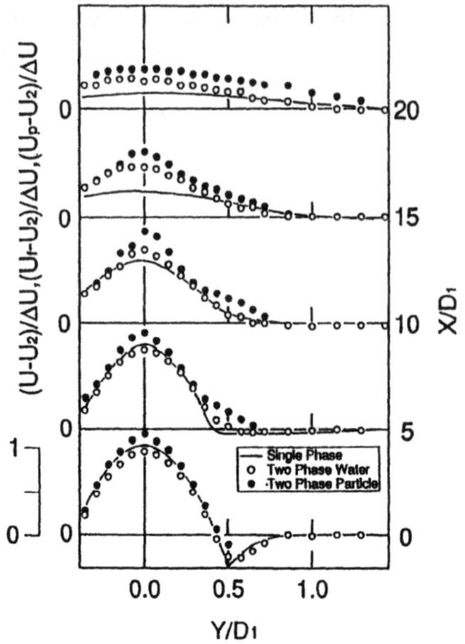

Figure 8. Cross sectional distributions of mean velocity for downstream locations.

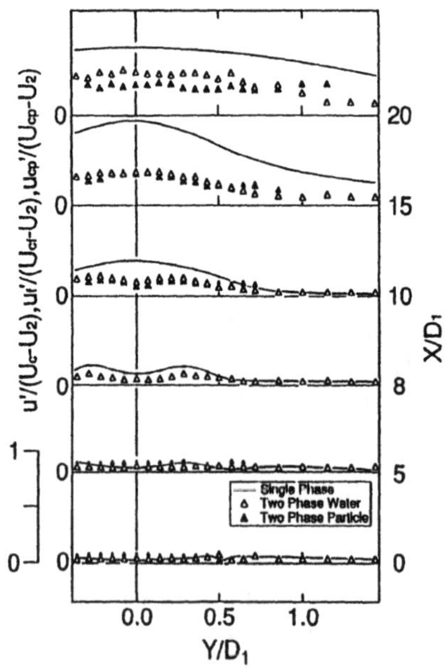

Figure 9. Cross sectional distributions of velocity fluctuation for downstream locations. (normalized by U_c-U_2)

has two maximum peaks in the shear layer of inner and outer flows where the disturbance is generated. The maximum value of velocity fluctuations located at the center of the jet increases at the downstream from X/D_1=10. In the two-phase flow, the velocity fluctuation of liquid phase at the outlet of the inner pipe becomes larger and the variation of the velocity fluctuation at $X/D_1=5$ and 8 is gradual in the radial direction and the value of the maximum peak which is located in the same region as in the single-phase flow decreases.

The variation of the half-width of both phases in the jet is shown in Fig.10. The half-width of the single-phase flow spreads faster than that of the two-phase flow and the development of jet is suppressed in the two-phase flow.

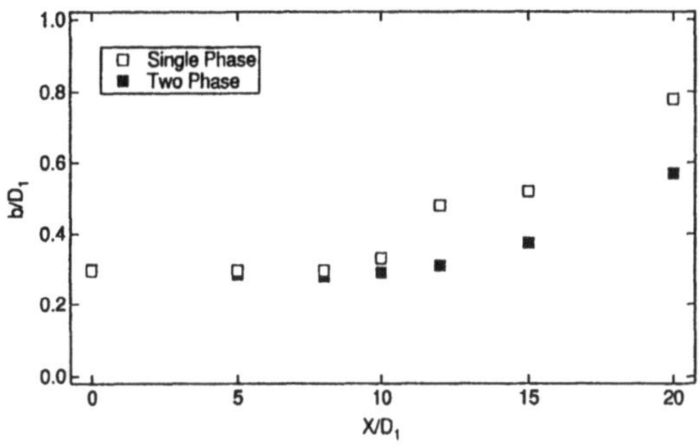

Figure 10. Streamwise distribution of half-width of jet.

Figure 11 indicates the Eulerian turbulent length scale L_e which is calculated by the equation,

$$L_e = U \cdot \tau_e \tag{2}$$

where τ_e is the Eulerian turbulent integral time scale, which is determined by auto-correlation coefficient obtained through the process to estimate the power spectrum by Blackman-Tukey method and defined as,

$$\tau_e = \int_0^\infty R(\tau)d\tau \tag{3}$$

The Eulerian scale is representative of eddy scale containing turbulent energy. In the single-phase flow, the length scale has a peak at $X/D_1=11$. The eddy of turbulence develops at $7<X/D_1<11$ and dissipates after $X/D_1=11$. In the two-phase flow, the length scale decays in the whole region compared with that of in the single-phase flow. That is to say, the growth of the turbulent eddy is suppressed due to the presence of the loaded particles.

Figure 12 shows the cross sectional distribution of the power spectra of velocity fluctuation which is estimated from time series data of velocity by using Blackman-Tukey method. In the single-phase flow, the naturally excited instability which has a frequency of about 3.5Hz appears around $Y/D_1=0.3$ at $X/D_1=5$, where the peak values of velocity fluctuation exist. After $X/D_1=5$, the instability of 3.5Hz expands all over the range of shear layer and its turbulent energy increases radically. Downstream after $X/D_1=10$, the second harmonic frequency of the natural instability seems to be raised. It is considered that the phenomenon is originated by the redistribution of turbulent energy from the natural instability to the second harmonic instability and this redistribution process begins by the trigger action of the peak energy of the natural instability. This process can be regarded as the transition process to turbulence. After $X/D_1=15$, the flow field tends to be homogeneous gradually. Although the same process observed in the single-phase flow appears in the two-phase flow, the value of peak energy of the natural instability decreases as compared with

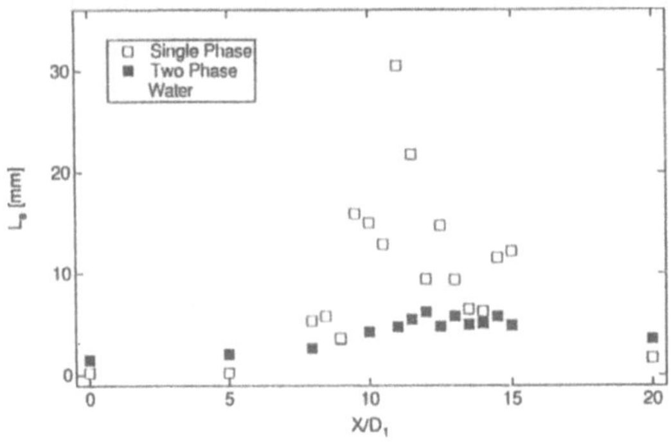

Figure 11. Streamwise distribution of Eulerian integral length scale L_e.

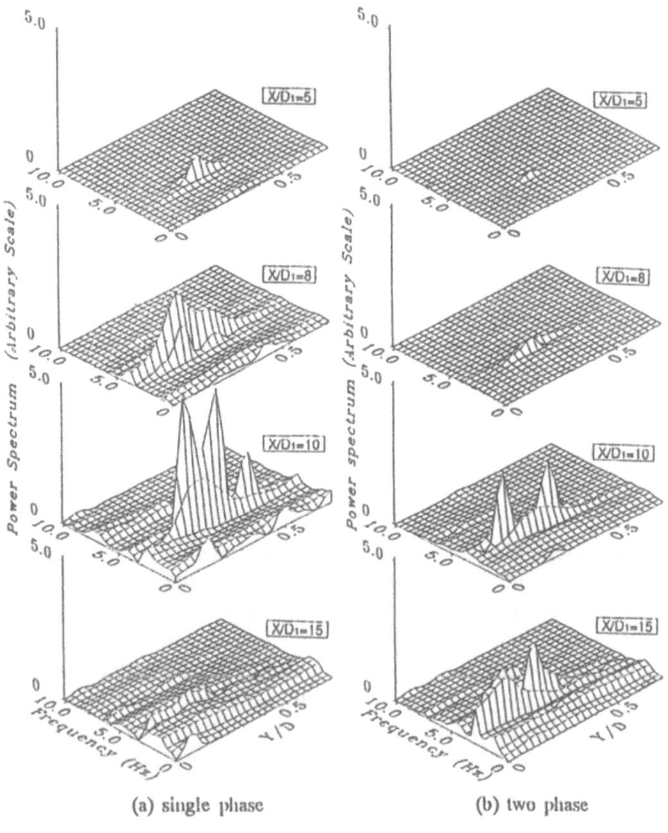

Figure 12. Distributions of power spectra in radial directions. (arbitrary scale)

that of in the single-phase flow due to the interaction between fluid and particles, therefore, the redistribution process by the peak energy hardly occurs and the transition process to turbulence is suppressed as a result.

CONCLUDING REMARKS

The effect of suspending particles on the laminar-turbulence transition process of the jet flow has been experimentally investigated. The following concluding remarks are derived from the present study;

The ratio of velocity fluctuation to mean flow reduces over the whole range in the test section for the two-phase flow. In the presence of particles, the disturbance generated in the shear layer is slowly transferred to the core region of jet. The transition process seems to begin by the trigger action of the energy of the naturally excited instability in the single-phase flow. The peak value of energy of the naturally excited instability in the two-phase flow decreases due to the energetic interaction between liquid and particles and the redistribution of turbulent energy among fluctuations is suppressed, therefore, the transition to fully developed turbulence of the jet is suppressed by the presence of solid particles.

Acknowledgment

The author would like to gratefully acknowledge the support of this experimental research by Mr. A.Hoshide.

References

Bendat, J.S., and Piersol, A.G., 1971, "RANDOM DATA: Analysis and Measurement Procedures", John Wiley & Sons Inc.

Fleckhaus, D., Hishida, K., and Maeda, M., 1987, Effect of laden solid particles on the turbulent flow structure of a round free jet, *Experiments in Fluids* **5**: 323-333.

Hishida, K., Nakano, H., Fujishiro, T., and Maeda, M., 1989, Turbulent characteristics of liquid-solids two-phase circular confined jet (in Japanese), *Trans. JSME* **511**: 648-654.

Hishida, K., Nakano, H., and Maeda, M., 1989, Turbulent Flow Characteristics of Liquid-Solid Particle Confined Jet, *International Conference on Mechanics of Two-Phase Flows*, Jun.12-15, Taiwan: 209-214.

Kobashi, K., Hishida, K., and Maeda, M., 1990, Measurement of Fuel Injector Spray Flow of I.C.Engine by FFT Based Phase Doppler Anemometer - An Approach to the Time Series Measurement of Size and Velocity, *5th International Symposium*, 9-12 July, Lisbon: 268-287.

Kourta, A., Boisson, H.C., Chassaing, P., and Minh, H.H., 1987, Nonlinear interaction and the transition to turbulence in the wake of a circular cylinder, *J. Fluid Mech.* **181**: 141-161.

Miksad, R.W., 1972, Experiments on the nonlinear stages of free-shear-layer transition, *J. Fluid Mech.* **56**: 695-719.

Miksad, R.W., 1973, Experiments on nonlinear interactions in the transition of a free shear layer, *J. Fluid Mech.* **59**: 1-21.

Thomas, F.O., and Chu, H.C., 1989, An experimental investigation of the transition of a planar jet: Subharmonic suppression and upstream feedback, *Phys. Fluids* **A1(9)**: 1566-1587.

Yang, Y., Chung, J.N., Troutt, T.R., and Crowe, C.T., 1991, The Effect of Particles on the Stability of a Two-phase Wake Flow, *Proceeding of The International Conference on Multiphase Flows*, 24-27 Sep., Tsukuba: 113-116.

TWO-PHASE HIGH TEMPERATURE MODELLING FOR GRANULATION

Iwanka P. Nikolova

Institute for Mechanics & Biomechanics
4 Acad. G. Bonchev Str.
Sofia 1113, Bulgaria

INTRODUCTION

Many kinds of industrial granulation processes exist, which defer each other by means of their realization methods. One of them is drop crystallization from a melt in air flux, which is frequently used by obtaining mineral fertilizers, e.g. NH_4NO_3-saltpetre ammonia. Granulation by drop crystallization includes three important stages: 1) jet-like melt flowing and drop formation after stability loss; 2) drop crystallization by cooling and granules forming; 3) granules moving in a granulation tower and collecting at the bottom in pseudo-boiling layer.

The first two stages are of special interest, but we concentrated on the third stage, ie, on the motion of granules in gas media by the presence of drag force and heat exchange between two phases and gas convection. All these factors make the motion of granules complicated. From a practical point of view it is quite important to know the trajectory of particles because of the necessity to calculate the granulation tower.

The numerical modelling of the falling granules' flow, whose height of flight is about 30-50m is the topic of the present paper. An assumption is made that granules are solid monosize particles (later we shall call them simply 'particles'), which provides an opportunity to apply the monodispersed media

Instabilities in Multiphase Flows, Edited by G. Gouesbet
and A. Berlemont, Plenum Press, New York, 1993

model. Particles are falling down from the upper wall of a channel with an arbitrary velocity. For simplicity we consider the one-dimensional case and consequently the one direction "closed" sedimentation is assumed.

Two approaches are applied for description of the gas and solid phase respectively. Navier-Stokes-equations are solved for the gas phase by taking into account the Stokes-resistance law. Solid particle motion is described by ordinary differential equations accordingly to the Newton's law for a so-called 'large particle'. In addition, some elements of Harlow-Welch PIC-method are used in order to determine the solid phase characteristics in the mesh points. Heat exchange between the solid and gas phases is taken into account.

LITERATURE REVIEW

A short review shows that there have been many studies treating gas- particle and gas-droplet behavior in a flow field or carrier gas. The distribution of velocity of gas-solid suspension flow in a vertical tube as well as the temperature are measured by Lodes and Mierka (1990). The dispersed phase was five individual fractions of polypropylene particles, which were separated in a cyclone and fed through a control and checking system into the gas flow. A numerical study of supersonic gas particle two-phase flow is performed by Ishii, Hatta, Umeda and Yuhi (1990) on the basis of so-called 'discrete particle-cloud model'. In this model, a whole particle cloud is divided into a large number of small subclouds. An important reason for applying such model when the mass-loading ratio is not negligible, are two opposite tendencies of particles behavior, namely their focusing and dispersion, which introduces serious diffities into the theoretical approach. It is obviously impossible to treat the particle phase as an ordinary continuum medium. Similar idea, which will be explained later, is explored in our paper for describing 'large particle' motion. On the contrary, for sufficiently dilute particles the possibility of the inter-particle collision will be small and so-called 'two-fluid model', which means that the particle phase as well as the gas phase can be treated as a continuum media, is simplified to a significant degree.

A combined experimental and numerical investigation for droplet trajectory is performed by Nguyen, Rangel and Dunn-Rankin (1991). A comparison of the measured trajectories with trajectories predicted by solving the equation of motion with different laws for droplet drag, e.g. Stokes, Renksizbulut, Chang et al laws, as well as investigation of the interaction effects occurring in a stream of droplets was carried out. Numerical calculations were based on a Runge-Kutta scheme.

GAS-PHASE EQUATIONS

Gas-phase equations (1-4) for viscous compressible heat-conductive carrier gas consist of discontinuity and momentum equations in a conservation form, taking into account interaction (friction) force and two-velocity assumption. Pressure and temperature equations include terms which express the difference between the temperature of the gas-phase and the solid phase and the heat exchange.

$$\frac{\partial \rho_1}{\partial \tau} + \frac{\partial(\rho_1 V_1)}{\partial y} = 0, \tag{1}$$

$$\frac{\partial V_1}{\partial \tau} + V_1 \frac{\partial V_1}{\partial y} = - \frac{1}{\rho_1 \gamma M^2} \frac{\partial P}{\partial y} + \frac{1}{\rho_1 Re} \frac{4}{3} \frac{\partial^2 V_1}{\partial y^2} - f - Fr, \tag{2}$$

$$\frac{\partial P}{\partial \tau} + V_1 \frac{\partial P}{\partial y} = - \gamma P \frac{\partial V_1}{\partial y} + \frac{\gamma}{RePr} \frac{\partial^2 T_1}{\partial y^2} - \frac{\alpha}{\gamma_1} \rho_2 (T_1 - T_p), \tag{3}$$

$$\frac{\partial T_1}{\partial \tau} + V_1 \frac{\partial T_1}{\partial y} = - (\gamma - 1)T \frac{\partial V_1}{\partial y^2} + \frac{\gamma}{\rho_1 RePr} \frac{\partial^2 T_1}{\partial y^2} \alpha \frac{\rho_2}{\rho_1} (T_1 - T_p) \tag{4}$$

$$Re = \frac{wL\rho_{10}}{\eta}, \quad Pr = \frac{\eta \, c_p}{\lambda}, \quad Fr = \frac{gL}{w^2}, \quad M^2 = \frac{w^2}{\gamma R T_0}, \tag{5}$$

$$\alpha = \frac{6\varepsilon\gamma\gamma_1 Nu}{\delta \, Re}, \qquad Nu = 2 + 0.6 \, Pr^{1/3} Re_p^{\,1/2}, \tag{6}$$

$$w = \rho_2^0 \, d^2 \, g/18\eta, \quad M_{21} = n_0 \Pi \, d^3 \rho_2^0 \, / \, 6 \, \rho_1^0, \quad T_p^0 = T_s + 1 \tag{7}$$

Here τ, y are the time and y-axial distance; ρ_1, V_1, P, T_1 are the density, the velocity, pressure and temperature for the gas phase. L is the characteristic length of the flow field and here the channel width is taken as L. The subscripts $_p$ and $_1$ denote particles and gas conditions, respectively, and sub and superscripts zero denote initial quantities. c_v and c_p are the specific heats at constant volume and pressure of the gas, c_2 is the specific heat of the particle material, γ is the ratio of the specific heats of the gas, γ_1 is equal to c_2/c_p. ε is determined below. T_s is the difference between the temperature of the gas and the particles.

Following the paper of Makhviladze et al (1987), non-dimensional parameters are introduced for M, Fr, Re, Pr, Nu - Mach, Fruid, Reynolds, Prandtl and Nusselt numbers. Further investigations were carried out as nondimensional analysis between these parameters.

PARTICLE PHASE MOTION

Particle assembly equations include particle motion and temperature equations (8,9) for 'large particles'. They are 'artificial' one and introducing by dividing the whole region into the definite numbers of similar subregions, which consist of an arbitrary number of physical particles with appropriate density (mass), velocity, temperature and so on. Each 'large particle' mass is assuming to be equal to the sum of masses of assembling physical particles. By labelling the 'large particles' by subscripts k ($k=1,2,3...$), the particle motion obeying Newton's law and heat exchange are described by

$$\frac{dV_k}{d\tau} = - Fr + Fr\left[V_1 - V_k \right], \tag{8}$$

$$\frac{dT_k}{d\tau} = \alpha\left[T_1 - T_k \right]. \tag{9}$$

It is assumed that high temperature particles are fed through a small region l_{in} on the top wall of the vessel (in a two-dimensional case) or through the whole wall in a one-dimensional case. The mass loading ratio M_{21} is small enough and particle-particle collisions are not taken onto account. Overwritten w, V_k and T_k are the steady state drop velocity of a single descending particle, the velocity and temperature of the k-th 'large particle' respectively; δ is the dimensionless diameter of the particle d/L; ε is ratio of the real densities of the gas and particles ρ_{10}/ρ_2^0 one; n_0 is the number of particles per unit volume.

RESULTS AND DISCUSSION

Numerical techniques

In a previous paper of Melikhov and Nikolova (1987) 'pure hydrodynamics'results for flow field and 'particle jet' focusing dependence on various parameters (Re, Fr, M_{21}, l_{in}) were obtained. Here a similar computational method was used to the one-dimensional case, but including the 'temperature' equations. The momentum equation (2) was divided into physical processes as follows: first, the homogeneous equation was solved; second, obtained velocity was improved by means of including the viscous terms and those responsible for the interphase interaction. Equation (2) written in discretization form is the following:

$$V_1^{n+1} = \frac{1}{1+\hat{\tau}\mathrm{Fr}\rho_2^0}\left[V_1^n + \tau\mathrm{Fr}(\rho_2^0 V_k^n - 1)\right], \tag{10}$$

where $\hat{\tau}$ is the time step and n denotes the number of time layer. Numerical results were performed for the following values of parameters: $M^2=0.05$, Re=30, Pr=1, $\hat{\tau} = 5.59\ 10^{-4}$, Ts=0-3. The rest of the parameters are variable.

Gas and particle-phase characteristics

As shown by the analytical expressions (8,10) by 'one-dimensional' sedimentation assumption, the particle velocity does not depend on the particle density ρ_2^0 but the gas velocity does. Numerical calculations play a decisive role in solving this problem. So it is important to know the velocity dependence $\rho V(\tau)$ on various densities ρ_2^0, (mass loading ratio M_{21}, respectively), on Fr and Re numbers in two regimes, isothermal and heat.

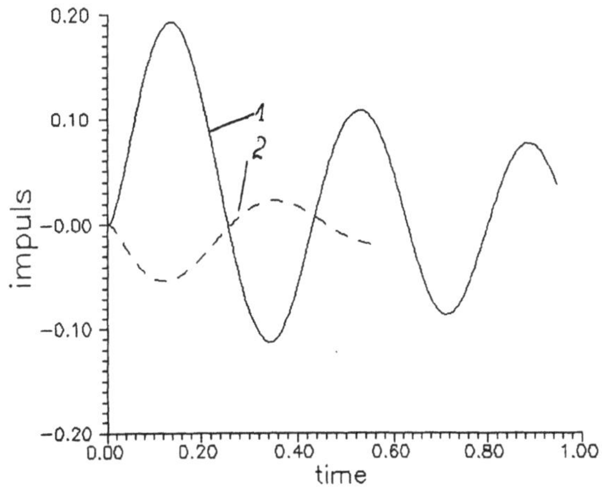

Figure 1. Time dependence of the integral impuls: 1 -Ts=3, 2-Ts=0; Re=30, Fr=40, α=34.

Due to the gravity force, particles fall down as a 'particle jet' and, because of the friction, carry gas in motion. For a certain characteristic time period $\tau_* = M$, gas stays at rest on the channel bottom. A high pressure area forms and starts moving upward, reducing gas velocity and afterwards changes its direction. A compressible wave, reflected from the top wall of channel moves downward. This process is illustrated in Fig.1., where vertical integral impulse $J(\tau) = \int \rho_1 V_1 dy$ is plotted as a function of time for isothermal ($T_s = 0$, dash line) and heat (solid curve) cases. Results are compared with those of Makhviladze et al (1987). Damping gas oscillations (standing wave) begins and influences precipitation of particles. In a case when 'heat' particles fall down buoyancy seems to be stronger than the friction force which causes upward gas moving initially (the first oscillation mode increases).

Fig.2 shows nonisothermal case ($T_s=3$) characteristics. Besides the gas-phase motion do to the sedimentation of particles, other kinds of effects, connected with gas-heating expansion and developing of natural heat convection can be observed. Initially the difference between gas-phase (curve-1) and particle-phase (curve-2) temperature exists. As a result of the temperatures of the two phases

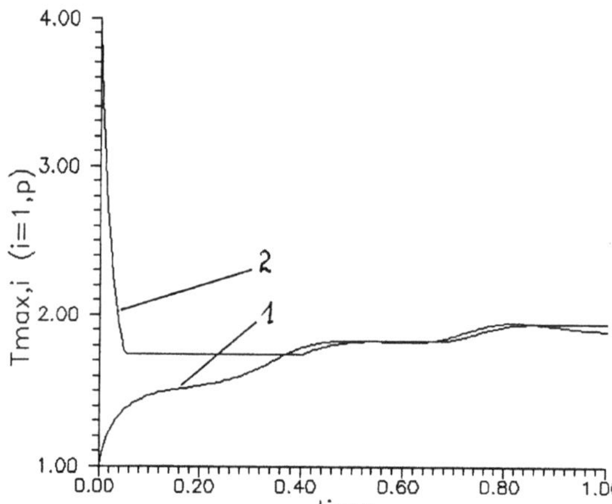

Figure 2. Maximum temperature of the gas-1 and solid phase-2, Re=30, Fr=40,α=34, M_{21}=0.1.

becoming even, gas is getting warm (see Fig.2, curve-1). Results are qualitatively compared with those of Makhviladze et al (1983).

Other important characteristics are the mass center y_c of the particle assembly and the K-ratio of already precipitated particles. Time dependence of vertical coordinate of nonsedimented particles is shown in Fig.3. The self-accelerated regime of descending particles in the isothermal case ($T_s = 0$, dash line) changes with oscillations of curves for the heat regime (curves 1- 4). Convection initially causes upward gas motion, which involves particles, and mass center rises, so that the picture significantly defers from the case T_s=0. Moreover, rising of the mass center depends on mass loading ratio. It is stronger by increasing of M_{21} parameter (see curves 1-4) in Fig.3. On the contrary Fig.4 shows slight dependence of precipitated particles on M_{21} parameter. It can

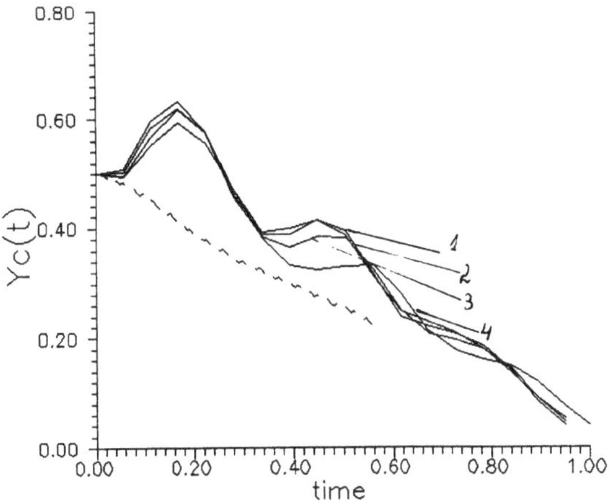

Figure 3. Time dependence of particle assembly mass center on the values of M21 parameter: 1-0.5, 2-0.2, 3-0.1, 4-0.01, $T_s=2$.

be shown that convection due to heating causes remarkable delay of precipitation in comparison with the isothermal case (see dash line).

Many important characteristics could be obtained from this program, but they rest outside the scope of this paper; they include the maximum and mean velocity for two phases as a function of time, vertical distance and mass loading ratio, as well as K-ratio dependence on Fr, Re and numbers.

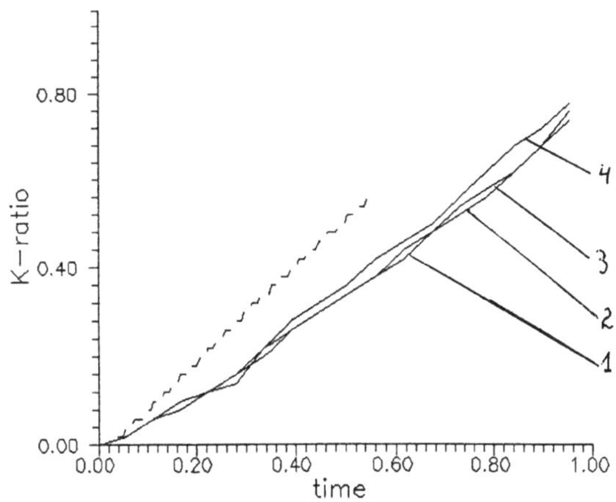

Figure 4. Time dependence of K-ratio on the values of parameter M21: 1-0.5, 2-0.2, 3-0.1, 4-0.01

CONCLUSION

Obtained characteristics provide the opportunity to follow gas and particle-phase behavior in a channel in conditions of heating and compare with the isothermal case. Mathematical programming is rather common and could be applied for a two-dimensional case, which is next step of study.

ACKNOWLEDGEMENT

Paper was sponsored by Grant 623 of the Ministry of Science and Higher Education during the time period 07.1990- 09.1991.

REFERENCES

Ishii, R., Hatta, N., Umeda, Y. and Yuhi, M., 1990, Supersonic gas-particle two-phase flow around a sphere, *J. Fluid Mech.* 221: 453.

Lodes, A. and Mierka, O., 1990, The velocity field in a vertical gas-solid suspension flow, *Int.J. Multiphase Flow* 16: 201.

Melikhov, O.I. Nikolova, I.P., 1987, Convection in closed volume due to the action of particle source (russ.), *Theor. and Appl.* Mech. 4: 53.

Makhviladze, G.M., Melikhov, O.I., 1983, Heat particle cloud motion above horizontal surface in a field of external force, *Journal Prikl. Mech.i Techn.Phyziki* 5: 115.

Makhviladze, G.M., Melikhov, O.I. and Soboleva, E.B., 1988, Gas-solid sedimentation in closed vessel, *Journal Prikl. Mech.i Techn.Phyziki* 6: 133.

Nguyen,Q., Rangel, R.H. and Dunn - Rankin, D., 1991, Measurement and collision of droplets, *Int. J. Multiphase flow* 17: 159.

CONTRIBUTORS

INDEX

MIX
Papier aus verantwortungsvollen Quellen
Paper from responsible sources
FSC® C105338

If you have any concerns about our products,
you can contact us on
ProductSafety@springernature.com

In case Publisher is established outside the EU,
the EU authorized representative is:
Springer Nature Customer Service Center GmbH
Europaplatz 3, 69115 Heidelberg, Germany

Printed by Libri Plureos GmbH
in Hamburg, Germany